Fish Respiration and Environment

Fish Respiration and Environment

Editors

Marisa N. Fernandes
Department of Physiological Sciences
Federal University of São Carlos
São Carlos, Brazil

Francisco T. Rantin
Department of Physiological Sciences
Federal University of São Carlos
São Carlos, Brazil

Mogens L. Glass
Department of Physiology
University of São Paulo
Ribeirão Prêto, Brazil

B.G. Kapoor
Formerly Professor of Zoology
The University of Jodhpur
Jodhpur, India

CRC Press
Taylor & Francis Group
Boca Raton London New York

CRC Press is an imprint of the
Taylor & Francis Group, an **informa** business
A SCIENCE PUBLISHERS BOOK

CIP data will be provided on request.

First published 2007 by Science Publishers Inc.

Published 2019 by CRC Press
Taylor & Francis Group
6000 Broken Sound Parkway NW, Suite 300
Boca Raton, FL 33487-2742

© 2007, Copyright reserved
CRC Press is an imprint of Taylor & Francis Group, an Informa business

First issued in paperback 2019

No claim to original U.S. Government works

ISBN 13: 978-0-367-45320-6 (pbk)
ISBN 13: 978-1-57808-357-2 (hbk)

Visit the Taylor & Francis Web site at
http://www.taylorandfrancis.com

and the CRC Press Web site at
http://www.crcpress.com

Preface

The diverse and ever-changing aquatic environment has a major impact on the organization of various organ-systems of fishes. This book contains seventeen chapters covering bony fishes which are focal to the current study. The chapters primarily cover fish respiration but also include osmoregulation, these being the two main functions of gills. Concurrently, cardio-respiratory synchronization has been well addressed.

Gills of healthy fishes are their life-line to meet the challenges arising from their changing environment: oxygen gradient, alkalinity, temperature fluctuations and overall the added pollutants.

Since the gills are considered multi-potent tissues, their architecture with bizarre modifications, morphometrics, O_2^- and mechans – receptors, Chloride (MR-rich) cells fulfill the varied functions; this functional plasticity is quite specific in nature. The counter-current blood flow is an added advantage.

What are the different views on the homology of swim (air-bladder, gas-bladder)? One of chapters presents these views. How do the dipnoans (lungfishes), with modified cardiac structure and functions, survive in different situations? This has received a lot of coverage.

Interest in amphibious fishes is no less. The causative factors for their emergence and related behavior are duly described. There is a focus on fishes with accessory respiratory organs, including skin. Information is given on how denizens of Magadi Lake (Kenya) with both high alkalinity and temperature – indeed a very hostile environment – efficiently perform the life-saving functions. The hypoxia – tolerant species (*Cyprinus carpio*) has been exemplified and functionally detailed. The performance of gills in fishes with diseases is also reported. An emphasis on osmo-respiratory compromises with related details can be obtained in specific chapters.

It is hoped that this book with its coverage in totality and well-supported with illustrations will not only infuse interest in readers but merit a permanent place on the shelves of ichthyological literature.

The Editors

Contents

Color Plate Section

List of Contributors

Almeida-Val Vera Maria de

Laboratory of Ecophysiology and Molecular Evolution, National Institute for Amazon Research, Manaus, Brazil.

E-mail: veraval@inpa.gov.br

Amin-Naves Jalile

Department of Physiology, Faculty of Medicine of Ribeirao Preto, University of São Paulo, Avenida Bandeirantes, 3900, 14049-900, Ribeirão Preto, SP, Brazil.

E-mail: jalile@rfi.fmrp.usp.br

Bassi Miriam

Department of Physiology, Faculty of Medicine of Ribeirão Preto, University of São Paulo, Avenida Bandeirantes 3900, 14049-900 Ribeirão Preto, SP, Brazil.

E-mail: mbassi@rfi.fmrp.usp.br

Brauner Colin J.

Department of Zoology, University of British Columbia, 6270 University Boulevard, Vancouver. BC V6T 1Z4, Canada.

E-mail: brauner@zoology.ubc.ca

Campbell Hamish

School of Biosciences, University of Birmingham, Edgbaston, Birmingham B15 2TT, UK.

E-mail: dr.hamish.Campbell@gmail.com

Chapman Lauren J.

Department of Biology, McGill University, 1205 Avenue Docteur Penfield, Montreal, PQ, Canada H3A 1B1 & Wildlife Conservation Society, 2300 Southern Boulevard, Bronx, New York, USA 10460.
E-mail: lauren.chapman@mcgill.ca

Etou Aya

Department of Animal and Marine Bioresource Science, Faculty of Agriculture, Kyushu University, Fukuoka 812-8581, Japan.
E-mail: etay1378@yahoo.co.jp

Fernandes Marisa Narciso

Laboratory of Zoophysiology and Comparative Biochemistry, Department of Physiological Sciences, Federal University of São Carlos, via Washington Luis, km 235, 13565-905 – São Carlos, SP, Brazil.
E-mail: dmnf@power.ufscar.br

Fernandes-Castilho Marisa

Research Center on Animal Welfare - RECAW & Laboratory of Studies on Animal Stress, Dept. of Physiology, Universidade Federal do Paraná, Curitiba, Brazil.
E-mail: mafernandes@ufpr.br

Freitas Eliane Gonçalves de

Research Center on Animal Welfare - RECAW & Laboratory of Animal Behavior, Dept. of Zoology and Botany, Universidade Estadual Paulista, São José do Rio Preto, Brazil.
E-mail: elianeg@ibilce.unesp.br

Giaquinto Percilia Cardoso

Research Center on Animal Welfare - RECAW & Laboratory of Physiology and Animal Behavior, Dept. of Physiology, Faculty of Medicine, Universidade de São Paulo, Ribeirão Preto, Brazil.
E-mail: percilia@rfi.fmrp.usp.br

Giusti Humberto

Department of Physiology, Faculty of Medicine of Ribeirao Preto, University of São Paulo, Avenida Bandeirantes 3900, 14049-900 Ribeirão Preto, SP, Brazil.
E-mail: hg@rfi.fmrp.usp.br

Glass Mogens L.

Department of Physiology, Faculty of Medicine of Ribeirão Preto, University of São Paulo. Avenida Bandeirantes 3900, 14049-900, Ribeirão Preto, SP, Brazil.

E-mail: mlglass@rfi.fmrp.usp.br

Graham Jeffrey B.

Center for Marine Biotechnology and Biomedicine and Marine Biology Research Division, Scripps Institution of Oceanography, University of California, San Diego, La Jolla, CA 92093-0204, USA.

E-mail: jgraham@ucsd.edu

Ishimatsu Atsushi

Institute for East China Sea Research, Nagasaki University, Tairamachi, Nagasaki 851-2213, Japan.

E-mail: a-ishima@net.nagasaki-u.ac.jp

Kalinin Ana Lúcia

Laboratory of Zoophysiology and Comparative Biochemistry, Department of Physiological Sciences, Federal University of São Carlos, Via Washington Luis, km 235, 13565-905 – São Carlos, SP, Brazil.

E-mail: akalinin@power.ufscar.br

Kisia S.M.

Department of Veterinary Anatomy, University of Nairobi, P.O. Box 30197, Nairobi, Kenya.

E-mail: kisiasm@yahoo.co.uk

Lee Heather J.

Center for Marine Biotechnology and Biomedicine and Marine Biology Research Division, Scripps Institution of Oceanography, University of California, San Diego, La Jolla, CA 92093-0204, USA.

E-mail: hboyle58@hotmail.com

Leite Cleo

Department of Physiological Sciences, Federal University of São Carlos, 13565-905, Sao Carlos, SP, Brasil.

E-mail: cleo.leite@gmail.com

Moron Sandro Estevan

Department of Physiological Sciences, Federal University of São Carlos, via Washington Luís km 235, 13565-905 São Carlos, SP, Brazil.

E-mail: sandromoron@uft.edu.br

Nelson Jay A.

Department of Biological Sciences, Towson University, Towson Maryland 21252-0001, USA.

E-mail: jnelson@towson.edu

Oliveira Christiane Patricia Feitosa de

Ph.D. student in Biotechnology, Universidade Federal do Amazonas, Manaus, Brazil.

E-mail: chris@inpa.gov.br

Onyango D.W.

Department of Veterinary Anatomy, University of Nairobi, P. O. Box 30197, Nairobi, Kenya.

E-mail: dwo@uonbi.ac.ke; dwonyango@yahoo.com

Perry Steven F.

Institut für Zoologie, Universität Bonn, Poppelsdorfer Schloss, 53115 Bonn, Germany.

E-mail: perry@uni_bonn.de

Powell Mark D.

School of Aquaculture, Tasmanian Aquaculture and Fisheries Institute, University of Tasmania, Locked Bag 1370 Launceston 7250 Tasmania Australia.

E-mail: Mark.Powell@utas.edu.au

Prati Mariangela

Department of Biotechnologies and Molecular Sciences, University of Insubria, via J.H. Dunant 3, 21100, Varese, Italy.

E-mail: mariangela.prati@uninsubria.it

Rantin Francisco Tadeu

Laboratory of Zoophysiology and Comparative Biochemistry, Department of Physiological Sciences, Federal University of São Carlos. Via Washington Luis, km 235, 13565-905 – São Carlos, SP, Brazil.

E-mail: ftrantin@power.ufscar.br

Rios Flavia Sant'Anna

Department of Cell Biology, Federal University of Paraná, Curitiba, PR, Brazil.

E-mail: flaviasrios@ufpr.br

Sakuragui Marise Margareth

Department of Physiological Sciences, Federal University of São Carlos, via Washington Luís km 235, 13565-905 São Carlos, SP, Brazil.

E-mail: msakuragui@yahoo.com

Sanches José Roberto

Department of Physiological Sciences, Federal University of São Carlos, Via Washington Luis, km 235, 13565-905 – São Carlos, SP, Brazil.

E-mail: fjrobs@power.ufscar.br

Sanchez Adriana Paula

Department of Physiology, Faculty of Medicine of Ribeirão Preto, Universidade de São Paulo, Avenida Bandeirantes 3900, 14049-900 Ribeirão Preto, SP, Brazil.

E-mail: drischiageto@yahoo.com.br

Sardella Brian A.

Department of Zoology, University of British Columbia, 6270 University Boulevard, Vancouver BC V6T 1Z4, Canada.

E-mail: sardella@zoology.ubc.ca

Saroglia Marco

Department of Biotechnologies and Molecular Sciences, University of Insubria, via J.H. Dunant 3, 21100 Varese, Italy.

E-mail: marco.saroglia@uninsubria.it

Soncini Roseli

Department of Biological Science, Federal University of Alfenas, Rua Gabriel Monteiro da Silva 714, 37130-000 Alfenas, MG, Brazil.

E-mail: soncinir@yahoo.com.br

Takeda Tatsusuke

Department of Animal and Marine Bioresource Science, Faculty of Agriculture, Kyushu University, Fukuoka 812-8581, Japan.

E-mail: takeda@agr.kyushu-u.ac.jp

Taylor Edwin W.

School of Biosciences, University of Birmingham, Edgbaston, Birmingham B152TT, UK.
E-mail: E.W. TAYLOR@bham.ac.uk

Terova Genciana

Department of Biotechnologies and Molecular Sciences, University of Insubria, via J.H. Dunant 3, 21100 Varese, Italy.
E-mail: genciana.terova@uninsubria.it

Val Adalberto Luís

Laboratory of Ecophysiology and Molecular Evolution, National Institute for Amazon Research, Manaus, Brazil.
E-mail: dalval@inpa.gov.br

Wang Tobias

Department of Zoophysiology, Aarhus University, 8000 Aarhus Denmark.
E-mail: tobias.wang@biol.au.dk

Wegner Nicholas C.

Center for Marine Biotechnology and Biomedicine and Marine Biology Research Division, Scripps Institution of Oceanography, University of California, San Diego, La Jolla, CA 92093-0204, USA.
E-mail: nwerner@ucsd.edu

Yoshida Yu

Institute for East China Sea Research, Nagasaki University, Tairamachi, Nagasaki 851-2213, Japan.
E-mail: mudskipper400@hotmail.com

Adaptation of Gas Exchange Systems in Fish Living in Different Environments

S.M. Kisia* and D.W. Onyango

INTRODUCTION

The survival of fish depends largely on its morphological, physiological and biochemical adaptations to the environment. One important system to the fish's survival is that of gas exchange that should be able to efficiently obtain oxygen from the external environment in order to meet the metabolic needs of fish and eliminate gaseous wastes, chiefly carbon dioxide. The evolution of different structures used in respiration by fish has been taking place for the hundreds of millions of years that fish have been in existence and is likely to continue with changing environmental conditions and metabolic demands of fish. The evolution of diverse respiratory systems has enabled fish survive and reproduce in most of the aquatic habitats on earth. The aquatic environment normally contains a

Authors' address: Department of Veterinary Anatomy, University of Nairobi, P.O. Box 30197, Nairobi, Kenya.
Corresponding author: E-mail: kisiasm@yahoo.co.uk

lower oxygen content than the terrestrial environment. Warmer waters have even less oxygen tension in comparison to cold waters. This is a constraint faced by tropical water fish when compared to those of temperate waters. Hypoxia and anoxia have been common constraints during the evolutionary history of tropical water fishes. Similarly, stagnant waters compared to moving and stirred water will contain less oxygen. The presence of organic matter or pollutants in water can further reduce the quantity of oxygen, thereby creating oxygen-poor zones as bacteria that oxidize the organic matter, in turn, reduce the available oxygen due to their respiratory activity.

To face the challenges of an aquatic environment, fish have to migrate to suitable environments, adjust their metabolic requirements, evolve a system that will enable them to get enough oxygen so as to meet metabolic needs or become endangered. For example, a genus of the South American electric eels, *Electrophorus* has developed a network of blood vessels in its mouth as a result of low concentrations of oxygen in the water they occupy in the Amazon basin. These fish periodically rise to the surface to gulp air into their mouths. The African lungfish (*Protopterus*) has to occasionally gulp air for survival and, during dry periods, it has to cocoon in the mud of dried up river beds, breathing through a small opening that leads to the surface.

Based on the diverse environments they live in, fish have evolved various respiratory structures they use mainly or in combination with others to meet their oxygen requirements. The purpose of this chapter is to examine the various gas exchange structures that have evolved in fish including gills, lungs, swim (gas) bladders and other accessory air-breathing organs and relate their use as respiratory structures to the different environments occupied by fish.

Gills

Gills are the primary organs of respiration in fish. Apart from performing the chief function of gaseous exchange, gills play important roles in osmoregulation and excretion of nitrogenous metabolic wastes. Although external gills are found in the larvae of many fish such as the torpedo (*Torpedo marmorata*) and the spiny dogfish (*Squalus acanthians*), internal gills are found in most fishes. Internal gills have undergone various structural and functional modifications depending on environmental conditions, metabolic activity and dependence of fish on these structures for respiration.

The living agnathans (jawless fish) such as lampreys and hagfishes that suck blood from living fishes or eat the flesh of dead or dying fishes have evolved a pharynx which is divided longitudinally into a dorsal food passage (esophagus) and a ventral blind ending respiratory tube. Internal gill slits arise from the ventral tube so that when the fish are feeding, water is continuously pumped in and out of the branchial pouches. Oxygen then diffuses into the filaments (primary gill lamellae) that line the branchial pouches.

The evolution of a countercurrent system of gas exchange whereby blood in the gill secondary lamellae and water flow in opposite directions enables the fish to obtain the maximal possible amount of oxygen from the water. The amount of oxygen that diffuses into the gills can be as high as 80% of the oxygen dissolved in water ventilating the gills as compared to about 25% of oxygen utilized by mammals from inspired air. Fish spend a lot of energy in ventilating the gills with water and this can account for up to 10 to 25% of their oxygen uptake, when compared to 1-2% spent by man in ventilating the lungs (Hughes, 1965). During ram ventilation in fast-swimming fish such as lamnid sharks and tunas, forward motion (when their mouths are open) is used to enhance water flow across the gill filaments. Most fast-swimming sharks have lost or reduced spiracles, whereas bottom-dwelling skates and rays have enlarged spiracles as most of the water that ventilates the gills enters the pharynx through these structures.

Gill Surface Area and Oxygen Diffusion

The surface area and oxygen diffusion distances of the gill secondary lamellae (gas exchange sites) vary in different fish, depending on their metabolic needs. Active pelagic tuna and menhaden have secondary lamellar surface areas of 3151 mm^2 (Muir and Hughes, 1969) and 1241 mm^2 (Hughes, 1984), respectively, as compared to 18 mm^2 in the much less active coelacanth (*Latimeria chalumnae*) (Hughes, 1980), 117 mm^2 in the torpedo (Hughes, 1978) and 151 mm^2 in the sluggish, bottom-dwelling toadfish (Hughes, 1984). Active fish such as the yellowfin tunny and mackerel have short diffusion distances over their gill respiratory surfac area of 0.17-1.13 μ and 0.60-3.63 μ, respectively (Hughes, 1970). This distance has been estimated at 5-6 μ in the coelacanth (Hughes, 1972).

Since water and salts are lost or gained through the gills depending on the concentration of solutes in the water and body fluids, fish are faced with the problem of ventilating or perfusing their gills to a level that is only necessary to meet their oxygen metabolic requirements. Such a mechanism is applied in fish such as the Lake Magadi tilapia *Alcolapia grahami* (Maina *et al.*, 1996) at the peak of daytime when water is supersaturated with oxygen as compared to almost anoxia later in the night (Narahara *et al.*, 1996). The process of preferential perfusion of gills—hence, partial participation in diffusion—has been noted in other fish (Randall *et al.*, 1972; Booth, 1978; Nilsson, 1986). The surface area of secondary lamellae that is ventilated can also be adjusted by contraction and relaxation of adductor and abductor muscles of the gills (Bijtel, 1949) as well as vascular shunts that divert some blood directly from afarent to efferent branchial arteries. Such blood bypasses the secondary lamellae. When the adductor muscles of a gill contract, water flows between the filaments of adjacent gills thus avoiding the secondary lamellae that are the respiratory gas exchange surface areas. The gill surface area that is exposed to ionic and water gain or loss is thus reduced. Contraction of the abductor and relaxation of the adductor muscles exposes a greater gill surface area to water.

ACCESSORY RESPIRATORY ORGANS

Accessory respiratory organs have enabled bony fishes belonging to more than 50 genera access oxygen directly from the air. These structures could have evolved as a result of the inability of gills to meet the gaseous matabolic demands of fish in water with low oxygen tension such as warm pools, swamps with a lot of decaying matter or stagnant water. Several changes have occurred with transition from aquatic to aerial respiration, including morphological and physiological changes in gas exchange, regulation of ions, excretion of nitrogenous wastes and acid-base balance. An obligate air-breathing fish such as *Arapaima gigas* of river Amazon that undergoes transition from water to air breathing during development and the accompanying changes in gill development offers a good model system in understanding the water to air transition in respiration (Brauner *et al.*, 2004).

Lungs

The three surving genera of the order Dipnoi and freshwater bichirs *Polypterus* (primitive chondrosteans) have developed a pair of long hollow sacs (lungs) in the dorsal part of the abdominal cavity. These lungs originate as ventral evaginations from the floor of the anterior part of the digestive system. The dipnoan lungfishes include *Neoceratodus* of Australia, *Protopterus* of Africa and *Lepidosiren* of South America and live in tropical rivers and ponds. The lungs are thought to have evolved in the Devonian when bodies of water become stagnant and hypoxic and would periodically dry up; hence, fish needed an accessory respiratory organ to supplement the gills in any gaseous exchange. *Protopterus* and *Lepidosiren* can also survive the dry season, during which they undergo aestivation by forming cocoons that are surrounded by mucus, leaving small tubes that lead to the surface of a dried up river bed or pond for breathing. The fish also lower metabolic rates and excrete urea instead of the more toxic ammonia at such times (Mommsen and Walsh, 1989).

Neoceratodus is a facultative air breather that uses its lungs to supplement the gills when the water becomes hypoxic. During hypoxic conditions, the fish shows an increase in gill-breathing frequency, that is followed by an increase in the number of air breaths together with increased pulmonary blood flow (Fritsche *et al.*, 1993) as well as increased haemoglobin oxygen affinity (Kind *et al.*, 2002). *Protopterus* and *Lepidosiren* are obligatory air-breathers due to the thickness of their gill epithelia and reduced gill respiratory surface areas (Laurent, 1996). These fish rely on their lungs for about 90% of their oxygen supply (Johansen, 1970). The internal lung septae are more elaborate in *Protopterus* which has lost its first two gills. The frequency of lung ventilation in lungfishes is low, leading to accumulation of carbon dioxide which is eliminated by gills, since it is extremely soluble in water.

The presence of both branchial and pulmonary circulatory systems in lungfishes differs from that of fish that entirely rely on their gills for respiration and represents a transition from which the tetrapod mode of respiration has evolved. The ductus arteriosus, pulmonary artery vasomotor segments and gill shunts (Fig. 1.1) are useful structures in the preferential perfusion of the lungs or gills in dipnoan fish (Laurent *et al.*, 1978). The ductus arteriosus connects anterior branches of the dorsal aorta with efferent branchial arteries that eventually give rise to pulmonary arteries. Pulmonary artery vasomotor segments are a thickening of the middle layer in the artery starting from the ductus arteriosus caudally over one-third of the artery. Gill shunts are large arteries of arches II and III (lack gills) with thick walls and connect the afferent and efferent arteries. Gill shunts greatly reduce the amount of

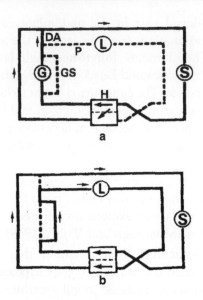

Fig. 1.1 Schematic representation of the ductus arteriosus (DA), pulmonary artery vasomotor segments (P), and gill shunts (GS) when fish are either using their gills or lungs for breathing. In Figure 1.1a, the lungfish is using its gills and in Fig. 1.1b its lungs exclusively for breathing. G—gills, L—lungs, S—systemic circulation, H—heart (lower part—left side and upper part—right side). Continuous line—normal blood flow, broken line—reduced blood flow and arrow—direction of blood flow. Adapted after Laurent (1996).

blood that flows into the gill filaments and the secondary lammellae. During air and aquatic respiration, the pattern of circulation in the lungfish alternates between that of a tetrapod and a fish (Fishman *et al.*, 1985). Figure 1.1 shows how the two systems operate when lungfish use aquatic and aerial respiration. When a lungfish relies on its gills for gaseus exchange exclusively, the pulmonary artery vasomotor segments (P) and the gill shunts (GS) are closed. The ductus arteriosus (DA) opens in the process. Blood is able to flow through the gill filaments and the efferent branchial arteries to the dorsal aorta. The reduction in pulmonary blood flow will result in reduced pulmonary venous return to the heart, which is compensated for by a right-to-left intracardiac shunt-through the incomplete partitioning of the ventricle. During pulmonary respiration, the pulmonary artery vasomotor segments and gill shunts are open, whereas the ductus arteriosus is closed. There is decreased flow of blood to the gill filaments and an increase in flow to the lungs. The right-to-left intracardiac shunt is reduced in the process. Burggren and Johansen (1986) have reviewed the cardiorespiratory morphology and physiology during aerial and aquatic respiration particularly those features that distinguish *Neoceratodus* from both *Protopterus* and *Lepidosiren* and lungfishes in general from aquatic vertebrates.

Swim (Air) Bladder

The swim (air) bladder in most fish is thought to have evolved from the lungs of ancestral bony fishes. Such fish have lived in oxygen-rich waters and conversion of the swimbladder to a hydrostatic organ was favored. Also known as the air bladder or ballast organ, this organ also plays other roles in different fish such as an air-breathing organ, a sound producer and a resonator in sound production. In species where it is used as an air-breathing organ (physostomous species), a pneumatic duct connects the caudal end of the esophagus to the swimbladder. In these species the organ has undergone structural and physiological modification to be used as an air-breathing organ. The organ supplements the gills in their function of gaseous exchange.

Maina *et al.* (1996) found that the swimbladder of the Lake Magadi Tilapia, *Oreochromis alcalicus (Alcolapia) grahami* was in a collapsed state during the day when fish were unstressed. The water is usually well aerated at such times due to photosynthetic activity of *Cyanobacteria,* though it is almost anoxic at night due to respiration by the same bacteria (Narahara *et al.,* 1996; Maina, 2000). The gills are adequate in meeting the oxygen needs of the fish during the daytime. At night, when the oxygen in water is not enough to meet the metabolic needs of the fish, the fish resorts to gulping air from the surface of the water into the buccal cavity and possibly the swimbladder. The swimbladder of the fish is well vascularized with blood capillaries bulging into the air space, exposing them to air. The swimbladder barrier diffusion distance can be as thin as 0.5 μ in the superficial venular capillaries (Maina *et al.,* 1996).

Physoclistous fish lack a pneumatic duct and posses a well-developed gas gland and a rete mirabile that consists of parallel capillaries that are located before and are in contact with the gland. Carbonic acid in the blood and lactic acid secreted by the gas gland favor the release of bound oxygen by hemoglobin that will accumulate in blood and the gas gland until its tension is higher than its partial pressure in the swimbladder. Oxygen will then diffuse from the gas gland into the swimbladder. Since 80% of air in the swimbladder of most fish species is oxygen, the organ can act as a storage organ of the gas (Black, 1940). Oxygen does not diffuse back into the blood through all parts of the swimbladder in these fish (contain a barrier of guanine plates) but by means of the oval which is a vascularized part of the swimbladder. Since the partial pressure of air in the oval is high, it diffuses into blood rapidly.

The rate of adjustments to the swimbladder vary in different fish. For example, physoclists that possess well-developed gas glands and retia mirabilia make great adjustments to the organ in a short time. Fish with poorly developed glandular tissue in the gas gland or lacking a rete mirabile such as the salmonids (Jones and Marshall, 1953) are unable to replace air lost from the swimbladder when denied access to surface air. *A. grahami* (a physostomous fish) has a well-developed rete mirabile (Maina *et al.*, 1996). The fish does not have to come to the surface of water in order to fill its bladder with air under normoxic conditions. The bladder thus also acts as a hydrostatic organ in this fish, which enables the fish to avoid exposing itself to the dangers of predation by birds.

OTHER ACCESSORY RESPIRATORY ORGANS

These structures have developed in the from of extensions of the pharyngeal, branchial or opercular chambers. The pharyngeal chambers have undergone diverse modifications in various species of fish. The air-breathing organs of various species of the snakehead of Asia, *Channa*, have developed on the dorsal part of the pharynx as a pair of extensions known as suprapharyngeal chambers (Hughes and Munshi, 1973; Munshi, 1985; Olson *et al.*, 1994). These chambers participate in both air and at times water breathing. The surface of the chamber has numerous vascular papillae that are also found in the buccopharynx, palate and even the tongue. The dome-shaped papillae (which increase the surface area) have infraepithelial capillaries that run in a spiral or wavelike fashion (Olson *et al.*, 1994). An air sac structure is found along the lateral sides of the head that is covered by opercula in *Monopterus cuchia* (Munshi and Singh, 1968). These air sacs are extensions of the pharynx.

The branchial chamber has undergone modification into air-breathing organs in various species such as the air sac catfish of Asia *Heteropneustes fossilis*, the climbing perch *Anabas testudineus*, *Clarias batrachus* and the African air-breathing catfish, *C. mossambicus*. *H. fossilis* has evolved four pairs of gill fans (formed by fusion of gill filaments) and two air sacs that are extensions of the suprabranchial chambers into the body trunk and embedded in myotomes (Munshi and Choudhary, 1994). The gill fans have lamellae that bear microvilli and the air tubes have many folds and ridges which also bear lamellae. C. *mossambicus* inhabits rivers and swamps in Africa that are liable to drying periodically and has evolved into an obligate air-breather. The accessory air-breathing organs (labyrinthine and suprabranchial chamber) contribute to 85% of the overall oxygen diffusing capacity compared to that of 15% by the gills (Maina and Maloiy, 1986). In the study—although the respiratory surface area of the gills was greater than that of the accessory air-breathing organs—the water-blood barrier diffusion distance was 1.97 µm in the gills and 0.30 µm in the accessory respiratory organs. C. *batrachus* has evolved four pairs of gill fans, two pairs of dendritic organs and a pair of suprabranchial chamber (Munshi, 1961). The climbing perch, A. *testudineus*—that is also an obligate air breather—has labyrinthine organs consisting of many plates covered by the suprabranchial chamber.

The opercular chamber has also evolved into accessory air-breathing organs in certain estuarine fishes of the family Gobiidae, for example, the mudskipper, *Periophthalmus*. The opercular bones are usually elastic with a thin epithelium that is richly vascularized (Munshi, 1985). These fish usually expand their opercular chambers with air. The snakeheads (*Channa*) have developed diverticula of the mouth and pharyngeal cavities which are well vascularized. *Electrophorus*, an air-breathing fish of the Amazon, has a highly vascularized mouth. Parts of the gastro-intestinal tract (GIT) such as the posterior intestine in the catfish *Corydorus aeneus* (Padkowa and Goniakowska-Witalinska, 2002) and modified stomach in the armoured catfish (de Oliveira *et al.*, 2001) have been modified into respiratory organs. Loaches (Cobitidae) use the middle and posterior parts of their intestines for gaseous exchange. The segment of the GIT used in gaseous exchange is normally modified into a thin-walled structure that is highly vascularized. Such fish normally suspend the digestive functions of the GIT and swallow air when using this structure for respiratory purposes. Air is then passed out of the GIT through the mouth or anus.

The skin of some fish functions as an accessory respiratory organ to complement gills and other accessory air-breathing organs as seen in *Saccobranchus* (*Heteropneustes*) *fossilis* (Hughes *et al.*, 1974a) and *Amphipnous* (*Monopterus*) *cuchia* (Hughes *et al.*, 1974b). Such fish normally lack scales on their skin which is highly vascularized. Such skin plays an important role in the elimination of carbon dioxide from the body.

CONCLUSION

Evolution of the fish respiratory system has undergone several changes in extant fish. The changes have enabled fish colonize most bodies of water on earth. Changes in environmental conditions have led to adaptive changes in the respiratory system, favoured colonization of such environments by fish that have adapted or migration of fish to suitable environments in which their gas exchange systems can effectively perform their functions.

Although most fish species rely on their gills entirely for their oxygen needs, some regions of the earth that have experienced hypoxia in water periodically have contributed to the evolution of structures that extract oxygen from the air as the gills are not suited for this function. Such structures supplement the function of gills in respiration to varying degrees and in some fish play the major role of gaseous exchange. Where the gills are not the main site of oxygen uptake in fish, it still plays a major role in elimination of gases such as carbon dioxide that are much more soluble in water than oxygen. The accessory air-breathing organs such as lungs and the swimbladder (as a respiratory organ) are found mainly in tropical fish whose water environment, at higher temperatures and with lower oxygen tensions, sometimes experience hypoxia.

References

Bijtel, H.J. 1949. The structure and the mechanism of the movement of the gill filaments in the Teleostei. *Archives Neerlandaises Zoologie* 8: 1-22.

Black, E.C. 1940. The transport of oxygen by the blood of freshwater fish. *Biological Bulletin,* (Woods Hole) 79: 215-224.

Booth, J.H. 1978. The distribution of blood flow in the gills of fish: application of a new technique to rainbow trout (*Salmo gairdneri*). *Journal of Experimental Biology* 83: 31-39.

Brauner, C.J., V. Matey, J.M. Wilson, N.J. Bernier and A.L. Val. 2004. Transition in organ function during the evolution of air-breathing insights from *Arapaima gigas*, an obligate air-breathing teleost from the Amazon. *Journal of Experimental Biology* 207: 1433-1438.

Burggren, W. and K. Johansen. 1986. Circulation and respiration in lungfishes (Dipnoi). *Journal of Morphology* 1: 217-236.

de Oliveira, C., S.R. Taboga, A.L. Smarra and G.O. Bonilla-Rodriguez. 2001. Microscopical aspects of accessory air breathing through a modified stomach in the armored catfish *Liposarcus anisitsi* (Siluriformes, Loricariidae). *Cytobios* 103: 153-162.

Fishman, A.P., R.G. DeLaney and P. Laurent. 1985. Circulatory adaptation to bimodal respiration in the dipnoan lungfish. *Journal of Applied Physiology* 59: 285-294.

Fritsche, R., M. Axelsson, C.E. Franklin, G.G. Grigg, S. Holmgren and S. Nilsson. 1993. Respiratory and cardiovascular responses to hypoxia in the Australian lungfish. *Respiration Physiology* 94: 173-187.

Hughes, G.M. 1965. *Comparative Physiology of Vertebrate Respiration*. Heinemann: London.

Hughes, G.M. 1970. Morphological measurements on the gills of fishes in relation to their respiratory function. *Folia Morphologia*, (Prague) 18: 78-95.

Hughes, G.M. 1972. Aspects of the respiration of *Latimeria chalumnae* and a comparison with the gills of associated fishes. *International Congress of Physiological Sciences, Sydney*, No. 36.

Hughes, G.M. 1978. On the respiration of *Torpedo marmorata*. *Journal of Experimental Biology* 73: 85-105.

Hughes, G.M. 1980. Ultrastructure and morphometry of the gills of *Latimeria chalumnae* and a comparison with the gills of associated fishes. *Proceedings of the Royal Society of London* B208: 309-328.

Hughes, G.M. 1984. General Anatomy of the gills. In: *Fish Physiology*, W.S. Hoar and D.J. Randall (eds.), Academic Press, New York, Vol. 10, pp. 1-72.

Hughes, G.M. and J.S.D. Munshi. 1973. Nature of the air-breathing organs of the Indian fishes, *Channa*, *Amphipnous*, *Clarias* and *Saccobranchus* as shown by electron microscopy. *Journal of Zoology* (London) 170: 245-270.

Hughes, G.M., B.R. Singh, G. Guha, S.C. Dube and J.S.D. Munshi. 1974a. Respiratory surface area of an air-breathing siluroid fish, *Saccobranchus* (=*Heteropneustes*) *fossilis* in relation to body size. *Journal of Zoology* (London) 172: 215-232.

Hughes, G.M., B.R. Singh, R.N. Thakur and J.S.D. Munshi. 1974b. Areas of the air-breathing surfaces of *Amphipnous cuchia* (Ham.). *Proceedings of the Indian National Science Academy* B40: 379-392.

Johansen, K. 1970. Air-breathing in fishes: In: *Fish Physiology*, W.S. Hoar and D.J. Randall (eds.). Academic Press, New York, Vol. 4, pp. 361-411.

Jones, F.R.H. and N.B. Marshall. 1953. The structure and functions of the teleostean swimbladder. *Biological Reviews* 28: 16-83.

Kind, P.K., G.C. Grigg and D.T. Booth. 2002. Physiological responses to prolonged aquatic hypoxia in the Queensland lungfish *Neoceratodus fosteri*. *Respiration Physiology and Neurobiology* 132: 179-190.

Laurent, P. 1996. Vascular organization of lungfish, a Landmark in ontogeny and phylogeny of air-breathers. In: *Fish Morphology, Horizons of New Research*, J.S. Datta Munshi and H.M. Dutta (eds.). Science Publishers, Enfield (NH), USA, pp. 47-56.

Laurent, P., R.G. Dehaney and A.P. Fishman. 1978. The vasculature of the gills in the aquatic and aestivating lungfish (*Protopterus aethiopicus*). *Journal of Morphology* 156: 173-208.

Maina, J.N. 2000. Functional morphology of the gas-gland cells of the air-bladder of *Oreochromis alcalicus grahami* (Teleostei: Cichlidae): An ultrastructural study of a fish adapted to a severe, highly alkaline environment. *Tissue and Cell* 32: 117-132.

Maina, J.N. and G.M.O. Maloiy. 1986. The morphology of the respiratory organs of the African air-breathing catfish (*Clarias mossambicus*): A light, election and scanning microscope study, with morphometric observations. *Journal of Zoology* (London) 209: 421-445.

Maina, J.N., C.M. Wood, A. Narahara, H.L. Bergman, P. Laurent and P. J. Walsh. 1996. Morphology of the swim (air) bladder of a cichlid teleost: *Oreochromis alcalicus grahami* (Trewavas, 1983), a fish adapted to a hyperosmotic, alkaline and hypoxic environment: A brief outline of the structure and function of the swimbladder. In: *Fish Morphology, Horizons of New Research*, J.S. Datta Munshi and H.M. Dutta (eds.). Science Publishers, Inc. Enfield (NH), USA, pp. 179-192.

Mommsen, T.P. and P.J. Walsh. 1989. Evolution of urea synthesis in vertebrates: the piscine connection. *Science* 243: 72-75.

Muir, B.S. and G.M. Hughes. 1969. Gill dimensions for three species of tunny. *Journal of Experimental Biology* 51: 271-285.

Munshi, J.S.D. 1961. The accessory respiratory organs of *Clarias batrachus* (Linn.). *Journal of Morphology* 109: 115-139.

Munshi, J.S.D. 1985. The structure, function and evolution of the accessory respiratory organs of air-breathing fishes of India. In: *Fortschritte der Zoologie: Functional Morphology in Vertebrates*, H.R. Dunker and G. Fleischer (eds.). Gustav Fischer-Verlag, Stuttgart, pp. 353-366

Munshi, J.S.D. and S. Choudhary. 1994. Ecology of *Heteropneustes fossilis* (Bloch): An air-breathing catfish of Southeast Asia. Monograph series. Freshwater Biological Association of India, Bhagalpur.

Munshi, J.S.D. and B.N. Singh. 1968. On the respiratory organs of *Amphipnous cuchia* (Ham.). *Journal of Morphology* 124: 423-444.

Narahara, A.B., H.L. Bergman, P. Laurent, J.N. Maina, P.J. Walsh and C.M. Wood. 1996. Respiratory physiology of the lake Magadi tilapia (*Oreochromis alcalicus grahami*), a fish adapted to a hot, alkaline and frequently hypoxic environment. *Physiological Zoology* 69: 1114-1136.

Nilsson, S. 1986. Conrol of gill blood flow. In: *Fish Physiology: Recent Advances*, S. Nilsson and S. Holmgren (eds.). Croom Helm, London, pp. 87-101.

Olson, K.R., P.K. Roy, T.K. Ghosh and J.S.D. Munshi. 1994. Microcirculation of gills and accessory respiratory organs from the air-breathing snakehead fish, *Channa punctata, C. gachua* and *C. marulius*. *Anatomical Record* 238: 92-107.

Padkowa, D. and L. Goniakowska-Witalinska. 2002. Adaptations to the air breathing in the posterior intestine of the catfish (*Corydoras aeneus*, Callichthyidae). A histological and ultrastructural study. *Folia Biologica* (Krakow) 50: 69-82.

Randall, D.J., D. Baumgarten and M. Malyusz. 1972. The relationship between gas transfer across the gills of fishes. *Comparative Biochemistry and Physiology* A 41: 629-637.

Morpho-physiological Divergence Across Aquatic Oxygen Gradients in Fishes

Lauren J. Chapman

INTRODUCTION

A central theme in ecology is understanding the patterns of distribution and abundance in organisms, the selective pressures underlying these patterns, and their evolutionary consequences. The abiotic environment has a major influence on the ecology of organisms; and for fishes, the availability of dissolved oxygen (DO) is one factor that can limit habitat quality and dispersal pathways (Kramer, 1983; Saint-Paul and Soares, 1987; Chapman and Liem, 1995; McKinsey and Chapman, 1998; Chapman *et al.*, 1999). All fish require oxygen for long-term survival. However, the physical properties of water can make oxygen uptake a challenge for non-air-breathing fishes even at high DO levels. In addition, without adequate mixing of water or light for photosynthetic oxygen

Author's address: Department of Biology, McGill University, 1205 Avenue Docteur Penfield, Montreal, PQ, Canada H3A 1B1 & Wildlife Conservation Society, 2300 Southern Boulevard, Bronx, New York, USA 10460.
E-mail: Lauren.chapman@mcgill.ca

production, respiration of inhabitants can deplete a water body of dissolved oxygen. Oxygen scarcity (hypoxia) occurs naturally in waters characterized by low light and low mixing (e.g., heavily vegetated swamps and flooded forests). This is particularly common in (but not limited to) tropical waters where high temperatures elevate decomposition rates and reduce oxygen tensions. Unfortunately, environmental degradation is increasing the occurrence of hypoxia as the influx of municipal wastes and fertilizer runoff accelerate eutrophication and pollution of water bodies (Prepas and Charette, 2003).

Fishes have evolved a variety of solutions to hypoxia including, as examples, the development of air-breathing organs, large gill surface area, change in oxygen carrying capacity of the blood, anaerobic metabolism, metabolic depression, and morphological specializations for exploitation of the oxygen-rich surface layer (Lewis, 1970; Galis and Barel, 1980; Liem, 1980; Hochachka, 1982; Kramer, 1983, 1987; Perry and McDonald, 1993; Chapman *et al.*, 1995; Graham, 1997). Although a great deal of effort has been directed towards describing the physiological, biochemical, and morphological adaptations of fishes to deoxygenation, the role of dissolved oxygen in the maintenance of fish faunal structure and diversity remains relatively unexplored. Hypoxic waters may act as barriers to fish movement, biofilters, refugia, or dispersal corridors, depending on the relative tolerance of fishes. Given the widespread occurrence of oxygen scarcity in aquatic systems and increasing levels of hypoxia associated with anthropogenic influence, microevolution in response to hypoxic stress may be a frequent phenomenon in nature and play an important role in evolutionary diversification.

There is a rich body of literature on intraspecific variation in fishes, much of which has focused on ecomorphological or life-history traits (Robinson and Wilson, 1994; Smith and Skulason, 1996; Robinson and Parsons, 2003) through such comparisons as benthic, limnetic, and littoral habitats, hard and soft prey, lotic and lentic systems, and the presence or absence of predators. Our studies of East African fishes have demonstrated that alternative dissolved oxygen environments provide another strong predictor of intraspecific variation, particularly with respect to respiratory traits (e.g., gill size) and associated characters. The objective of this chapter is to review the existing patterns of morpho-physiological divergence across oxygen gradients in fishes and to highlight the value of aquatic oxygen gradients as novel systems to study mechanisms generating and maintaining phenotypic diversity. To meet this goal, I shall: (a) describe the patterns of interdemic variation in respiratory traits of fishes with a focus on gill morphology; (b) discuss the potential role of environmental effects on interdemic variation in gill traits across oxygen gradients; and (c) explore performance trade-offs that may contribute to the maintenance of respiratory phenotypes in the field. I shall focus on our studies of East African fishes and expand the geographic and phylogenetic breadth through existing literature data.

INTERDEMIC VARIATION IN GILL MORPHOMETRICS

Hypoxia and Non-air-breathing Fishes in African Wetlands

In East Africa, hypoxia is prevalent in the extensive wetlands dominated by papyrus (*Cyperus papyrus*) and *Miscanthidium violaceum* (Carter, 1955; Beadle, 1981; Chapman et al., 1998). For example, in the Rwembaita Swamp of Kibale National Park of Uganda, DO levels in the dense papyrus averaged only 1.2 mg l^{-1} over a one-year study and were less than 0.7 mg l^{-1} for most of the year (Chapman et al., 1998). Carter (1955) reported DO values averaging less than 0.1 mg l^{-1} for the near shore areas of littoral papyrus swamps in Lake Victoria, and average values of 2.5 mg l^{-1} for the interface between papyrus and pelagic waters. Similarly, in the dense papyrus at the mouth of the Chambura River in Uganda, Beadle (1932) found no detectable oxygen within a few centimeters of the mat surface.

The few accounts of fish faunas in African swamps include a number of air breathers (e.g., *Protopterus aethiopicus*, *Clarias* spp., *Ctenopoma muriei*; Carter, 1955; Beadle, 1981; Chapman, 1995, Chapman et al., 1996a,b, 2002). However, some non-air-breathing fishes also occur in these dense swamps including species of cichlids, cyprinids, killifishes, and mormyrids (Chapman et al., 1996a,b, 2002). Several of these species have broad habitat ranges that include hypoxic wetland waters as also well-oxygenated open waters of lakes and rivers. This provides an excellent system for exploring the role of dissolved oxygen as a divergent selective factor contributing to phenotypic diversity in respiratory traits.

Gill Morphometry

Several studies based on interspecific comparisons have suggested that large gill respiratory surface in non-air-breathing fishes may relate to hypoxic conditions in their environment (Gibbs and Hurwitz, 1967; Galis and Barel, 1980; Fernandes *et al.*, 1994; Mazon *et al.*, 1998). It is of interest, therefore, to look for similar patterns of variation within species across dissolved oxygen gradients. We began our study of interdemic variation by quantifying the relationship between total gill filament length and dissolved oxygen (DO) concentration for the cyprinid *Barbus neumayeri* from six sites in the Mpanga River drainage of western Uganda. These sites varied from the dense interior of a papyrus swamp—where dissolved oxygen averaged 1.2 mg l^{-1} over 3 years—to the well-oxygenated waters of an ever flowing river where DO averaged 7.4 mg l^{-1} (Chapman *et al.*, 1999). The morphological parameters most easily and accurately measured for large numbers of small fish are those related to gill filament length. Total gill filament length was measured for 10 fishes from each of the six populations using standard methods (Muir and Hughes, 1969; Hughes, 1984a; Chapman and Liem, 1995; Chapman *et al.*, 1999). Although the total gill surface area is a better indicator of oxygen uptake capacity, the total gill filament length is generally correlated with the area of respiratory surface across a range of freshwater fishes (Palzenberger and Pohla, 1992), and we assumed that a longer total gill filament length in one population as compared to another of the same species reflects a greater capacity to extract oxygen from the water. Across six populations of *B. neumayeri*, we found that total gill filament length increased with a corresponding decrease of DO (Chapman *et al.*, 1999). In a more recent investigation, we quantified variation in the total gill filament across 10 populations of *B. neumayeri* to increase the rigor of our earlier analyses. DO varied from a mean monthly average of 1.4 mg l^{-1} at a site in the dense interior of a papyrus swamp to 7.3 mg l^{-1} at a fast-flowing river site. All sites were within the Mpanga River drainage of western Uganda with the exception of one fast-flowing site from the adjacent Dura River drainage. Again, DO was a very significant predictor of log_{10} total gill filament length across populations (F=11.59, P=0.009), explaining 59% percent of the variation in this character (Fig. 2.1, Chapman, DeWitt, and Langerhans, unpublished data). In a stepwise regression that included dissolved oxygen, water current, water depth, and water temperature, DO was the only significant predictor of total gill filament length. These studies of *B. neumayeri* indicate significant variation in a morpho-respiratory trait over a small geographical scale that correlates with DO availability.

To increase the phylogenetic breadth of this analysis, we have measured gill metrics on four additional species of non-air breathers. All

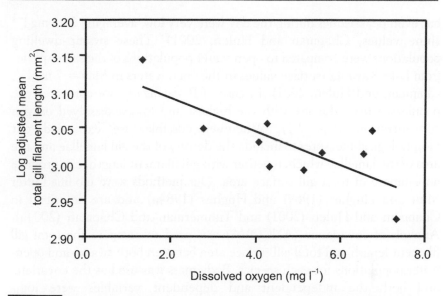

Fig. 2.1 Relationship between mean mass-adjusted total gill filament length ($\log_{10}$ transformed) and mean dissolved oxygen concentration of the environment for 10 populations of *Barbus neumayeri* from the Mpanga and Dura River drainages of western Uganda. Data were adjusted to a mean body mass of 3.5 g. Sites ranged from the dense interior of a papyrus swamp to well-oxygenated ever flowing river habitats. (Chapman, DeWitt and Langerhans, unpubl. data).

species were from East Africa, with the exception of *Poecilia latipinna* (F. Poeciliidae, sailfin molly) that was collected from a periodically hypoxic salt marsh and a well-oxygenated river site in Florida (Timmerman and Chapman, 2004a). At the Florida marsh site, dissolved oxygen levels were highly variable ($0.2–10.8$ mg l^{-1}); at the nearby river site, DO levels were high and showed little variation (mean$=6.8$ mg l^{-1}, Timmerman and Chapman, 2004a). The African cichlid *Pseudocrenilabrus multicolor victoriae* was collected from the dense interior of a papyrus swamp surrounding Lake Manywa, Uganda where DO levels were extremely low (0.4 mg l^{-1} at the time of fish collection, Chapman *et al.*, 2000). This population was compared to *P. multicolor* collected from well-oxygenated ecotonal waters of nearby Lake Kayanja, where DO at the collection site averaged 6.1 mg l^{-1} over the period of one year (Chapman *et al.*, 2000). Two mormyrid species (*Petrocephalus catostoma* and *Gnathonemus victoriae*) were collected from hypoxic lagoons in the Lwamunda Swamp that surrounds Lake Nabugabo, Uganda, a wetland dominated by the grass *Miscanthidium violaceum*. At this site, DO showed nocturnal reduction, but even peak values during the day were very low, averaging 0.93 mg l^{-1} (June values, Chapman and Hulen, 2001). These swamp-dwelling populations were compared to open-water populations of the same species from Lake Kayanja (surface values in the open waters in May = 7 mg l^{-1}, Chapman and Hulen, 2001). In case of B. *neumayeri*, we compared the population from the site with the highest and lowest dissolved oxygen concentration (Fig. 2.1). Wherever possible, we expanded our morphological measures to include the density of the gill lamellae and the area of the lamellae which, together with gill filament length, can produce an estimate of total gill surface area. Our methods were modified after Muir and Hughes (1969) and Hughes (1984a) and are described in Chapman and Hulen (2001) and Timmerman and Chapman (2004a). Analysis of covariance (ANCOVA) was used to compare the total gill filament length and total gill surface area between both swamp and open-water populations for each species. Body mass was used as the covariate, and both the independent and dependent variables were log_{10} transformed. Total gill surface area was significantly larger in wetland-dwelling populations of the African mormyrid *Gnathonemus victoriae* (Chapman and Hulen, 2001), the African cichlid *Pseudocrenilabrus multicolor* (Chapman et al., 2000), and the North American poeciliid *Poecilia latipinna* (Timmerman and Chapman, 2004a) relative to open-water populations (Fig. 2.2). In case of *Barbus neumayeri* and the mormyrid *Petrocephalus catostoma*, we did not determine total gill surface area. However, total gill filament length was significantly larger in the wetland-dwelling population of each species (B. *neumayeri* – slopes: F=0.527, P=0.482, intercepts: F=91.107, P<0.001; P. *catostoma* – slopes: F=1.98, P=0.190, intercepts: F=13.55, P=0.004).

Palzenberger and Pohla (1992) reviewed the existing literature on gill morphometry of fishes. From their data set for 28 non-air-breathing freshwater species (with multiple populations for eight species), they extracted the mean slope of significant regressions for gill morphometric parameters and body weight. They set the lowest and highest mean values within each parameter range to 0% and 100% respectively, in order to create a range of values for each gill character. This permitted them to express the values of a species as a percentage within the range of values for freshwater fishes. We used their parameter estimates for total gill surface area to estimate this character for each population expressed as a percentage of freshwater fishes. The total gill surface area of wetland-dwelling populations, expressed as a range of freshwater fishes, averaged

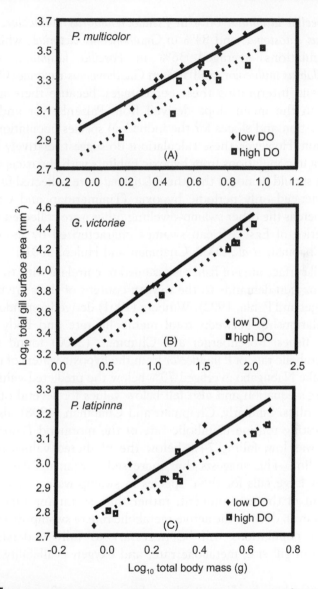

Fig. 2.2 Bilogarithmic relationship between total gill surface area (mm²) and body mass (g) for three species of teleost fishes collected from a hypoxic swamp or marsh environment and well-oxygenated lake or river habitat. **(A)** *Pseudocrenilabrus multicolor*: slopes – F=0.149, P=0.703; intercepts – F=149.871, P<0.001. Adapted from Chapman *et al.*, (2000), **(B)** *Gnathonemus victoriae*: slopes – F=0.510, P=0.491. intercepts – F=9.123, P=0.012. Adapted from Chapman and Hulen, (2001) and **(C)** *Poecilia latipinna*: slopes – F=0.436, P=0.522, intercepts – F=5.653, P=0.033. Adapted from Timmerman and Chapman (2004).

44% in *Poecilia latipinna*, 66% in *Pseudocrenilabrus multicolor*, 74% in *Petrocephalus catostoma*, and 84% in *Gnathonemus victoriae*; while open-water populations averaged 36% in *Poecilia latipinna*, 45% in *Pseudocrenilabrus multicolor*, and 56% in *Gnathonemus victoriae*. One must be cautious in interpreting these percentages, because there may be a difference in the mean slope derived from Palzenberger and Pohla's general equation and slopes for the individual species/populations under consideration. However, these calculations do suggest relatively large gill surface area in populations from hypoxic habitats with the exception of *P. latipinna*. It should be noted that the *P. latipinna* were collected from a site that experienced only periodic hypoxia (Timmerman and Chapman, 2004a), whereas the other swamp-dwelling fishes were collected from the dense interior of East African swamps characterized by more severe hypoxia (Chapman *et al.*, 2000; Chapman and Hulen, 2001).

The gill surface area of fishes is assumed to correlate with the ratio of metabolic oxygen demands to the oxygen content of their environment (Palzenberger and Pohla, 1992). Winberg (1961) derived a standard curve for the relationship between total metabolic rate and body size for freshwater fishes. Rosenberger and Chapman (2000) found that the routine metabolic rate of *P. multicolor* from the hypoxic waters of the Juma River of Lake Nabugabo averaged 70% below the predicted values based on Winberg's equation and also fell below values for several other East African cichlids. Similarly, Chapman and Chapman (1998) also found that the resting routine metabolic rate of the mormyrid *Petrocephalous catastoma* was low, falling 74% below the predicted values based on Winberg's line. This suggests that increased oxygen uptake efficiency afforded by large gills for these species in swamp waters may reflect a requirement of the environment, rather than sustaining high activity levels. However, data on the active metabolic rate of swamp-dwelling and open-water populations will be necessary to fully understand the interaction of gill size, metabolic rate, and oxygen availability in these species.

Although there has been much work on interspecific variation in gill size among species that differ in ecological context, studies of interdemic variation in gill metrics across oxygen gradients in the field are rare. However, a recent study by Bouton and colleagues (2002) correlates the head shape for six species (one to three populations) of rock-dwelling haplochromine cichlids to a series of ecological variables. They estimated the volumes of three compartments of the head (oral, suspensorial, and opercular). Although their analyses combined population-level and species-level variation using principal component analysis (PCA), they do provide evidence for interdemic variation in opercular volume that relates to dissolved oxygen concentration. Their first principal component (PC1) divided populations mainly according to geographic origins, and they found that DO levels explained the largest part of the variation in the PC1 and the largest part of the variation in the five measures of the opercular compartment that correlate highly with PC1. They further suggested that gill volume varies with oxygen levels and influences the outer head shape.

Variation in gill metrics has also been observed between historical and contemporary populations of fishes and is related to changes in oxygen levels of their habitat. In Lake Victoria, East Africa, dramatic changes in the fish fauna coincided with an explosive increase in the population of introduced Nile perch in the 1980s. Most notably, over 50% of the non-littoral haplochromines—or about 40% of the endemic haplochromine cichlid assemblage—in the lake disappeared (Kaufman, 1992; Witte et al., 1992; Seehausen et al., 1997). Predation by the Nile perch is thought to have been a major cause of the faunal collapse, although other factors such as eutrophication and deoxygenation of the deeper waters of the lake (Hecky, 1993; Hecky et al., 1994), may have also played a role (Kaufman and Ochumba, 1993; Balirwa et al., 2003). Despite dramatic declines in populations of many native fishes, some species have persisted with Nile perch. In contrast to the cichlids, the zooplanktivorous cyprinid *Rastrineobola argentea* strongly increased in numbers, despite being prey for Nile perch (Wanink, 1999; Wanink and Witte, 2000). This increase has been attributed to a relaxation of competitive pressure by haplochromines subsequent to the Nile perch boom (Wanink and Witte, 2000), during which time, *R. argentea* began to explore the bottom zone of the lake during the daytime (a habitat previously occupied by haplochromines) and include benthic macroinvertebrates in its diet. Wanink and Witte (2000) found evidence for rapid morphological change in *R. argentea* between samples collected in 1988, one year after bottom dwelling became common, and samples collected in 1983 prior to the habitat shift. With respect to gill metrics, *R. argentea* collected in 1988 were characterized by a higher number of gill filaments than the conspecifics collected in 1983. Wanink and Witte postulate that this change may have improved oxygen uptake capacity that would allow exploitation of hypoxic deeper waters. Preliminary observations also revealed an ~25% increase in the average number of secondary gill lamellae in the zooplanktivorous *Yssichromis pyrrhocephalus*, a haplochromine cichlid from Lake Victoria (Witte *et al.*, 2000). It was speculated that this may be a response to increased hypoxia in the lake (Witte *et al.*, 2000), but it is as yet unknown whether this is due to heritable response to selection or due to environmentally induced phenotypic plasticity. Schaack and Chapman (2003) also reported a change in the gill size of swamp-dwelling *Barbus neumayeri* between samples collected in the early 1990s and samples collected several years later in the late 1990s. The findings suggested that this may reflect the differences in the developmental history and selection environment experienced by fish during the years previous to each study (e.g., higher levels of rainfall and DO during the later part of the decade).

Shared and Unique Features of Gill Proliferation

When multiple species of fishes face a similar environmental gradient, their patterns of divergence might exhibit both shared and unique features (Langerhans and DeWitt, 2004). Although our list of species is still modest, we can begin to look at this question. Future studies that quantify the response of replicate populations for multiple species across DO gradients will be an important follow up to these initial observations. There are many ways in which fish might increase the surface area of the gill, irrespective of lamellar perfusion. However, there exist three basic strategies that can be easily compared among species. The fish can increase the length or number of gill filaments to increase its total gill filament length; it can increase the density of lamellae along the filaments; or it can increase the area of the individual lamellae. A comparison of the gill morphometrics across low- and high-oxygen populations for our focal species suggests both shared and unique elements of diversification in gill characters (Table 2.1). In four of the five species for which data are available on interdemic variation in gill metrics, gill filament length was larger in the wetland-dwelling fish than in open-water conspecifics. In two of the three populations for which we were able to measure lamellar area, it was also larger in wetland fishes; however, lamellar density, or the packing of lamellae, was lower in the swamp-dwelling populations or showed no significant difference.

In their review of gill morphometrics, Palzenberger and Pohla (1992) reported a positive relationship between the percent gill lamellar density and that of total gill surface area (where each character was expressed as a percentage of the range of freshwater fishes) and a positive relationship

Table 2.1   Shared and unique elements of gill morphometric variation between wetland (W) and open-water (O) populations of five species of non-air-breathing fishes. The wetland sites were characterized by chronic or periodic hypoxia. W=O indicates no significant difference between wetland and open-water fish; W>O indicates that the trait was larger in wetland fish; and W<O indicates that the trait was smaller in the wetland fish. (—— no data).

Family	Species	Location	Total gill Filament length	Lamellar density	Lamellar area
Mormyridae	*Gnathonemus victoriae*	Uganda	W=O	W=O	W>O
	Petrocephalus catostoma	Uganda	W>O	----	----
Cichlidae	*Pseudocrenilabrus multicolor*	Uganda	W>O	W<O	W>O
Cyprinidae	*Barbus neumayeri*	Uganda	W>O	W<O	----
Poeciliidae	*Poecilia latipinna*	Florida	W>O	W=O	W=O

between the percent total gill filament length and that of total gill surface area. In general, for the suite of species considered, a large total gill filament length tended to be combined with dense spacing of the lamellae and with smaller lamellar area. Hughes (1966) argued that for highly active species with high oxygen-uptake demands, the best strategy for increasing the gill surface area without compromising the streamline of the head would be to have densely packed, small lamellae, on longer filaments. According to Hughes, increasing gill surface area by increasing the length of the filaments would minimize resistance to flow. Of course, such an increase would require greater space in the head, which may constrain other non-functionally related characters (see trade-offs between respiratory and trophic structures below). This idea was supported by data on fish with active life styles and high gill surfaces area like tuna that use ram ventilation and may require rigid lamellae (Hughes, 1966; Muir and Hughes, 1969; Hughes and Morgan, 1973).

In our comparisons of swamp-dwelling and open-water fish populations, we find that the most consistent interdemic trend in response to hypoxia is an increase in the total gill filament length. With respect to lamellar density, interpopulational variation that occurs across oxygen gradients does not support the general interspecific trend reported by Palzenberger and Pohla (i.e., a positive correlation between lamellar density and gill surface area). Swamp-dwelling populations that we have considered to date either show no increase or a decrease in lamellar packing. However, these swamp-dwelling species tend to have low resting routine metabolic rates (Chapman and Chapman, 1998; Rosenberger and Chapman, 2000; Timmerman and Chapman, 2004b), and rely primarily on buccal pumping rather than ram ventilation. So, it is possible that greater spacing of lamellae in these ecotypes may minimize the resistance to water flow through the gill sieve. Hughes (1984b) noted that sluggish fish expected to have low rates of oxygen consumption exhibit lamellae that are relatively large in area and characterized by wider spacing than other species. Galis and Barel (1980) compared the gill morphology of the gills of more than 80 species of African lacustrine cichlids. They found that, in general, closely spaced lamellae were related to the fast flow of water over the gills and suggested that three conditions could lead to increase water flow: an active pelagic life style, hypoxic water due to the need to increase ventilation frequency and water flow over the gills, and a limited space for the gill apparatus that may be compensated for by increased water flow. Galis and Barel found that cichlids that frequent oxygen-poor waters were generally characterized by a high lamellar frequency. However, they did report one exception, *Limnochromis permaxillaris*, a zooplanktivore from Lake Tanganyika. This species was characterized by a lower lamellar density than other cichlids known to occur regularly in oxygen-poor waters. They noted that their feeding mode is likely to preclude fast flow of water over the gills, which may favor a lower lamellar density to minimize resistance.

In the future, studies of interdemic variation that include measurement of morphometric diffusing capacity (Kisia and Hughes, 1992), which is a good metric for summarizing the overall adaptation of a gas exchange surface, will be very useful in more fully understanding the response of species to divergent aquatic oxygen environments.

Variation in other Respiratory Traits

The variation in gill morphometrics across oxygen gradients is accompanied by variations in other physiological and biochemical characters. Martinez *et al.* (2004) evaluated whether *B. neumayeri* from low and high oxygen environments in the Kibale National Park, Uganda differed in traits related to aerobic and anaerobic metabolic potential. Hematocrit was measured as an index of blood oxygen-carrying capacity, and tissue activities and isozyme composition of lactate dehydrogenase (LDH) were measured as indices of tissue anaerobic capacity. They measured these traits in fish both sampled shortly after collection and after several months in the lab held under normoxic conditions. They found that *B. neumayeri* collected from the hypoxic swamp site had higher hematocrit (Fig. 2.3A), but this difference disappeared after long-term normoxia exposure. Thus, the higher hematocrit of swamp-dwelling *B. neumayeri* appeared to be an acute compensatory response to environmental hypoxia. An increase in hematocrit is a common feature of the short-term hypoxic response in a variety of fishes (Jensen *et al.*, 1993; Gallaugher and Farrell, 1998).

In contrast to the hematocrit response, Martinez *et al.* (2004) found that *B. neumayeri* from the hypoxic swamp site had higher liver LDH activities than fish from the high-oxygen site (Fig. 2.3B), and this difference persisted after long-term normoxia exposure, suggesting that higher LDH activity is not a temporary adjustment to the hypoxic conditions of the swamp habitat, but may represent environmentally induced developmental response or genetic differentiation between sites.

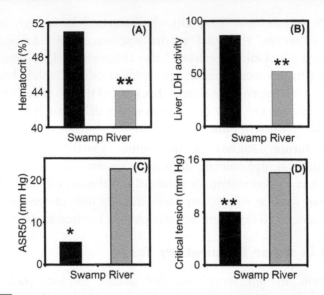

Fig. 2.3   Mean values for **(A)** hematocrit (%, adapted from Martinez *et al.*, 2004), **(B)** liver LDH activity (U g⁻¹ wet mass. Adapted from Martinez *et al.* (2004), **(C)** oxygen tension (mm Hg) at which fish spend 50% of their time skimming the surface film (aquatic surface respiration, ASR). Adapted from Olowo and Chapman (1996) and **(D)** critical oxygen tension (mm Hg; L. Chapman, *unpubl. data*) for *Barbus neumayeri* from a river and swamp site. * — different at P<0.05, ** — different at P<0.001.

They also detected polymorphism at both the LDH-A and LDH-B loci, and found that genotype frequencies for LDH-B differed significantly between collection sites, supporting the notion that there is limited gene flow between these populations.

RESPIRATORY PERFORMANCE ACROSS OXYGEN GRADIENTS

If the observed variation in respiratory traits across oxygen gradients is adaptive, we should see higher respiratory performance in swamp-dwelling fish. Although we have not demonstrated a direct link between larger gill size in swamp-populations and higher oxygen uptake capacity, we have carried out a series studies on the respiratory performance of large- and small-gilled conspecifics from field populations. Two integrative measures of hypoxia tolerance include the threshold for aquatic surface respiration (ASR) and critical oxygen tension. ASR thresholds can be measured by exposing fish to progressive hypoxia and estimating the oxygen partial pressure at which 10% (ASR_{10}), 50% (ASR_{50}), and/or 90% (ASR_{90}) of the time is spent skimming water from the surface film (Chapman *et al.*, 1995; Rosenberger and Chapman, 2000). Critical oxygen tension (P_c) is the oxygen partial pressure where metabolic rate becomes dependent on oxygen concentration (Ultsch *et al.*, 1978, 1980). For *B. neumayeri*, swamp-dwelling fish characterized by a large total gill filament length exhibited a lower frequency of ASR (Olowo and Chapman, 1996) and a lower critical oxygen tension than conspecifics from well-oxygenated waters (Fig. 2.3C,D). Timmerman and Chapman (2004a) reported a lower critical oxygen tension and lower ASR thresholds in sailfin mollies from a marsh population characterized by a relatively large gill surface area (see Fig. 2.3) than in conspecifics from a nearby river population characterized by a smaller gill surface area. Similarly, *P. multicolor* from a swamp-dwelling population with a large gill surface area exhibited a lower critical oxygen tension than conspecifics from a well-oxygenated lake site (Chapman *et al.*, 2002).

It is possible that gill proliferation in response to hypoxic stress may also have negative effects. Comparing two populations of *Gambusia affinis* known to differ in pesticide resistance, McCorkle *et al.* (1979) found that the pesticide resistant population was more vulnerable to hypoxic stress. It is possible that traits leading to superior tolerance to hypoxic stress— such as an increase in the surface area of the gill—increase pesticide susceptibility due to increased uptake, thus selecting against more hypoxia-tolerant individuals.

DEVELOPMENTAL PLASTICITY IN FISH GILLS IN RESPONSE TO DIVERGENT OXYGEN ENVIRONMENTS

Interdemic variation in respiratory traits could be genetically based, fixed by environmental pressures at a critical period of ontogeny, and/or simply be a phenotypic response that remains labile throughout an individual's lifetime. The degree to which interdemic variation in gill morphometry represents phenotypic plastic and/or genetic differences in fishes remains unknown. However, proliferation of gills in response to rearing under hypoxia does occur. In a study of the cichlid *P. multicolor*, we compared the gill size of a population from a stable hypoxic habitat with one of a stable well-oxygenated habitat (Chapman *et al.*, 2000; Fig. 2.2A). In addition, we compared siblings (split-brood) raised under hypoxic or well-oxygenated circumstances. The response to hypoxia was an increase in gill area, both in the field (29%, Fig. 2.2A) and in the plasticity experiment (18%,

Fig. 2.4A). In the field, the increase was due to longer filaments and larger secondary lamellae. In the experiment, the increase was due to more and longer filaments (Chapman *et al.*, 2000). We suggest that the differences in the response to hypoxia between field and experimental fishes may be due to both differences in selection pressures between populations, and a combination of inherited changes and plasticity that may allow for a finer tuned response to severe hypoxic conditions in the field population (Chapman *et al.*, 2000; also see functional morphological trade-offs below).

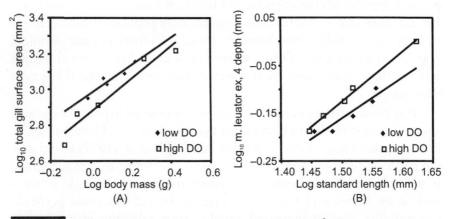

Fig. 2.4   Bilogarithmic plots of **(A)** of total gill surface area (mm²) and body mass (g) for *Pseudocrenilabrus multicolor* from a lab-rearing experiment where fish for one split brood were grown under hypoxia (1 mg l⁻¹) and normoxia and **(B)** the depth of the m. levator ex. 4 muscle (mm) and standard length (mm) for the same individual fish. Adapted from Chapman *et al.* (2000).

Saroglia *et al.* (2002) explored gill morphometric response of sea bass (*Dicentrarchus labrax*) to different dissolved oxygen partial pressures. They reported a negative relationship between the gill surface area and the oxygen partial pressure of the water in which the bass were reared for three months, again providing evidence of phenotypic plasticity in fish gills in response to oxygen availability. An interesting example of gill plasticity was highlighted by Schwartz (1995) in a comparison of brooding and non-brooding *Oreochromis mossambicus*, a mouth-brooding tilapiine. Brooding fishes were characterized by more and heavier gill filaments, which may facilitate increase water flow over the gills, eggs and wrigglers (Schwartz, 1995).

Developmental plasticity in gill characters in response to hypoxic stress is not unique to fishes. Burggren and Mwalukoma (1983) documented a dramatic branchial hypertrophy associated with exposure to hypoxia in larval bullfrogs (*Rana catesbeiana*). The internal gills of the larvae exhibited an increase in the number of gill filaments and filament size upon exposure to 28 days of hypoxia. Bond (1960) reported changes in the size of the gills and the shape and structure of gill filaments in salamander larvae (two species of *Ambystoma* and one species of *Salamandra*) in response to exposure to hypoxia during development.

Together, this suite of studies suggests a strong element of developmental plasticity in explaining the variation in gill size across populations of gilled ectotherms.

However, there is also evidence to suggest heritable variation, as Chapman *et al.* (2000) indicated in their study of respiratory and trophic morphology of *P. multicolor*. An important question in evaluating the potential for genetic variation among populations across oxygen gradients is whether respiratory phenotypes co-occur in the field, i.e., whether there is movement of individuals among populations and therefore, the potential for genetic mixing. Using a 4.5-year mark and recapture study, we found evidence for ecological mixing of morphological variants (large- and small-gilled populations) of the cyprinid *Barbus neumayeri* across DO gradients (Chapman *et al.*, 1999). However, our studies also indicated genetic differentiation across these populations and no relationship between genetic differences and geographic distances among sites. Although this study did not address the source of variation in gill size among populations, the results do suggest habitat-specific selection pressures on dispersers and potential for immigrant inviability (Chapman *et al.*, 1999).

Future studies that quantify the degree to which interdemic variation in gill size and associated characters represent genetic differences among populations, environmentally induced phenotypic variation, and/or the interaction of genetic and environmental influences will be critical for clarifying the nature of the morphological response to hypoxic stress and the role of aquatic oxygen as a driver of diversification among populations.

INTERACTIONS BETWEEN RESPIRATORY AND TROPHIC MORPHOLOGY

An important step in understanding morphological divergence across aquatic oxygen gradients is to explain why alternative respiratory phenotypes persist in different environments. Performance of different phenotypes between alternative habitats can be used to detect potential fitness trade-offs, i.e., the phenotype with the highest performance or fitness in one habitat performs sub-optimally in the alternative habitat (Van Buskirk et al., 1997). The role of trade-offs in generating or maintaining variation has been explored in fishes [e.g., (Lepomis gibbosus) Robinson et al., 1996; Mittelbach et al., 1999; (Salvelinus alpinus) Adams, 1998; (Gasterosterus spp.) Schluter, 1993, 1995; Day and McPhail, 1996; (Cichlasoma citrinellum) Meyer, 1989); (Carassius carassius) Nilsson et al., 1995; Pettersson and Bronmark, 1999]. However, these studies have focused primarily on trade-offs associated with trophic specialization, interspecific competition, benthic vs limnetic habitat use, and predation. Trade-offs between feeding and respiratory structures may also be important in fishes because of their compact, laterally compressed head morphology (Barel, 1983; Cech and Massingill, 1995).

We have begun to explore functional-morphological trade-offs between respiratory and trophic morphology that may maintain respiratory phenotypes in the field. Recent work has demonstrated that adaptive change in gill size in swamp fish correlates with a reduction in the size of key trophic muscles and feeding performance. In the cyprinid B. neumayeri, Schaack and Chapman (2003) compared a series of non-respiratory structures between small- and large-gilled populations. The differences that emerged between high- and low-oxygen sites were in the head region of the fish (swamp fish had longer heads, shorter geniohyoideus, and less depth of the sternohyoideus muscles), suggesting a potential association with gill proliferation and low oxygen specifically (Fig. 2.5). In paired feeding trials on a novel prey type, we found that large-gilled fish spent more time feeding than small-gilled fish of the same body size without ingesting more food (Schaack and Chapman, 2003). This suggests less efficient food uptake in large-gilled fish from hypoxic habitats. Although these morphological differences may result from several selection pressures (e.g., differences in food type or availability among sites), the associations observed suggest that increased gill size may affect associated, but functionally unrelated, characters; and this could impose a liability in the alternative oxygen environment (e.g., reduced feeding efficiency).

The correlation between large gills and reduced trophic muscle size has also been documented in experiments with the cichlid P. multicolor (Chapman et al., 2000), in which food and dissolved oxygen were controlled. We compared the gill apparatus, the surrounding structural

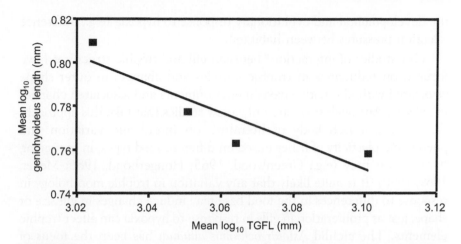

Fig. 2.5 The bilogarithmic relationship between the mean length of geniohyoideus muscle (mm) and the mean total gill filament length for four populations of the cyprinid *Barbus neumayeri* from the Rwembaita Swamp system of Kibale National Park, Uganda. Population values were adjusted to the common body mass of all populations using ANCOVA. Adapted from Schaack and Chapman (2003).

elements, and the outer shape of the fish for a population of a stable hypoxic habitat with one of a stable well-oxygenated habitat. In addition, we compared individuals raised under hypoxic or well-oxygenated circumstances (split-brood experiment). Swamp-dwelling fish and fish raised under hypoxia had a greater gill surface area than fish from normoxic habitats or fish raised under normoxia (Fig. 2.4A). The plastic change in gill size in this split-brood experiment allowed for a direct test of whether any change in gill size in response to growth under hypoxia leads to changes in non-functionally related structures. Three muscles displayed a reduction in size in response to growth under hypoxia (e.g., Fig. 2.4B). In addition, the lower pharyngeal jaw showed a reduced thickness (Chapman *et al.*, 2000). Interestingly, in the field population from the low-oxygen site, only one muscle was reduced in cross-sectional area relative to fish from the high-oxygen site. However, the upper pharyngeal jaws exhibited a decrease in width, length, and depth in the swamp-dwelling *P. multicolor*, which may result in a decrease in the maximum force of the jaw. A combination of inherited and plasticity changes may permit a finer-tuned response to hypoxic condition in the field, because in the swamp population, a larger increase in gill surface area was realized with a less detrimental effect on trophic functioning of the fish (Chapman *et al.*, 2000). Other explanations may also account for the discrepancy between the developmental and evolutionary response to hypoxia (e.g., difference selection pressures between habitats).

Our studies of interactions between gill and trophic structures have focused on reductions in trophic muscles and changes in outer shape associated with alternative oxygen environments and associated changes in gill size. Although there are only a few studies that take this approach, there exists a rich body of literature on interdemic variation and phenotypic plasticity of fishes raised on different food types, in particular, hard vs soft prey (e.g., Greenwood, 1965; Hougerhoud, 1986; Meyer, 1989, 1990). It is quite likely that any variation in trophic morphology in response to differences in the food base may induce changes in gill size or shape, just as proliferation in gills in response to hypoxia can affect trophic elements. The cichlid *Astatoreochromis alluaudi* has been the focus of several studies on trophic morphological variation in response to hard and soft foods. When fed on hard prey (e.g., mollusks), *A. alluaudi* will develop a massive pharyngeal mill with hypertrophied muscles, while a softer diet leads to reduction in pharyngeal jaw size and associated musculature (Greenwood, 1965; Huysseune *et al.*, 1994). Smits *et al.* (1996) found the total head volume in snail-eating *A. alluaudi* to be 31% larger than in fish from an insect-eating population. The researchers reported internal reallocations of the respiratory apparatus (change in the shape of the gills). However, they found that gill size did not differ between the two morphs. In this case, a trade-off between pharyngeal crushing and respiration was not supported; however, the changes in trophic morphology did result in a change in gill shape. It should be noted that Smits and colleagues did not measure the lamellar area, which may be important in evaluating potential implications of the shape change.

Galis and Barel (1980) studied the functional morphology of the gills of African lacustrine cichlids from a sample of more than 80 species. In some less active, algal scraping species they reported a high frequency (or density) of gill lamellae, suggesting that available space for the gill apparatus may be relatively small for species of this trophic guild, but might be compensated for by a high lamellar frequency and high water flow over the gills. Mazon *et al.* (1998) studied the functional morphology of the gills and respiratory area of two Brazilian fishes, a microphagous feeder and an opportunistic carnivore. They found that the total gill surface area did not differ between the species; however, there were morphological differences in the respiratory apparatus that seemed to relate to their feeding mechanisms (i.e., the large gill surface area observed in both species was achieved differently, reflecting a complex interaction between head shape and feeding habits). Thus, studies exploiting both interspecific and intraspecific comparisons suggest that non-respiratory factors may influence the extent of divergence in respiratory characters, and vice versa; though the implications for respiratory performance are not yet known. Future studies that explore the direct interactions between oxygen and food on trophic and respiratory morphology will be of great value in understanding the nature of the morphological response to hypoxic stress.

SUMMARY

The significance of variation in dissolved oxygen as a driver of phenotypic divergence is a largely unexplored aspect of aquatic biodiversity. However, there is now good evidence that alternative oxygen environments provide a strong predictor of intraspecific variation in fishes, particularly in respiratory traits (e.g., gill morphometry) and associated characters (e.g., trophic elements). Although studies are limited to date, developmental plasticity in gill morphometry occurs in response to aquatic oxygen availability, but there is also evidence to suggest an interaction between long-term selection and environmentally-induced phenotypic variation in response to hypoxic stress.

Trade-offs between trophic and respiratory morphology may contribute to the maintenance of respiratory phenotypes in nature. Our studies of East African fishes have demonstrated that adaptive change in gill size (large gills) in fish from hypoxic waters correlates with reduced size of key trophic muscles and feeding performance relative to small-gilled conspecifics. These trade-offs may lead to fitness costs in the field that impose habitat-specific selection pressures on dispersers and directly reduce gene flow.

There are many areas for future study that will be important to our understanding of the role of aquatic oxygen as an ecological source of divergent selection. These fields include studies that define and quantify the shared and unique aspects of morpho-physiological diversification in response to DO gradients, quantify the degree to which developmental plasticity and genetic differences contribute to observed patterns, compare gene flow vs physical immigration, and explore the potential for ecologically dependent reproductive isolation (i.e., mate choice experiments). Given the widespread nature of hypoxia in aquatic systems and potential connectivity of populations across oxygen gradients, alternative oxygen environments provide an ideal system to explore the nature and consequences of divergent selection.

Acknowledgments

Funding for the studies described in this chapter was provided from the University of Florida, the National Science Foundation (INT 93-08276, DEB-9622218, IBN-0094393), the Wildlife Conservation Society, the National Geographic Society, and McGill University. Permission to conduct research in Uganda was acquired from the Uganda National Council for Science and Technology and Makerere University. I thank the graduate students and field assistants at the Makerere University Biological Field Station and Lake Nabugabo, and our colleagues at the Fisheries Resources Research Institute of Uganda for assistance with various aspects of this project. I also thank Colin Chapman, Erin Reardon, Thom DeWitt, and Brian Langerhans for their valuable input on this chapter.

References

Adams, C.E. 1998. Trophic polymorphism amongst Arctic charr from Loc Rannoch, Scotland. *Journal of Fish Biology* 52: 1259-1271.

Balirwa, J.S., C.A. Chapman, L.J. Chapman, I.G. Cowx, K. Geheb, L. Kaufman, R.H. Lowe-McConnell, O. Seehausen, J.H. Wanink, R.L. Welcomme and F. Witte. 2003. Biodiversity and fishery sustainability in the Lake Victoria Basin: An unexpected marriage? *Bioscience* 53: 703-715.

Barel, C.D.N. 1983. Towards a constructional morphology of cichlid fishes (Teleostei, Perciformes). *Netherlands Journal of Zoology* 33: 357-424.

Beadle, L.C. 1932. Scientific results of the Cambridge Expedition to the East African Lakes, 1930-1. 3. Observations on the bionomics of some East African swamps. *Journal of the Linnean Society (Zoology)* 38: 135-155.

Beadle, L.C. 1981. *The Inland Waters of Tropical Africa*. Longman, London.

Bond, A.N. 1960. An analysis of the response of salamander gills to changes in the oxygen concentration of the medium. *Developmental Biology* 2: 1-20.

Bouton, N., J. de Visser and C.D.N. Barel. 2002. Correlating head shape with ecological variables in rock-dwelling haplochromines (Teleostei: Cichlidae) from Lake Victoria. *Biological Journal of the Linnean Society* 76: 39-48.

Burggren, W.W. and A. Mwalukoma. 1983. Respiration during chronic hypoxia and hyperoxia in larval and adult bullfrogs (*Rana catesbeiana*). I. Morphological responses of lungs, skin and gills. *Journal of Experimental Biology* 105: 191-203.

Carter, G.S. 1955. *The Papyrus Swamps of Uganda*. Heffer, Cambridge.

Cech, J.J. and M.J. Massingill. 1995. Tradeoffs between respiration and feeding in Sacramento blackfish, *Orthodon microlepidotus*. *Environmental Biology of Fishes* 44: 157-163.

Chapman, L.J. 1995. Seasonal dynamics of habitat use by an air-breathing catfish (*Clarias liocephalus*) in a papyrus swamp. *Ecology of Freshwater Fish* 4: 113-123.

Chapman, L.J. and C.A. Chapman. 1998. Hypoxia tolerance of the mormyrid *Petrocephalus catostoma*: Implications for persistence in swamp refugia. *Copeia* 1998: 762-768.

Chapman, L.J. and K. Hulen. 2001. Implications of hypoxia for the brain size and gill surface area of mormyrid fishes. *Journal of Zoology* 254: 461-472.

Chapman, L.J. and K.F. Liem. 1995. Papyrus swamps and the respiratory ecology of *Barbus neumayeri*. *Environmental Biology of Fishes* 44: 183-197.

Chapman, L.J., L.S. Kaufman, C.A. Chapman and F.E. McKenzie. 1995. Hypoxia tolerance in twelve species of East African cichlids: Potential for low oxygen refugia in Lake Victoria. *Conservation Biology* 9: 1274-1288.

Chapman, L.J., C.A. Chapman and M. Chandler. 1996a. Wetland ecotones as refugia for endangered fishes. *Biological Conservation* 78: 263-270.

Chapman, L.J., C.A. Chapman, R. Ogutu-Ohwayo, M. Chandler, L. Kaufman and A.E. Keiter. 1996b. Refugia for endangered fishes from an introduced predator in Lake Nabugabo, Uganda. *Conservation Biology* 10: 554-561.

Chapman, L.J., C.A. Chapman and T.L. Crisman. 1998. Limnological observations of a papyrus swamp in Uganda: Implications for fish faunal structure and diversity. *Verhandlungen Internationale Vereinigung Limnologie* 26: 1821-1826.

Chapman, L.J., C.A. Chapman, D. Brazeau, B. McGlaughlin and M. Jordan. 1999. Papyrus swamps and faunal diversification: Geographical variation among populations of the African cyprinid *Barbus neumayeri*. *Journal of Fish Biology* 54: 310-327.

Chapman, L.J., F. Galis and J. Shinn. 2000. Phenotypic plasticity and the possible role of genetic assimilation: Hypoxia-induced trade-offs in the morphological traits of an African cichlid. *Ecology Letters* 3: 387-393.

Chapman, L.J., C.A. Chapman, F.G. Nordlie and A.E. Rosenberger. 2002. Physiological refugia: Swamps, hypoxia tolerance, and maintenance of fish biodiversity in the Lake Victoria Region. *Comparative Biochemistry and Physiology* A133: 421-437.

Day, T. and J.D. McPhail. 1996. The effect of behavioural and morphological plasticity on foraging efficiency in the threespine stickleback (*Gasterosteus* sp.). *Oecologia* 108: 380-388.

Fernandes, M.N., F.T. Rantin, A.L. Kalinin and S.E. Moron. 1994. Comparative study of gill dimensions of three erythrinid species in relation to their respiratory function. *Canadian Journal of Zoology* 72: 160-165.

Galis, F. and C.D.N. Barel. 1980. Comparative functional morphology of the gills of African lacustrine Cichlidae (Pisces, Teleostei): An ecomorphological approach. *Netherlands Journal of Zoology* 30: 392-430.

Gallaugher, P. and A.P. Farrell. 1998. Hematocrit and blood oxygen carrying capacity. In: *Fish Physiology*, S.F. Perry and B.L. Tufts (eds.). Academic Press, London, Vol. 17, pp. 185-227.

Gibbs, R.H. and B.A. Hurwitz. 1967. Systematics and zoogeography of the stomiatoid fishes, *Chauliodus pammelas* and *C. sloani*, of the Indian Ocean. *Copeia* 1967: 798-805.

Graham, J.R. 1997. *Air-breathing Fishes: Evolution, Diversity and Adaptation*. Academic Press, San Diego.

Greenwood, P.H. 1965. Environmental effects on the pharyngeal mill of a cichlid fish, *Astatoreochromis alluaudi*, and their taxonomic implications. *Proceedings of the Linnean Society of London* 176: 1-10.

Hecky, R.E. 1993. The eutrophication of Lake Victoria. *Verhandlungen Internationale Vereinigung Limnologie* 25: 39-48.

Hecky, R.E., F.W.B. Bugenyi, P. Ochumba, J.F. Talling, R. Mugidde, M. Gophen and L. Kaufman. 1994. Deoxygenation of the deep water of Lake Victoria, East Africa. *Limnology and Oceanography* 39: 1476-1481.

Hochachka, P.W. 1982. Anaerobic metabolism: living without oxygen. In: *A Companion to Animal Physiology*, C.R. Taylor, K. Johansen and L. Bolis (eds.). Cambridge University Press, Cambridge, pp. 138-150.

Hougerhoud, R.J.C. 1986. Ecological morphology of some cichlid fishes. Ph.D. Dissertation, University of Leiden, Leiden, The Netherlands.

Hughes, G.M. 1966. The dimensions of fish gills in relation to their function. *Journal of Experimental Biology* 45: 177-195.

Hughes, G.M. 1984a. Measurement of gill area in fishes: Practices and problems. *Journal of Marine Biology Association of the United Kingdom* 64: 637-655.

Hughes, G.M. 1984b. General anatomy of the gills. In: *Fish Physiology*, W.S. Hoar and D.J. Randall (eds.). Academic Press, Orlando, Vol. 10A, pp. 1-72.

Hughes, G.M. and M. Morgan. 1973. The structure of fish gills in relation to their respiratory function. *Biological Reviews* 48: 419-475.

Huysseune, A., J.-Y. Sire and F.J. Meunier. 1994. Comparative study of lower pharyngeal jaw structure in two phenotypes of *Astatoreochromis alluaudi* (Teleostei: Cichlidae). *Journal of Morphology* 221: 25-43.

Kaufman, L.S. 1992. Catastrophic change in species-rich freshwater ecosystems: The lessons of Lake Victoria. *BioScience* 42: 846-858.

Kaufman, L. and P. Ochumba. 1993. Evolutionary and conservation biology of cichlid fishes as revealed by faunal remnants in northern Lake Victoria. *Conservation Biology* 7: 719-730.

Kisia, S.M. and G.M. Hughes. 1992. Estimation of oxygen-diffusing capacity in the gills of different sizes of tilapia, *Oreochromis niloticus*. *Journal of Zoology* (London) 227: 405-415.

Kramer, D.L. 1983. The evolutionary ecology of respiratory mode in fishes: an analysis based on the costs of breathing. *Environmental Biology of Fishes* 9: 145-158.

Kramer, D.L. 1987. Dissolved oxygen and fish behavior. *Environmental Biology of Fishes* 18: 81-92.

Langerhans, R.B. and T.J. DeWitt. 2004. Shared and unique features of evolutionary diversification. *American Naturalist* 164: 335-349.

Lewis, W.M. Jr. 1970. Morphological adaptations of cyprinodontoids for inhabiting oxygen deficient waters. *Copeia* 1970: 319-326.

Liem, K.F. 1980. Air ventilation in advanced teleosts: Biomechanical and evolutionary aspects. In: *Environmental Physiology of Fishes*, M.A. Ali (ed.). Plenum Press, New York, pp. 57-91.

Martinez, M.S., L.J. Chapman, J.M. Grady and B.B. Rees. 2004. Interdemic variation in hematocrit and lactate dehydrogenase in the African cyprinid *Barbus neumayeri*. *Journal of Fish Biology* 65: 1056-1069.

Mazon, A. de F., M.N. Fernandes, M.A. Nolasco and W. Severi. 1998. Functional morphology of gills and respiratory area of two active rheophilic fish species, *Plagioscion squamosissimus* and *Prochilodus scrofa*. *Journal of Fish Biology* 52: 50-61.

McCorkle, F.M., J.E. Chambers and J.D. Yarbrough. 1979. Tolerance of low oxygen stress in insecticide-resistant and susceptible populations of mosquito fish (*Gambusia affinis*). *Life Sciences* 25: 1513-1518.

McKinsey, D.M. and L.J. Chapman. 1998. Dissolved oxygen and fish distribution in a Florida spring. *Environmental Biology of Fishes* 53: 211-223.

Meyer, A. 1989. Cost of morphological specialization: Feeding performance of the two morphs in the trophically polymorphic cichlid fish, *Cichlasoma citrinellum*. *Oecologia* 80: 431-436.

Meyer, A. 1990. Morphometrics and allometry in the trophically polymorphic cichlid fish *Cichlasoma citrinellum*: Alternative adaptations and ontogenetic changes in shape. *Journal of Zoology* (London) 221: 237-260.

Mittelbach, G.G., C.W. Osenberg and P.C. Wainwright. 1999. Variation in feeding morphology between pumpkinseed populations: Phenotypic plasticity or evolution? *Evolution and Ecology Research* 1: 111-128.

Muir, B.S. and G.M. Hughes. 1969. Gill dimensions for three species of tunny. *Journal of Experimental Biology* 51: 271-285.

Nilsson, P.A., C. Bronmark and L.B. Pettersson. 1995. Benefits of a predator-induced morphology in crucian carp. *Oecologia* 104: 291-296.

Olowo, J.P. and L.J. Chapman. 1996. Papyrus swamps and variation in the respiratory behaviour of the African fish *Barbus neumayeri*. *African Journal of Ecology* 34: 211-222.

Palzenberger, M. and H. Pohla. 1992. Gill surface area of water-breathing freshwater fish. *Reviews in Fish Biology and Fisheries* 2: 187-216.

Perry, S.F. and G. McDonald. 1993. Gas exchange. In: *Fish Physiology*, D.H. Evans (ed.). CRC Press, Boca Raton, pp. 251-278.

Pettersson, L.B. and C. Bronmark. 1999. Energetic consequences of an inducible morphological defense in crucian carp. *Oecologia* 121: 12-18.

Prepas, E.E. and T. Charette. 2003. Worldwide eutrophication of water bodies: Causes, concerns, controls. In: *Treatise on Geochemistry*, H.D. Holland and K.K.Terekian (eds.). Elsevier, Amsterdam, Science Direct online version, Vol. 9, pp. 311-331.

Robinson, B.W. and K.J. Parsons. 2003. Changing times, spaces, and faces: Tests and implications of adaptive morphological plasticity in the fishes of northern postglacial lakes. *Canadian Journal of Fisheries and Aquatic Sciences* 59: 1819-1833.

Robinson, B.W. and D.S. Wilson. 1994. Character release and displacement in fishes: A neglected literature. *American Naturalist* 144: 596-627.

Robinson, B.W., D.S. Wilson and G.O. Shea. 1996. Trade-offs of ecological specialization: An intraspecific comparison of pumkinseed sunfish. *Ecology* 77: 170-178.

Rosenberger, A.E. and L.J. Chapman. 2000. Respiratory characters of three haplochromine cichlids: Implications for persistence in wetland refugia. *Journal of Fish Biology* 57: 483-501.

Saint-Paul, U. and B.M. Soares. 1987. Diurnal distribution and behavioral responses of fishes to extreme hypoxia in an Amazon floodplain lake. *Environmental Biology of Fishes* 20: 91-104.

Saroglia, M., G. Terova, A. De Stradis and A. Caputo. 2002. Morphometric adaptations of sea bass gills to different dissolved oxygen partial pressures. *Journal of Fish Biology* 60: 1423-1430.

Schaack, S.R. and L.J. Chapman. 2003. Interdemic variation in the African cyprinid *Barbus neumayeri*: Correlations among hypoxia, morphology, and feeding performance. *Canadian Journal of Zoology* 81: 430-440.

Schluter, D. 1993. Adaptive radiation in sticklebacks—Size, shape, and habitat use efficiency. *Ecology* 74: 699-709.

Schluter, D. 1995. Adaptive radiation in sticklebacks—Trade-offs in feeding performance and growth. *Ecology* 76: 82-90.

Schwartz, F.J. 1995. Gill filament responses and modifications during spawning by mouth brooder and substratum brooder (Tilapiine, Pisces) cichlids. *Acta Universitatis Carolinae Biologica* 39: 231-242.

Seehausen, O., F. Witte, E.J. Katunzi, J. Smits and N. Bouton. 1997. Patterns of the remnant cichlid fauna in southern Lake Victoria. *Conservation Biology* 11: 890-904.

Smith, T.B. and S. Skulason. 1996. Evolutionary significance of resource polymorphisms in fishes, amphibians, and birds. *Annual Review of Ecology and Systematics* 27: 111-133.

Smits, J.D., F. Witte and F.G. Van Veen. 1996. Functional changes in the anatomy of the pharyngeal jaw apparatus of *Astatoreochromis alluaudi* (Pisces, Cichlidae), and their effects on adjacent structures. *Biological Journal of the Linnean Society* 59: 389-409.

Timmerman, C.M. and L.J. Chapman. 2004a. Hypoxia and interdemic variation in the sailfin molly (*Poecilia latipinna*). *Journal of Fish Biology* 65: 635-650.

Timmerman, C.M. and L.J. Chapman. 2004b. Behavioral and physiological compensation for chronic hypoxia in the live-bearing sailfin molly (*Poecilia latipinna*). *Physiological and Biochemical Zoology* 77: 601-610.

Ultsch, G.R., M.E. Ott and N. Heisler. 1980. Standard metabolic rate, critical oxygen tension, and aerobic scope for spontaneous activity of trout (*Salmo gairdneri*) and carp (*Cyprinus carpio*) in acidified water. *Comparative Biochemistry and Physiology* A67: 329-335.

Ultsch, G.R., H. Boschung and M.J. Ross. 1978. Metabolism, critical oxygen tension, and habitat selection in darters (*Etheostoma*). *Ecology* 59: 99-107.

Van Buskirk, J., S.A. McCollum and E.E. Werner. 1997. Natural selection for environmentally induced phenotypes in tadpoles. *Evolution* 51: 1983-1992.

Wanink, J.H. 1999. Prospects for the fishery on the small pelagic *Rastrineobola argentea* in Lake Victoria. *Hydrobiologia* 407: 183-189.

Wanink, J.H. and F. Witte. 2000. Rapid morphological changes following niche shift in the zooplanktivorous cyprinid *Rastrineobola argentea* from Lake Victoria. *Netherlands Journal of Zoology* 50: 365-372.

Winberg, G.G. 1961. New information on metabolic rate in fishes. Transactions Series No. 362. *Fisheries Research Board Canada.* Nanaimo, British Columbia. (Translated from Russian).

Witte, F., T. Goldschmidt, J.H. Wanink, M.J.P. van Oijen, P.C. Goudswaard, E.L.M. Witte-Maas and N. Bouton. 1992. The destruction of an endemic species flock: Quantitative data on the decline of the haplochromine cichlids of Lake Victoria. *Environmental Biology of Fishes* 34: 1-28.

Witte, F., B.S. Msuku, J.H. Wanink, O. Seehausen, E.F.B. Katunzi, P.C. Goudswaard and T. Goldschmidt. 2000. Recovery of cichlid species in Lake Victoria: an examination of factors leading to differential extinction. *Reviews in Fish Biology and Fisheries* 10: 233-241.

Swimbladder-Lung Homology in Basal Osteichthyes Revisited

Steven F. Perry

INTRODUCTION

In spite of more than a century of investigation, the origin of the actinopterygian swimbladder remains unclear. The phylogenetically ancient actinopterygian group Polypteriformes (Cladistia), like lungfish, possesses a ventral air-breathing organ (ABO), and lung-like structures have also been described in the Devonian placoderm, *Bothriolepis* (Denison, 1941; Mayer, 1942). Consequently, recent reviews and textbooks (e.g., Liem, 1988; Kardong, 2002; Roux, 2003; Bartsch, 2004; Mickoleit, 2004) state that lungs were present in the earliest gnathostomes, were lost in the Chondrichthyes and migrated dorsally to form the swimbladder during the evolution of actinopterygians (Fig. 3.1A, entries in parentheses).

The above conclusions, however, are only valid provided all lungs are homologous and if the swimbladder is homologous to them. In addition,

Author's address: Institut für Zoologie, Universität Bonn, Poppelsdorfer Schloss, 53115 Bonn, Germany.

E-mail: perry@uni_bonn.de

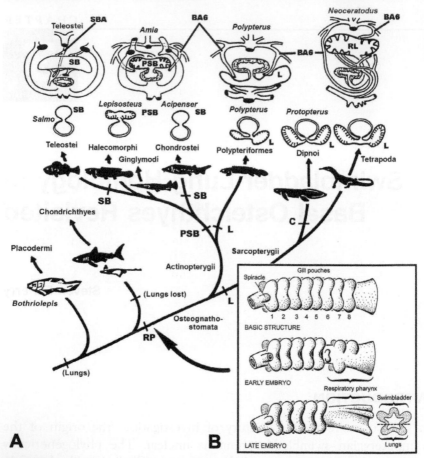

Fig. 3.1 Cladogram of jawed vertebrates illustrating two different scenarios for the origin of lungs and swimbladders (Part A) and the respiratory pharynx (Part B). Curved arrow indicates the location of part B in the cladogram. Entries in parentheses are according to Liem (1988) and show the lungs as plesiomorphic in jawed vertebrates, becoming lost in cartilaginous fish, and being maintained in all other lines. Chondrostei (Acipenseriformes) and Teleostei, separately evolve a true swimbladder (SB) from lungs as in the preferred scenario below, but the pulmonoid swimbladder (PSB) is considered to be a dorsal lung. **Boldface** entries represent the preferred scenario. Placoderm 'lungs' are considered a homeoplasy because of their anterior pharyngeal origin (Denison, 1941): ancestors of cartilaginous fish never had lungs. The respiratory pharynx (RP and Part B) evolved in air-breathing early bony fish (Osteognathostomata) and gave rise to lungs (L) in basal lobe-finned fish (Sarcopterygii). In ray-finned fish, the ventral part gave rise to lungs in Polypteriformes and the dorsal part formed the pulmonoid swimbladder (PSB) in all others. The true swimbladder (SB) evolved not from lungs but the PSB, thus eliminating the necessity of explaining the dorsal migration of lungs. In the Sarcopterygii, lungfish (Dipnoi) and possibly coelacanths (Actinistia, not shown) developed lungs from an unpaired ventral rudiment: in lungfish, the vascular and nerve supply is contralateral (C). Schemata of lungs and pharynx and of vascular supply are viewed caudally: i.e., right lung to the right. BA6 is the sixth branchial artery; SBA, swimbladder artery; RL, right lung. (Modified after Perry and Sander, 2004).

the question of lung or swimbladder homology and that of the origin of the swimbladder from lungs must be treated separately, because one does not necessarily follow from the other. In this chapter, the homology question will be critically evaluated in light of largely forgotten ontogenetic and comparative anatomical studies and a plausible explanation for the origin of the actinopterygian swimbladder will be sought.

HISTORICAL BACKGROUND

As summarized by Marcus (1937), Goette suggested in 1875—based on studies of lung ontogeny in anurans—that tetrapod lungs developed ontogenetically and evolved phylogenetically from modified gill pouches in the posterior pharynx. In 1882, Boas (cited in Marcus, 1937) proposed that tetrapod lungs were homologous to the swimbladder, and originated from the division and ventral migration of the latter. Bashford Dean then reversed this hypothesis, placing adult *Polypterus, Neoceratodus, Erythrinus, Lepisosteus/Amia* in a sequence that has no relation to the phylogenetic position of the groups (Liem, 1988), but presumably demonstrated that the pneumatic duct (and, therefore, the lungs) migrates dorsally around the pharynx: that paired ventral lungs could evolve into an unpaired, dorsal swimbladder.

Alternatively, Greil (1905, 1914) proposed that the lungs and swimbladder were one and the same: 'Swimbladder and lungs originate in corresponding locations, develop in the same general way and also show similarities in their further differentiation. We can therefore rightfully conclude that these structures of the same origin are homologous.' Greil (1905) supported his position by citing a 1902 study by Piper of swimbladder ontogeny in *Amia*. Piper had demonstrated that the primordial swimbladder, which originates in the pharyngeal dorsal midline, later assumes a lateral position due to the presence of a large ventral yolk mass. Moser (1904) had made similar observations in developing teleosts. Greil (1905) concluded that this change constituted lateral relocation of the swimbladder, and made the connection to amphibians, citing similar anatomical constraints that forced their lungs to develop laterally. He interpreted the lung-like organs of *Polypterus* to be paired swimbladders.

In 1913, Makuschok pointed out that the apparent lateralization of the swimbladder observed by Moser (1904) was temporary, since the ostium remained in or near the dorsal midline in the genera she studied (*Rhodeus, Cyprinus*), or, in amphibians, took place posterior to the lung rudiments (Moser, 1902). Also, in *Lepisosteus* (Makuschok, 1913), the swimbladder forms as a dorsomedial pharyngeal ridge parallel to the vertebral axis (Fig. 3.4). When the gut rotates and swings to the left, forming the stomach, the primordial swimbladder separates and extends caudally, remaining in the dorsal midline. Later, Neumayer (1930) and Wassnetzov (1932), working separately, demonstrated a similar dorsal origin of the non-respiratory swimbladder in *Acipenser* (Fig. 3.5). In this genus, however, paired rudiments are formed but only the right-hand one develops.

There is no convincing embryological support of the migration theory, but numerous studies demonstrate the separate ontogenetic origin of lungs and swimbladder. Nevertheless, as Liem (1988) laments, the migration theory has persisted until today.

RE-EVALUATION

Homology Criteria

The homology concept was introduced in 1843 by Owen to designate the 'same' structure in different species. With the advent of cladistics, molecular systematics and modern developmental biology, several sets of definitions of homology have come into common use. These definitions depend both on the properties of the data (morphological, molecular; phenotypical, DNA-based; developmental, adult-based; behavioral, static) and on the use to which the knowledge of homology will be put (Butler and Saidel, 2000; Brigandt, 2003).

Since we are interested here in comparative anatomy and evolutionary biology, homologous structures are defined by three criteria: (1) same location relative to other structures in the adult, (2) same embryological origin, and (3) continuity, i.e., the possibility of constructing an ontogenetic or phylogenetic sequence to explain different expressions of the attribute in related organisms (Butler and Saidel, 2000). Our present goals are to identify the 'same' regions in different organisms and to describe them, explaining the adaptive modifications observed (Brigandt, 2003). Problems lie in the explicit definition of the regions to be compared and in different conceptions of 'sameness'.

If the criteria are all met, the organs must be homologous. For two structures in different organisms to be homologous, one structure does not have to be derived from the other. They could be deemed homologous by virtue of being part of larger homologous structures. Thus, for example, the homology of the pharynx in all jawed vertebrates is unquestioned. Most authors (e.g., Gegenbaur, 1901; Moser, 1902; Greil, 1905; Wiedersheim, 1909; Makuschok, 1913, 1914; Marcus, 1923, 1937; Bertin 1958) also concur that the lungs of amphibians and fish originate in the posterior pharyngeal region. Similarly, there is consensus that the swimbladder also originates here (Makuschok, 1913; Neumayer, 1930; Wassnetzov, 1932). Thus, lungs and swimbladder are homologous as 'posterior pharynx', just as the first and fifth digits in related species are homologous as 'hand'. As Wassnetzov (1932) recognized, the answer to the homology question also depends on our designation of the region concerned.

If one questions whether lungs and swimbladders are homologous because they are derived from the same **part** of the pharynx, one must apply the homology criteria to that part. Although it is possible that the lung became unpaired and came to occupy a dorsal position without leaving any ontogenetic or phylogenetic trace, in this case the hypothesis of its homology with the swimbladder would have to be rejected because none of the three criteria can be verified.

Lung Homology

Polypteryformes and amphibians show a ventrolateral origin of paired lung buds (Marcus, 1937), but the unpaired glottis develops differently in each group. In the polypteriform *Polypterus* and *Erpetoichthyes* (Fig. 3.2 A, B) the left lung connects with the right, which enters the pharynx ventromedially. In recent amphibians, on the other hand, the ostia of the bilaterally symmetrical, ventromedial lung buds meet in the ventral pharyngeal midline (Fig. 3.2 E) through differential growth of the pharynx in Anura and Urodela or through elongation of the posterior pharynx to a pseudo-trachea in Gymnophiona (Makuschok, 1913, 1914; Marcus, 1923).

In lungfish, the lungs originate ontogenetically as unpaired ventral swellings which give rise to the paired lung rudiments (Kerr, 1910). Probably, material from the left side crosses to form the right lung and vice versa (Fig. 3.2C,D), since both the blood supply and the vagal innervation in the Lepidosirenidae (*Protopterus* and *Lepidosiren*), in which both lungs are well developed (Fig. 3.3), are contralateral. The left pharyngeal duct degenerates and the two lungs join cranially, connecting to the pharynx

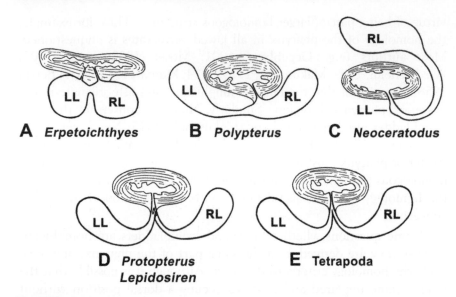

A *Erpetoichthyes*   B *Polypterus*   C *Neoceratodus*

D *Protopterus*   E Tetrapoda
   Lepidosiren

Fig. 3.2 Schematic crosssections through the posterior pharyngeal region of polypteriform (A, B) and dipnoan (C, D) fish, and tetrapods (E) in caudal view. Note degeneration (A) or lack of formation (B, D) of left lung duct. C, junction of left lung with right in Lepidosirinid lungfish. LL, left lung; RL, right lung. Modified after Marcus (1937).

only via the right duct (Marcus, 1937). Only the right lung develops in *Neoceradotus*, and it is supplied by both pulmonary arteries (Figs. 3.1A and 3.2C) (Bertin, 1958). In *Polypterus*, the predominant right lung is supplied by both the right and left vagus; the left lung, only by the left vagus (Fig. 3.6) (Bertin, 1958). The blood supply, however, is ipsilateral as is the case in tetrapods.

In summary, the ABOs in Polypteriformes, lungfish, amphibians and amniotes have in common their ventral, posterior pharyngeal origin and their plesiomorphically paired structure. Some disagreement, however, remains whether lungs are derived from gill pouches 6, 7 or 8 (Wiedersheim, 1909; Makuschok, 1913; Marcus, 1923), or from pouches at all (Greil, 1914). In addition, since the lungs of amniotes form by the branching of the medioventral laryngotracheal tube (not unlike the situation in lungfish, but remaining ipsilateral), it is not possible to trace their origin to specific gill pouches (Moser, 1902; Hesser, 1905; Corliss, 1976). The vascular supply from the sixth branchial artery and vagal innervation is plesiomorphic for the posterior pharynx of gnathostomes and is therefore of no value in determining the homology of lungs and swimbladders.

Homology of the Swimbladder

The swimbladder originates from the pharynx as a single ridge (*Acipenser, Lepisosteus*) (Figs. 3.4 and 3.5) or tube (*Cyprinus, Salmo*) in or near the dorsal midline, and invades the dorsal mesentery (Moser, 1904; Makuschok, 1913; Neumayer, 1930; Wassnetzov, 1932). Thus, although the swimbladder lies in the dorsal midline, the pneumatic duct may enter the pharynx somewhat laterally. In *Lepisosteus* (Wiedersheim, 1909) and the basal teleost *Arapaima* (personal observation), the respiratory swimbladder opens through a slit-like glottis directly into the pharynx. In *Acipenser,* the ostium intestino-vesicale secondarily connects the anterior part of the stomach with the swimbladder (Marinelli and Strenger, 1973). As in teleosts, a swimbladder artery supplies the non-respiratory organ in this group (Fig. 3.1A) (Marinelli and Strenger, 1973).

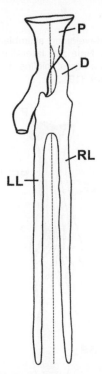

Fig. 3.3 Outline of lungs and pharxnx of a 6-cm larval *Lepidosiren paradoxa* in dorsal view, showing right-side duct entrance. D indicates duct; LL, left lung; P, pharynx; RL, right lung. Adapted from Marcus (1937).

Fig. 3.4 Artistic reconstruction of developing swimbladder in an 8.5 mm embryo of *Lepisosteus osseus* in dorsal view. Note the lateral displacement of gut, while the developing swimbladder remains in the dorsal midline. Li indicates liver; P, pharynx; SB, swimbladder. Adapted from Makuschok (1913).

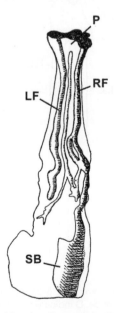

Fig. 3.5 Early stage of swimbladder development in *Acipenser* in dorsal view, showing right (RF) and left (LF) dorsal folds. The right fold extends to form the swimbladder. Adapted after Neumayer (1932, in Marcus, 1937).

Modern molecular developmental biology could reveal congruence in gene expression during ontogeny in lungs and swimbladder. Such results, however, would prove homology only of developmental pathways (viz., fingers and toes; consecutive vertebrae) in conformity with the definition used in molecular developmental biology. It would not, however, demonstrate that one organ is derived from the other and cannot satisfy the homology criteria of descriptive comparative anatomy and evolutionary biology posed here. Since we have defined the dorsal and ventral parts of the pharynx as separate regions, however, and no evidence exists that the lungs have migrated dorsally to form a swimbladder in lower teleosts and non-teleost actinopterygians, the homology of lungs and swimbladders must be rejected.

Extant Phylogenetic Bracketing (EPB) and Functional Morphological Approximation (FMA)

The validity of EPB depends on the accuracy of the cladogram on which it is based. Among the Actinopterygii, most treatments of higher fish systematics (Lauder and Liem, 1983; Nelson, 1994; Grande and Bemis, 1996; Mickoleit, 2004) show the Polypteriformes either as an unresolved trichotomy with the Acipenseriformes and all other ray-finned fish or as the sister group to all others, with the Acipenseriformes lying between them and the Neopterygii. Thus, the basal position of the Polypteriformes among the Actinopterygii is not questioned. The situation among the Sarcopterygii, however, remains unclear. Recent lines of evidence point to the Dipnoi, the Actinistia = Coelacanthiformes, or both as the sister group to tetrapods (Clack, 2002).

EPB using Fig. 3.1A reveals that the posterior pharynx of basal Osteichthys was an ABO, supplied by the right and left sixth branchial arteries. It contained paired, ventral sacs (lungs) and also gave rise to the swimbladder, including the pulmonoid swimbladder of basal neopterygian fish. Since the Polypteriformes, lungfish and tetrapods differ in the combinations of vascular supply and innervation of the lungs, as described above, it is not possible to further specify the plesiomorphic state of these organs.

FMA attempts to provide plausible explanations for the attributes observed. In light of variations in lung structure and the lack of plausible ontogenetic transitions between lungs and swimbladder, it appears probable that the ABO of basal bony fish had both dorsal and ventral components, and that one may have developed at the expense of the other during evolution of the separate groups (Wiedersheim, 1905).

If this is true, it is puzzling why lungs and a swimbladder never appear in the same individual, not even as ephemeral structures during ontogeny. Wiedersheim's (1903) attempts to recognize dorsal and ventral glottis rudiments in *Protopterus* and *Amia* were unsuccessful. The lack of both components in modern fish could be explained if the primitive ABO remained pharyngeal. Analogs are found today in *Clarias* (Moussa, 1956; Munshi, 1961; Hughes and Munshi, 1973a; Maina and Maloyi, 1986) and *Anabas* (Munshi, 1968; Hughes and Munshi, 1973b; Peters, 1978; Liem, 1987).

A Plausible Alternative

Recently, the idea of the respiratory pharynx (Fig. 3.1B) (Neumayer, 1930; Wassnetzov, 1932) as the primitive vertebrate ABO has been re-introduced (Perry et al., 2001). A scenario based on this concept is presented below.

The plesiomorphic state in basal bony fish before the separation of Sarcopterygii and Actinopterygii was that both the dorsal and the ventral parts of the posterior foregut had modified gill pouches that were capable of air breathing, and were supplied by the sixth branchial artery (Fig. 3.1). These structures were not inflatable and thus provided no significant buoyancy. Among the Sarcopterygii, the ventral parts must have evolved to form inflatable lungs. Assuming ipsilateral innervation and blood supply to be plesiomorphic, lungs further evolved separately among the ancestors of modern lungfish. The basal Actinopterygii retained the respiratory pharynx. The Polypteriformes developed the ventral part to form lung-like organs, while in the remaining actinopterygians the pulmonoid (Bartsch, 2004) swimbladder formed from the dorsal rudiment. Both the Sarcopterygii and Actinopterygii retained the plesiomorphic arterial supply (Perry and Sander, 2004). The origin of a separate swimbladder artery in the non-respiratory organ of Aciperseriformes is easier to justify than the same development in teleosts, in which the swimbladder of basal groups retained (or regained?) respiratory function.

For an air-breathing fish without a flotation device, life in the water column would have been energy intensive. Thus, the above scenario is consistent with the presence of flexible, sturdy appendages (archaepterygium) to support weight and provide propulsion in basal bony

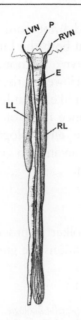

Fig. 3.6 Lungs of adult *Polypterus* sp. in dorsal view, showing vagal innervation. Note that the left vagus nerve (LVN), after innervating the left lung (LL), crosses the esophagus (E) and joins the right vagus nerve (RVN) in innervating the right lung (RL). P indicates pharynx. Adapted from Kerr (1910).

fish. For those that moved to open water, short, highly mobile appendages (neopterygium) and a dorsal, pulmonoid swimbladder, that in addition to providing oxygen also gave neutral buoyancy and allowed them to keep the dorsal side up, would have been advantageous. On the other hand, for those inhabiting shallow or tangled shorelines, paddle-like appendages and ventrolateral lungs would allow buoyancy combined with maximum maneuverability. These scenarios suggest plausible arguments for the repeated evolution of lungs and swimbladders from a relatively undifferentiated respiratory pharynx without postulating the conversion one highly differentiated organ into another.

CONCLUSION

The hypothesis based on extant phylogenetic bracketing that Osteichthyes originally possessed paired ventrolateral pouches that served as air-breathing organs is supported by embryological data. This does not eliminate the possibility of simultaneous presence of a dorsal rudiment that could give rise to the unpaired swimbladder, particularly if the rudimentary pharyngeal ABO did not store air. Regarding the homology question, there is no doubt that both the lungs and the swimbladder are derivatives of the posterior pharynx, the homology of which among jawed vertebrates is not questioned. However, there is no evidence to support the migration hypothesis. It is unlikely that lungs and the swimbladder are derived one from the other, since none of the three homology criteria (same relative location, same embryological origin, ontogenetic or phylogenetic continuum) relevant to comparative anatomy and evolutionary biology is satisfied.

Acknowledgements

The contributions of Klaus Völker in procuring literature and Marion Schlich in preparing the illustrations are gratefully acknowledged.

References

Bartsch, P. 2004. Actinopterygii. In: *Spezielle Zoologie*, W. Westheide and R. Rieger (eds.) Spektrum Akademischer Verlag, Heidelberg, *Teil 2: Wirbel-oder Schädeltiere*, pp. 226-287.

Bertin, L. 1958. Organes de la respiration aérienne. In: *Traité de Zoologie. Anatomie, systematique, biologie. Aganthes et poissons. Anatomie, éthologie, systématique*, P.P. Grassé (ed.), Masson et Cie, Paris, Vol. 13, pp. 1363-1398.

Brigandt, I. 2003. Homology in comparative, molecular, and evolutionary developmental biology: The radiation of a concept. *Journal of Experimental Zoology* B299: 9-17.

Butler, A. and W.M. Saidel. 2000. Defining sameness: Historical, biological and generative homology. *BioEssays* 22: 846-853.

Clack, J.A. 2002. *Gaining Ground. The Origin and Evolution of Tetrapods*. Indiana University Press, Bloomington, Indiana.

Corliss, D.E. 1976. *Patten's Human Embryology*. McGraw-Hill, New York.

Denison, R.H. 1941. The soft anatomy of *Bothriolepis. Journal of Paleontology* 15: 553-561.

Gegenbaur, C. 1901. *Vergleichende Anatomie der Wirbelthiere mit Berücksichtigung der Wirbellosen. Zweiter Band*. Verlag Wilhelm Engelmann, Leipzig.

Goodrich, E.S. 1930. *Studies on the Structure and Development of Vertebrates*. MacMillan, London, New York.

Grande, L. and W.E. Bemis. 1996. Interrelationships of acipenseriformes with comments on "Chondrostei". In: *Interrelationships of Fishes*, M.L.J. Stiassny, L.R. Parenti and G.D. Johnson (eds.). Academic Press, San Diego, pp. 85-115.

Greil, A. 1905. Bemerkungen zur Frage nach dem Ursprunge der Lungen. *Anatomischer Anzeiger* 26: 625-632.

Greil, A. 1914. Zur Frage der Phylogenese der Lungen bei den Wirbeltieren. Erwiderung an Herrn M. Makuschok. *Anatomischer Anzeiger* 39: 202-206.

Hesser, K. 1905. Entwicklung der Reptilienlunge. *Anatomische Hefte* 29: 215-310.

Kardong, K.V. 2002. *Vertebrates: Comparative Anatomy, Function, Evolution*. McGraw-Hill, New York.

Kerr, J.G. 1910. On certain features of the development of the alimentary canal in *Lepidosiren* and *Protopterus*. *Quarterly Journal of Microscopical Sciences* 54: 483-518.

Hughes, G.M. and J.S.D. Munshi. 1973a. The nature of the air-breathing organs of Indian fishes, *Channa, Amphipnous, Clarias* and *Saccobronchus* as shown by electron microscopy. *Journal of the Zoological Society* (London) 170: 245-270.

Hughes, G.M. and J.S.D. Munshi. 1973b. The fin structure of the respiratory organs of the climbing perch, *Anabas testudineus* (Pisces, Anabantidae). *Journal of the Zoological Society* (London) 170: 201-225.

Lauder, G.V. and K.F. Liem. 1983. The evolution and interrelationships of the actinopterygian fishes. *Bulletin of the Museum of Comparative Zoology* 150: 95-197.

Liem, K.F. 1987. Functional design of the air ventilation apparatus and overland excursions by teleosts. *Fieldiana: Zoology* 37: 1-29.

Liem, K.F. 1988. Form and function of lungs: the evolution of air breathing mechanisms. *American Zoologist* 28: 739-759.

Maina, J.N. and G.M.O. Maloiy. 1986. The morphology of the African air-breathing catfish (*Clarias mossambicus*): A light, electron and scanning microscopic study, with morphometric observations. *Journal of the Zoological Society* (London) A209: 421-445.

Makuschok, M. 1913. Über genetische Beziehung zwischen Schwimmblase und Lungen. *Anatomischer Anzeiger* 44: 33-55.

Makuschok, M. 1914. Zur Frage der phylogenetischen Entwicklung der Lungen bei den Wirbeltieren. *Anatomischer Anzeiger* 46: 495-514.

Marcus, H. 1923. Beitrag zur Kenntnis der Gymnophionen. VI. Über den Übergang von der Wasser-zur Luftatmung mit besonderer Berücksichtigung des Atemmechanismus von Hypogeophis. *Zeitschschrift für Anatomie und Entwicklungsgeschichte* 69: 328-343.

Marcus, H. 1937. Lungen. In: *Handbuch der Vergleichenden Anatomie der Wirbertiere*, L. Bolk, E. Göppert, E. Kallius and W. Lubosch (eds.). Urban and Schwarzenberg, Berlin, pp. 909-988.

Marinelli, W. and A. Strenger. 1973. *Vergleichende Anatomie und Morphologie der Tiere. IV Liefertung. Acipenser ruthenus*. Verlag Franz Deuticke, Wien.

Mickoleit, G. 2004. *Phylogenetische Systematik der Wirbeltiere*. Dr. Friedrich Pfeil-Verlag, Munich.

Moser, F. 1902. Beiträge zur vergleichenden Entwicklungsgeschichte der Wirbeltiere. *Archive der Mikroskopie, Anatomie und Entwicklungsgeschichte* 60: 587-668.

Moser, F. 1904. Beiträge zur vergleichenden Entwicklungsgeschichte der Schwimmblase. *Archive der Mikroskopie, Anatomie und Entwicklungsgeschichte* 63: 532-574.

Moussa, T.A. 1956. Morphology of the accessory air-breathing organs of the teleost, *Clarias lazera* (C. and V.). *Journal of Morphology* 98: 125-160.

Munshi, J.S.D. 1961. The accessory respiratory organs of *Clarias batrachus* (Linn.). *Journal of Morphology* 109: 115-139.

Munshi, J.S.D. 1968. The accessory respiratory organs of *Anabas testudineus* (Bloch). (Anabantidae, Pisces). *Proceedings of the Linnean Society* (London) (Zoology) 179: 107-126.

Myers, G.S. 1942. The "lungs" of *Bothriolepis*. *Ichthyological Bulletin* 2: 134-136.

Nelson, J.S. 1994. *Fishes of the World*. 3rd edition. John Wiley and Sons, New York.

Neumayer, L. 1930. Die Entwicklung des Darms von *Acipenser*. *Acta Zoologica* (Stockholm) 39: 1-151.

Owen, R. 1843. *Lectures on the comparative anatomy and physiology of vertebrate animals delivered at the Royal College of Surgeons, in 1843*. Longman, Brown, Green and Longmans, London.

Perry, S.F. and M. Sander. 2004. Reconstructing the evolution of the respiratory apparatus in tetrapods. *Respiratory Physiology and Neurobiology* 144: 125-139.

Perry, S.F., R.J.A. Wilson, C. Straus, M.B. Harris and J.E. Remmers. 2001. Which came first, the lung or the breath? *Comparative Biochemistry and Physiology* A129: 37-47.

Peters, H.M. 1978. On the mechanism of air ventilation in anabantoids (Pisces:Teleostei). *Zoomorphologie* 89: 93-123.

Roux, E. 2002. Origine et évolution de l'appareil respiratoire aérien des Vertébrés. *Revue des Maladies Respiratoires* 19: 601-615.

Schmalhausen, I.I. 1968. *The Origin of Terrestrial Vertebrates*. Academic Press, New York, London.

Wassnetzov, W. 1932. Über die Morphologie der Schwimmblase. *Zoologische Jahrbücher Abteilung Anatomie und Ontogenie der Tiere* 56: 1-36.

Wiedersheim, R. 1904. Über das Vorkommen eines Kehlkopfes bei Ganoiden und Dipnoërn sowie über die Phylogenie der Lunge. *Zoologische Jahrbücher. Supplement* 7. Festschrift zum 7. Geburtstag A. Weissmann's.

Wiedersheim, R. 1909. *Vergleichende Anatomie der Wirbeltiere. Siebente, vielfach umgearbeitete und stark vermehrte Auflage des "Grundriss der Vergleichenden Anatomie der Wirbeltiere."* Verlag Gustav Fischer, Jena.

Witmer, L.M. 1995. The extant phylogenetic bracket and the importance of reconstructing soft tissues in fossils. In: *Functional Morphology in Vertebrate Paleontology*, J.J. Thompson (ed.). Cambridge University Press, Cambridge, pp. 19-33.

The Effects of Temperature on Respiratory and Cardiac Function of Teleost Fish

Francisco Tadeu Rantin[1],*, Ana Lúcia Kalinin[1] and Mogens L. Glass[2]

INTRODUCTION

Some animals adapt to darkness (e.g., cave or deep sea dwellers) or to continuous light (e.g., polar animals during the summer). Animals can also adapt to large variations of humidity. By contrast, most organisms are adapted to a very limited species-specific temperature range.

In the aquatic environment, smaller animals are unable to maintain body temperatures far from the water temperature. The reason is that water has a high thermal conductivity and, moreover, a large heat capacity (Schmidt-Nielsen, 1975; Stevens and Sutterlin, 1976). Most teleost fish must cope with ambient temperature changes, which has a large impact on

Authors' addresses: [1]Laboratory of Zoophysiology and Comparative Biochemistry, Department of Physiological Sciences, Federal University of São Carlos. Via Washington Luis, km 235, 13565-905 – São Carlos, SP, Brazil. E-mail: akalinin@power.ufscar.br

[2]Department of Physiology, Faculty of Medicine of Ribeirão Preto, University of São Paulo. 14049-900 – Ribeirão Preto, SP–Brazil. E-mail: mlglass@rfi.fmrp.usp.br

Corresponding author: E-mail: ftrantin@power.ufscar.br

their physiology. Some highly active species of large aerobic scope such as blue fin tuna, skipjack tuna and big eyed tuna are, however, able to maintain an elevated body temperature in relation to their environment (Stevens *et al.*, 1974; Stevens and Neil, 1978).

Any teleost fish exhibits a preference for a specific temperature range, which usually coincides with optimal growth (Jobling, 1981) and/or with a high aerobic capacity (Kelsch and Neil, 1990). A simple behavioral response is to move to waters of favorable temperatures. The preferred temperature range of a given species may, however, vary with season, diurnal rhythms, and growth (Reynolds and Casterlin, 1979; Roberts, 1979).

Additionally, O_2-availability exerts an influence, since fish prefer lower temperatures when challenged by low O_2 availability. This behavior decreases metabolic needs and, thereby, prolongs survival under adverse ambient conditions (Schurmann and Schou-Christiansen, 1994).

Some habitats do not allow an efficient behavioral thermoregulation. As an example, shallow and small lakes and minor rivers change temperature on a diurnal or seasonal basis without any major temperature gradient from surface to bottom (Harder *et al.*, 1999). In this chapter we discuss what happens to respiratory and cardiovascular function in teleost fish exposed to unavoidable changes of body temperature. Some aspects have been little explored, but the few available data are rather consistent. Therefore, we find it useful to review the data at this point, hoping to provoke much further research in this field, that call for interactions between ecology, ethology, physiology and aquaculture.

OXYGEN UPTAKE, GILL VENTILATION AND RESPIRATORY GASES

Gill Ventilation and Oxygen Uptake

In ectothermic animals, O_2-uptake and CO_2 production increase with temperature, often in an exponential fashion (Krogh, 1914, 1968). In teleost fish, a higher temperature provokes an increased gill ventilation and perfusion, which serves to maintain an adequate delivery of oxygen to tissues. This relationship becomes understandable when O_2 transport equations are considered. According to Dejours (1981), a simple and fundamental equation in respiratory physiology is:

$$\dot{V}O_2 = \dot{V} \cdot C_1O_2 \cdot EO_2, \tag{1}$$

where $\dot{V}O_2$ = oxygen uptake; $\dot{V}$ = ventilation (= irrigation) of the gas exchanger; C_1O_2 = oxygen concentration of the inspired medium (water or air) and EO_2 = extraction of oxygen from the inspired water, i.e., the fraction of O_2 that entered the circulation divided by the total amount of inspired O_2. This equation is equally valid for pulmonary and gill respiration. The effects of temperature on these variables are shown in Figs. 4.1 and 4.2.

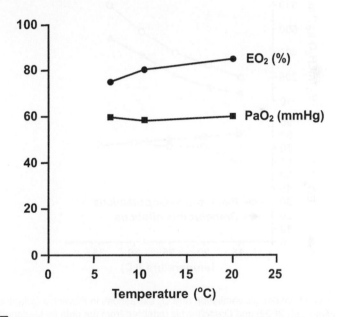

Fig. 4.1 Arterial PO_2 and percentage of O_2 extraction by the gills of rainbow trout. Based on data by Randall and Cameron (1973).

In teleost fish, the degree of O_2-extraction by the gills is usually very high (Randall and Cameron, 1973; Lomholt and Johansen, 1979). About 85% of total inspired O_2 is removed by the blood flow through the gills (Fig. 4.1). The extraction may, however, decrease with higher ventilatory stroke volume and increased respiratory frequency as occurs during ambient hypoxia (for a classic study, see Saunders, 1962). The high degree of O_2-extraction is possible due to the countercurrent gas exchange system of teleost fish. The blood flow of the secondary lamellae (the gas

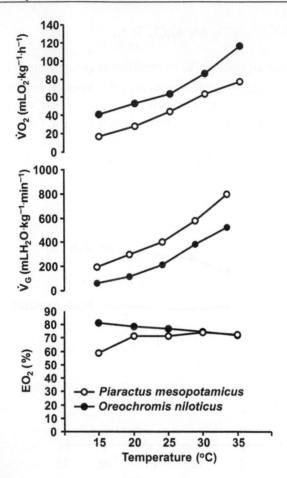

Fig. 4.2   Oxygen uptake, gill ventilation and O$_2$ extractions in *Piaractus* (adapted from the data by Aguiar *et al.*, 2002) and *Oreochromis* (adapted from the data by Maricondi-Massari *et al.*, 1998) in relation to temperature. Notice that extraction is relative constant with temperature, because gill ventilation is adjusted to meet temperature-dependent demands for oxygen.

exchange units) moves in the opposite direction of the inspired water. Ideally, the system permits PaO$_2$ (arterial PO$_2$) to equilibrate with the P$_I$O$_2$ (PO$_2$ of the inspired water), but this condition is not reached in any known species (Piiper and Scheid, 1984).

This degree of extraction in teleost fish is little modified by temperature (Randall and Cameron, 1973; Maricondi-Massari *et al.*, 1998). In relation to equation (1), this implies that ventilation of the gills must increase in proportion to temperature-induced elevation of metabolism. Moreover, ventilation must increase whenever C_IO_2 (O_2 content of the inspired water) becomes reduced.

The concentration of O_2 obeys Henry's law:

$$[O_2] = \alpha O_2 \cdot PO_2, \tag{2}$$

where αO_2 ($= O_2$ solubility) decreases with increases of temperature and salinity. In distilled water the solubility (αO_2) decreases from about 2.5 $\mu mol \cdot L^{-1}$ $mmHg^{-1}$ at 5°C to about 1.4 μ $mol \cdot L^{-1}$ $mmHg^{-1}$ at 35°C (Dejours, 1981). For O_2-saturated water, this implies (1) that increases of ventilation with temperature must compensate decreased amounts of O_2 in the inspired water, and (2) must match increases of metabolism (Maricondi-Massari *et al.*, 1998).

Control of Gill Ventilation

The level of gill ventilation is predominantly linked to the variable O_2 levels of the aquatic environment rather than to CO_2 levels in the water (Dejours, 1981). In most teleost fish some principal O_2 sensing receptors are located on the first gill arch. Blood screening O_2 receptors have been identified along with units screening the inspired water (Milsom and Brill, 1986; Soncini and Glass, 2000). The effects of temperature on ventilatory responses to hypoxia have only been studied in few species. Fernandes and Rantin (1989) evaluated the respiratory responses of the Nile tilapia, *Oreochromis niloticus*, to environmental hypoxia under different thermal conditions, while Glass *et al.* (1990) studied these responses in carp, *Cyprinus carpio*. The ventilatory responses of the two species are consistent, since low temperatures did not abolish ventilatory responses to ambient hypoxia. This emphasizes the importance of an O_2-oriented regulation that alleviates the effects of changing temperatures and O_2-availability of the aquatic environment. Much by contrast, amphibians and turtles considerably reduce, or even loose, ventilatory responses to hypoxia when exposed to low temperatures (Jackson, 1973; Kruhøffer *et al.*, 1987). Their ventilation is, however, maintained due to a predominantly acid-base oriented drive to ventilation (cf. Dejours, 1981). This pH-dependent manner of respiratory control characterizes amphibians, reptiles, birds and mammals (Figs. 4.3 and 4.4).

Blood Gases

Some species of Antarctic teleosts lack hemoglobin, and O_2 is transported to the tissues only in the dissolved form. The small amount delivered to

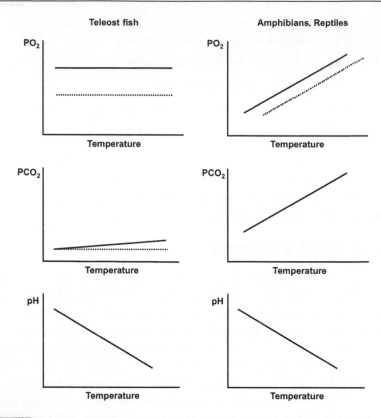

Fig. 4.3 Schematic drawing of the relationships between arterial blood status (PO$_2$, PCO$_2$, pH) and blood temperature in teleost fish and in ectothermic land vertebrates (amphibians and reptiles). Combinations of continuous and broken lines indicate that the magnitude and/or the effect of temperature on variable are species dependent.

tissues is sufficient for two reasons: (1) high solubility of O$_2$ at temperatures close to 0°C and (2) a low metabolism due to such low temperatures (Zummo *et al.*, 1995).

The position of the Hb-O$_2$ dissociation curve of teleost fish often reflects the habitat and mode of life of the species. This becomes evident, comparing carp (*Cyprinus carpio*) to rainbow trout (*Oncorhynchus mykiss*) and other fast swimming teleosts: Carp is a stationary species with a high tolerance to severely hypoxic ambient conditions. By contrast, trout is a fast swimming species of flowing, well-oxygenated waters. These adaptations are reflected in their relative Hb-O$_2$ affinities, i.e., high in carp and low in trout (Cameron, 1971; Albers *et al.*, 1983; Souza *et al.*, 2001). The high Hb-O$_2$ affinity of carp permits survival in environments

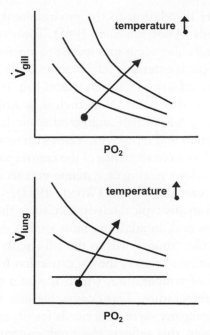

Fig. 4.4 Gill ventilation of teleost fish exposed to changes of temperature and O_2 levels (normoxia/hypoxia) in the water (upper drawing). Lung ventilation in ectothermic land vertebrates in relation to the same environmental changes in the inspired gas phase. Notice that low temperature does not abolish ventilatory responses in teleost fish. By contrast, ectothermic land vertebrates at low body temperature do not respond to hypoxia or, alternatively, the responses are highly reduced. See text for further explanation.

of moderate to severe hypoxia, because branchial O_2 loading of the blood can take place in spite of a very low ambient PO_2. In turn, the low circulatory PO_2 limits O_2 delivery to tissues and, thereby, the aerobic scope (Dejours, 1981). Oppositely, the low affinity of trout blood limits O_2 loading under hypoxic conditions, while it favors a large aerobic scope as long as the fish remains in normoxic waters. Normally, the PaO_2 of carp ranges from 10 to 40 mmHg, whereas the rainbow trout operates within a range of 90 to 110 mmHg (Gilmour and Perry, 1994). Both species vigorously defend their respective PaO_2 levels when exposed to hypoxic waters. Furthermore, they maintain a constant PaO_2 over a wide range of temperatures as seen in Figs. 4.1 and 4.2 (Randall and Cameron, 1973; Glass *et al.*, 1990; Soncini and Glass, 1997)

The PO_2 loading O_2 of hemoglobin is most critical at high temperature. This is because the O_2-hemoglobin affinity becomes reduced with increases of temperature. Triphosphate-mediated modulation of

O_2-hemoglobin affinity in relation to the environment may, however, occur on a seasonal basis (Andersen et al., 1985). Triphosphate mediated adjustments of O_2-affinity also occur in response to ambient hypoxia, and the literature on the topic is extensive (cf. Nikinmaa, 1990). Data are few concerning temperature effects on PaO_2 in teleost fish. It may, therefore, be risky to generalize at this point. Nonetheless, a striking difference appears when comparing teleost fish and ectothermic land vertebrates. Arterial PO_2 of amphibians and reptiles increases with rising temperature, which can be explained as a consequence of the central vascular anatomy of these animals, that allows mixing of systemic venous and oxygenated returns, i.e., central vascular shunt (Wood, 1982). This particular situation will not occur in any typical teleost fish, since their myocardium consists of serial pumps and, in addition, most studies do not favor the presence of cardiovascular 'shunts' within the gill circulation (Glass and Soncini, 1999) As mentioned above, the O_2 extraction from the water is virtually independent of temperature, which is also a prerequisite for constant PaO_2 with temperature. This occurs, because increases of gill ventilation closely accompany increasing needs for O_2-uptake at higher temperature. Once again this reflects the predominantly O_2-oriented ventilatory control in teleost fish (Soncini and Glass, 2000).

Acid-base Status of the Blood

The effects of temperature on acid-base status have been studied in several teleost fish (cf. Heisler, 1984). In most species, arterial pH decreases with rising temperature by about 0.01 to 0.015 units/°C. The same change of pH with temperature has repeatedly been reported for ectothermic tetrapods (cf. Glass and Wood, 1981); see Fig. 4.3.

Reeves (1972, 1976) discussed early data on temperature dependent acid-base status in tetrapods. Based on the few data available at the time, he suggested that the key variable is constant dissociation of the imidazol group of histidine on proteins (constant α-imidazol). The corresponding equation,

$$pH = pK_{Im} + \log [Im/H^+Im], \qquad (3)$$

states that pH of the extracellular must change in parallel to pK_{Im} with temperature provided that $\alpha = Im/H^+Im$ is kept constant. Reeves (1972) argued that constant α-imidazol would protect enzyme function against changes of temperature (Hazel et al., 1978). The $\Delta pK_{Im}/\Delta t$ values for imidazol groups ranges between -0.18 and -0.24 units/$10°C$, but it turns out that most ectothermic animals regulate acid-base status to achieve only about half of the required decline of pH with rising temperature. On the other hand, it is striking that nearly all ectothermic vertebrates share a negative $\Delta pH/\Delta t$ with rising temperature (Fig. 4.3). To date the alphastat model remains disputed. Therefore, it seems too early to confirm why the set point for pH declines with rising temperature. We can, however, ask how a regulated change of pH with temperature can be achieved. According to the Henderson-Hasselbalch equation, the mechanisms are modulation of bicarbonate concentration and/or a change of CO_2 pressure. Studying amphibians, Reeves (1972) proposed that ventilatory modulation of PCO_2 accounted for control of pH in relation to temperature, assuming a key role of CO_2-receptors (alphastat receptors). Cameron and Randall (1972) and Randall and Cameron (1973), however, found that trout achieves a negative $\Delta pH/\Delta t$ by adjustments of bicarbonate levels and not by modulation of PCO_2. This indicated that teleost fish mainly adjust temperature-dependent pH through modulation of bicarbonate levels, while ectothermic tetrapods mainly achieve this regulation by ventilatory adjustments (Jackson, 1989).

Recently, this distinction between control mechanisms complicated, since carp and pacu (*Piaractus mesopotamicus*) increase $PaCO_2$ with rising temperature, which contributes to a concomitant fall of arterial pH (Glass *et al.*, 1990; Soncini and Glass, 1997). Therefore, a strict distinction is not justified when comparing ectothermic tetrapods and teleost fish in the present context (Fig. 4.3). The increased $PaCO_2$ at higher temperature is not necessarily a consequence of adjustments of gill ventilation, since limitations to CO_2 output might be caused by factors related to release from the blood, including concentrations of carbonic anhydrase (Desforges *et al.*, 2002).

CARDIAC FUNCTION

Temperature changes substantially affect the ability of the fish heart to maintain cardiac adaptations that permit adequate function under conditions crucial to survival. Acutely and seasonally, fish become exposed to large temperature changes. For example, as a fish moves vertically through the water column, its heart temperature may rapidly change, which requires intrinsic adjustment to maintain adequate force and power and, thereby, avoid cardiac arrhythmia. Fish also experience gradual and more predictable seasonal temperature changes that adjust tissues and organ function during thermal acclimatization. The specific strategy depends on the particular habitat and mode of life of the species (Vornanen et al., 2002).

CARDIAC OUTPUT

Few studies report on the effects of temperature on cardiac function in teleost fish (Fernandes and Rantin, 1989; Maricondi-Massari et al., 1998) and the data are inconsistent (compare Matakainen and Vornanen, 1992 and Morita and Tsukuda, 1995).

Temperature changes reveal an inverse relationship between heart rate (f_H) and stroke volume (V_S) temperature changes applied to in vitro fish hearts (Graham and Farrell, 1985; Yamamitsu and Itazawa, 1990; Korsmeyer et al., 1997). Temperature does not necessarily affect myocardial force development (Driedzic and Gesser, 1994; Coyne et al., 2000; Anelli-Jr. et al., 2004) as the opposing changes in V_S reflect alterations in filling time. In vivo cardiac responses to acute temperature change in teleosts document that f_H normally determine cardiac output, while stroke volume is rather constant (Cech et al., 1976; Farrell and Jones, 1992; Korsmeyer et al., 1997).

Moreover, high temperature increases tissue metabolism and O_2-consumption. According to the Fick principle,

$$\dot{V}O_2 = \dot{Q} \ ([O_2]a - [O_2]v), \tag{4}$$

where $\dot{V}O_2$ = O_2 uptake, $\dot{Q}$ = blood flow, and ($[O_2]a - [O_2]v$) = the O_2 content difference between arterial and mixed venous blood. According to this principle, an increased O_2-uptake would be accompanied by an increased cardiac frequency (f_H) as reported by Maricondi-Massari et al. (1998). An increase of f_H does not imply an increased cardiac output. Working on rainbow trout, Farrell et al. (1996) found a reduction of cardiac stroke volume with an increase from 15 to 22°C. Consequently, cardiac output rose only between 5 to 20°C to reach a plateau above that temperature. Importantly, the optimal cardiac performance coincided with the preferred temperature range. Different from high temperature, ambient hypoxia causes a bradycardia that is partly under neural control (Sundin et al., 1999, 2000).

Recently, Stecyk and Farrell (2002) evaluated cardiac function in carp (*Cyprinus carpio*) within the range of 6 to 15°C. In normoxic water, both cardiac output and f_H increased with temperature ($Q_{10} = 1.7$—output; 2.6—f_H). Severe hypoxia depressed both variables to minimum values that were independent of acclimation temperature. Further, Farrell (2002) reviewed the effects of temperature on swimming performance of salmonids and found that the best performance coincided with the largest aerobic and cardiac scopes (see also Farrell and Clutterham, 2003).

Acute temperature changes markedly influence the contraction of the heart. At elevated temperatures, the heart rate and the velocity of cardiac contraction increase, whereas the force contraction usually decreases (Vornanen, 1989; Møller-Nielsen and Gesser, 1992; Keen *et al.*, 1994). With these opposing changes in force and frequency, the pumping capacity (product of heart rate and force) of the fish heart is low at extreme temperatures, while its optimal performance is achieved within the preferred temperature range (Shiels and Farrell, 1997; Shiels *et al.*, 2002).

As in other vertebrates, the f_H of teleost fish is determined both by the intrinsic rate of pacemaker cells in the sinus venosus and by the extrinsic control by the autonomic nervous system and humoral factors (Laurent *et al.*, 1983). Temperature affects both intrinsic and extrinsic components. Thus, increased temperature raises the depolarization frequency of the pacemaker cells and, consequently, the heart rate (for a review, see Tibbits *et al.*, 1992). In addition, high temperatures reduce the duration of the ventricular action potential (V_{APD}) (Lennart and Hubbard, 1991; Maricondi-Massari *et al.*, 1998; Aguiar *et al.*, 2002). This effect is due to an increased permeability of the sarcolemma (SL), leading to an early final repolarization. Consistent with these temperature effects, several tropical species increase f_H when exposed to high temperature (Figs. 4.5 and 4.6).

Beyond the limits of modulation by temperature and/or activity, the absolute lower and upper limits for f_H in most teleost fish are $10 < f_H < 60$ bpm (Driedzic and Gesser, 1994). This range has been reported for tropical species, including Nile tilapia, *Oreochromis niloticus* (Costa *et al.*, 2000), traíra, *Hoplias malabaricus* (Olle, 2003), pacu, *Piaractus mesopotamicus* (Anelli-Jr. *et al.*, 2004), and muçum, *Symbrachus marmoratus* (Kalinin, unpublished data). As an exception, the frillfin goby, *Bathygobius soporator*, has a very high f_H that may border the lower limit for small mammals (Rantin *et al.*, 1998). Within the mentioned limits,

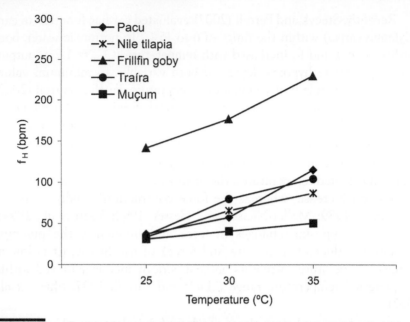

Fig. 4.5 The effects of temperature on the *in vivo* heart rate of several tropical teleost species. Frillfin goby (Rantin *et al.,* 1998), Nile tilapia (Costa *et al.,* 2000), traíra (Olle, 2003), Pacu (Anelli-Jr *et al.,* 2004) and muçum (= marbled swamp eel — unpublished data).

species of a wide range of climates show changes of f_H during acute temperature transitions, having a Q_{10} of about 2 (Bailey and Driedzic, 1990). This suggests that alteration of f_H is a main cardiovascular variable in relation to ambient temperature changes.

Contraction Force and Temperature

In mammalian heart muscle, L-type Ca^{2+} channels are activated during the cardiac action potential and Ca^{2+} enters the cell via Ca^{2+} current and, in addition, a much smaller amount enters via Na^+-Ca^{2+} exchange (NCX). Ca^{2+} influx triggers Ca^{2+} release from the sarcoplasmic reticulum (SR) and, to some extent, can also contribute directly to activation of the myofilaments. The Ca^{2+} entry plus the amount released from the SR via Ca^{2+}-induced Ca^{2+} release (CICR) raises cytosolic free calcium concentration ($[Ca^{2+}]i$), causing Ca^{2+} binding to the thin-filament protein troponin C (TnC), which activates contraction (Bers, 2000; 2002). The teleost myocardium contains all the cellular components of the mammalian system for E-C coupling. Nevertheless, the exact

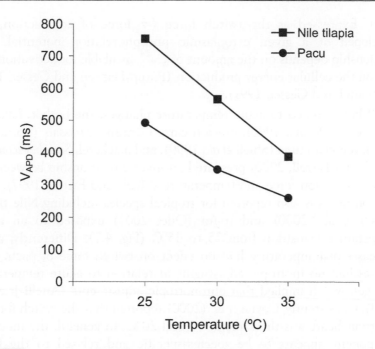

Fig. 4.6 The effects of temperature on the duration of the ventricular action potential (V$_{APD}$) of Nile tilapia (Maricondi-Massari *et al.*, 1998) and pacu (Aguiar *et al.*, 2002).

interactions in force generation still need further study. Particularly, more information is needed concerning the relative importance of the sources of calcium for contractile activation in teleosts, in particular as to transarcolemmal calcium flux vs mobilization from the SR. It should also be pointed out that the use of different procedures, test parameters and ranges of thermal acclimation may explain the discrepancies between studies (Coyne *et al.*, 2000).

Temperature affects development of myocardial twitch force in several ways. Higher temperatures increase: (1) the Ca^{2+} sensitivity of contractile proteins (Harrison and Bers, 1990) and (2) the interval that the cross bridges remain in maximal force development (Kuhn *et al.*, 1979).

Temperature also affects the energetic efficiency of the heart (Graham and Farrell, 1990) along with contraction and cellular Ca^{2+} balance (Bailey and Driedzic, 1990; Lennart and Huddard, 1991). In addition, the force-frequency relationships are highly temperature-dependent (Hove-Madsen, 1992). This indicates that temperature influences the relationships between contractility and cellular energy state of the teleost heart. Expressed as the twitch force (= force of contraction, Fc) developed at a given cytoplasmic phosphorylation potential, this relationship depends on the amount of Ca^{2+} available for activation and also on the cellular energy production (Purup-Hansen and Gesser, 1987; Hartmund and Gesser, 1991).

When exposed to acute temperature changes, the burbot, *Lota lota* (Tompe and Keurs, 1990), crucian carp, *Carassius carassius* (Vornanen, 1989), yellowfin tuna (Shiels *et al.*, 1999), and mackerel, *Scomber scombrus* (Shiels and Farrell, 2000) presented an inverse relationship between the peak contraction force and temperature (Shiels and Farrell, 1997). This relationship was also reported for tropical species including Nile tilapia (Costa *et al.*, 2000) and traíra (Olle, 2003) exposed to an acute temperature transition from 25 to 35°C (Fig. 4.7). Differently, acute increases of temperature had no effect on twitch force of pacu. This species had no inotropic adjustment in relation to acute temperature changes, which implied that chronotropic adjustments (Anelli-Jr *et al.*, 2004). Consistently, Coyne *et al.* (2000) reported that the twitch force of the trout heart was the same at 10 and 20°C. In general, the inotropic component appears to be species-specific and related to the Ca^{2+} sensitivity of the myofilaments.

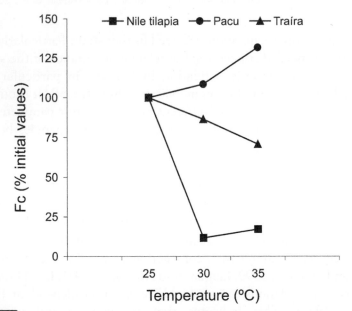

Fig. 4.7 The effect of temperature transition from 25° to 35°C and subsequent return to 25°C on F_c — % of initial values) of ventricle strips of tropical teleosts.

Stimulation Frequency and Force Development of the Myocardium

Considerable information is available for the effects of stimulation frequency on force development of vertebrate hearts. Information on stimulation and force development are performed *in vitro*, while the stimulation frequency is adjusted to desired levels and Fc is measured. These procedures serve to evaluate the excitation-contraction coupling capacity of the myocardium. Within the physiologically relevant range, an increased stimulation frequency increases Fc of the myocardium. This applies to most mammals (Koch-Weser and Blinks, 1963), turtles, amphibians (Driedzic and Gesser, 1985) and elasmobranchs (Maylie *et al.*, 1979; Driedzic and Gesser, 1988). By contrast, in most teleost fish Fc decreases with increased stimulation frequency (negative staircase). This applies to Atlantic cod, *Gadus morhua*, Atlantic mackerel, *Scomber scombrus* (Driedzic and Gesser, 1985), crucian carp, *Carassius carassius* (Vornanen, 1989), skipjack tuna, *Katsuwonus pelamis* (Keen *et al.*, 1992), frillfin goby, *Bathygobius soporator* (Rantin *et al.*, 1998), yellowfin tuna, *Thunnus albacares* (Shiels *et al.*, 1999), Nile tilapia, *Oreochromis niloticus* (Costa *et al.*, 2000), Pacific mackerel, *Scomber Japonicus* (Shiels and Farrell, 2000) and rainbow trout, *Onchorhynchus mykiss* (Driedzic and Gesser, 1985; Hove-Madsen and Gesser, 1989; Hove Madsen, 1992; Møller-Nielsen and Gesser, 1992; Keen *et al.*, 1994; Gesser, 1999; Hove-Madsen and Tort, 1998; Shiels *et al.*, 1998; Aho and Vornanen, 1999; Harwood *et al.*, 2000). The effects of increasing stimulation frequency on force development for a number of teleosts are shown in Fig. 4.8.

The high frequency-induced reduction in Fc could result from (1) a reduced Ca^{2+} influx to the SL or from (2) a reduced Na^+ efflux via a reverse-mode Na^+-Ca^{2+} exchanger or, finally, from (3) a reduction of Ca^{2+} release from the SR (SR). These three mechanisms are probably interrelated through a Ca^{2+}-induced release of Ca^{2+}—from the SR (Fabiato, 1983; Shiels *et al.*, 2002) (see Fig. 4.9).

As an exception, this negative Fc/f_H response is absent or reversed in some high performance species, including tuna (Keen *et al.*, 1992; Shiels *et al.*, 1999). Atlantic cod (Driedzic and Gesser, 1985) and mackerel (Driedzic and Gesser, 1988; Shiels and Farrell, 2000). This implies that high performance teleosts are able to maintain Fc, f_H and cardiac output and aerobic scope during exposure to high temperature and/or exercise.

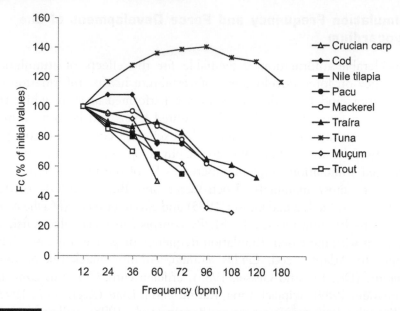

Fig. 4.8 The force-frequency relationship for a number of teleosts. Fc is normalized to the value for the lowest contraction frequency (0.2 Hz) for each species. Modified from Shiels *et al.* (2002).

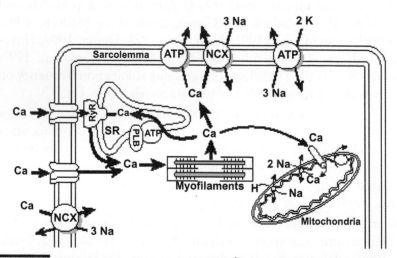

Fig. 4.9 The putative sources of activator Ca^{2+} during E-C coupling of fish cardiac myocytes. The teleost and mammalian myocardium contains similar cellular components for E-C coupling, i.e., SR calcium stores, L-type calcium channels, contractile elements, Na^+/Ca^{2+} exchanger (NCX), and sarcolemmal Ca^{2+}-ATPase. The relative importance of these sources of calcium for contractile activation need to be better defined in teleosts, i.e., transsarcolemmal calcium fluxes vs mobilization from the SR. Moreover, the sources of calcium for contraction are not well established for the fish heart. In addition, the Ca^{2+} source may be temperature-dependent (Coyne *et al.*, 2000). Modified from Bers (2002).

Sources of Ca^{2+}

Interruption of regular stimulation of heart muscle produces characteristic changes in Fc during the recovery phase. In this case the sequence is (1) Stimulation and contraction. (2) No stimulation and rest. (3) Re-initiation of stimulation ("post-rest period"). The post-rest response provides important information on basic cellular components underlying cardiac contraction. These include the relative contribution from the intracellular stores of activator Ca^{2+} during contraction (Hajdu, 1969) and the role of the NCX (Sutko *et al.*, 1986). Upon interruption of stimulation of the myocardium, the SL Ca^{2+}-pump and/or the NCX remove Ca^{2+} from the cell. Additionally, Ca^{2+} is stored at intracellular sites, principally within the SR (Mill *et al.*, 1992). As a net result of these Ca^{2+} transfers an amount of activator Ca^{2+} is stored and can be liberated as soon as further contraction is initiated (Bers *et al.*, 1993; Hove-Madsen *et al.*, 2000). In this situation, an increased amount of intracellular stored Ca^{2+} causes a 'post-rest potentiation' of Fc, which leads to a larger amplitude of the twitch signal when stimulation is reinitiated.

Conversely, Ca^{2+} efflux from the cell reduces the amount of stored Ca^{2+} and, consequently, Fc of the heart muscle (Bers, 2001). Undoubtedly, Ca^{2+} management in most fish depends both on SR development and on NCX activity, whereas the Ca^{2+}-pump of the SL is of secondary importance (Tibbits *et al.*, 1991).

In isometric force studies, SR calcium cycling is usually assessed indirectly with ryanodine, which specifically inhibits the SR calcium release channels. Any ryanodine-induced loss in force is attributed to the loss of the SR calcium cycling pathway and a reduction in available activator Ca, and assumes no compensation by other Ca cycling mechanisms (Rousseau *et al.*, 1987).

In most teleost fish, the calcium current across the L-type Ca^{2+} channels accounts for a significant fraction of the total Ca^{2+} flux during the ventricular contraction (Tibbits *et al.*, 1992; Vornanen 1996, 1997, 1998). This is due to small myocyte diameters, that facilitate a direct myofilament activation by the SL Ca^{2+} influx as confirmed by a negligible effect of ryanodine on Fc (Hove-Madsen, 1992; Møller-Nielsen and Gesser, 1992).

The SR is, nevertheless, the main source of cytosolic Ca^{2+} in high performance species, e.g., tuna (Keen *et al.*, 1992; Shiels *et al.*, 1999; Shiels and Farrell, 2000). By contrast, the SR was not inhibited by application of ryanodine in Atlantic cod, sea-raven (Driedzic and Gesser, 1988), crucian carp (Vornanen, 1998), tide pool goby (Rantin *et al.*, 1998) and Nile tilapia (Costa *et al.*, 2000). Consistently, ryanodine sensitivity of trout increased, when unphysiological stimulation frequencies and/or temperatures were applied (Keen *et al.*, 1994; Hove-Madsen *et al.*, 1998). Further, species-specific activity and the SR participation in the E-C coupling are probably related, since ryanodine reduced force development by 30 to 40% in some highly aerobic performance species, including the Pacific mackerel (Shiels and Farrell, 2000), skipjack tuna (Keen *et al.*, 1992) and yellowfin tuna (Shiels *et al.*, 1999), which indicates an important role of the SR for Ca^{2+} activation.

The L-type Ca^{2+} channel current was insufficient to fully activate contraction in isolated trout myocytes, which suggests that the SR contributed to the cytosolic Ca^{2+} regulation during the E-C coupling (Hove-Madsen and Tort, 1998; Hove-Madsen *et al.*, 1998). It should, however, be noted that the experimental temperature was 25°C, a high temperature for the species in question. This motivated the hypothesis that the performance of the SR is favored by temperature changes close to the upper ecologically relevant temperature range for the species (Shiels *et al.*, 2002). This may not apply to tropical species in which the upper temperature range is much higher. Thus, Rantin *et al.* (1998) and Costa *et al.* (2000) studied the frillfin goby and the Nile tilapia, species that are exposed to daily temperature changes by as much as15°C. These species did not show any sensibility to ryanodine, neither at 25°C (annual mean temperature of their environment) nor at 40°C (Fig. 4.10). This indicates a negligible role of SR-based Ca^{2+} for contraction.

As mentioned above, Fc of some tropical species depends on stored intracellular Ca^{2+}. Consequently, the post-rest potentiation of Fc could be abolished by application of ryanodine (Fig. 4.10). Evidence of a functional SR has been described for atrial muscle strips of a highly aerobic performer, the tuna (Keen *et al.*, 1992) and for ventricular muscle strips of pacu—a neotropical species exposed to high temperatures (Bers, 2001; Anelli-Jr. *et al.*, 2004).

It should be stressed that highly active species such as curimbatá, pacu and tuna possess a functional SR that busts Ca^{2+} delivery for myocardial contraction. This leads to a cardiac performance that matches high aerobic demand during intensive activity. This supports Keen *et al.* (1992), Aho and Vornanen (1998), Shiels *et al.* (1999), and Shiels and Farrell (2000), since they proposed that a functional SR characterizes fast

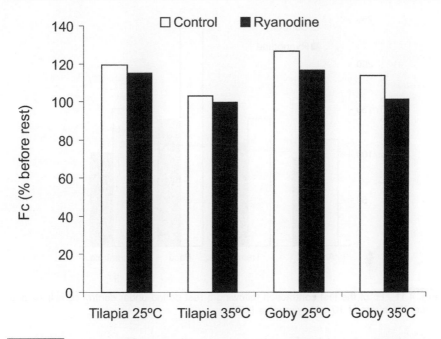

Fig. 4.10 Force of the first contraction following a rest period. The figure compares a control preparation and the contraction resulting from application of 10 mM ryanodine at 25 and 40°C.

swimming 'athletic fish' as expressed by the authors. As an exception, the armored catfish has a functional SR but is sedentary. A functional SR is essential for high performance, but nothing prevents that a sedentary fish has a functional SR (Fig. 4.11).

Phylogenetic background may explain why several species, including traíra, armored catfish, curimbata, carp and pacu, possess a functional SR. These species belong to the same superorder Ostariophysi (Chugun *et al.*, 1999; Anelli-Jr. *et al.*, 2004). Phylogenetic aspects could also explain the absence of a functional SR to deliver Ca^{2+} for the myocardial contraction. This applies to the superorder Acanthopterygii, including Nile tilapia, tide-pool goby, flounder and the sea raven. Likewise, the superorder Protoacanthopterygii includes cod and trout that do not have a functional SR.

As exceptions within the Acanthopterygii, a functional SR is present in the tunas *Katsuwonus pelamis* (Keen *et al.*, 1992) and *Thunnus albacares* (Shiels *et al.*, 1999) and also in the Pacific mackerel, *Scomber japonicus* (Shiels and Farrell, 2000). In these species, a functional RS is probably

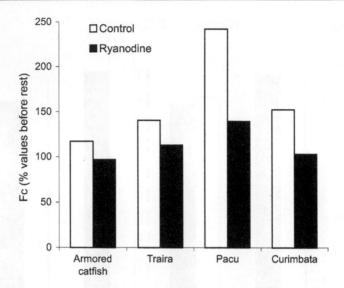

Fig. 4.11 Fc of the first contraction following a rest period under control conditions and under the effect of 10 mM ryanodine (25°C).

related to high levels of aerobic activity. More studies are, nevertheless, needed to evaluate the complex relationships between high temperature, high aerobic activity and the available sources for powerful cardiac performance.

To sum up: high performance teleosts possess large stores of intracellular activator Ca^{2+}, which permits the heart to provide sufficient O_2 to tissues at high temperature and/or exercise. In this aspect the heart of high performance fish (Keen *et al.*, 1992) resemble mammals rather than terrestrial ectothermic vertebrates Tibbits (1996) (Fig. 4.12). As a difference, the teleost fish also depend on extracellular sources of Ca^{2+} which, according to Keen *et al.* (1992), provides more flexibility to modulate the Fc.

GENERAL CONCLUSION

With the adaptation of cardio-respiratory functions to temperature, the teleost fish is able to alleviate the impact of changing temperatures on gas transport. As we have pointed out the adjustments take place both on a daily and on a seasonal basis. The adjustments involve the functions of whole organs and changes on cellular basis. As a key feature, the cardio-respiratory control of teleosts is highly O_2-oriented, reflecting a low and

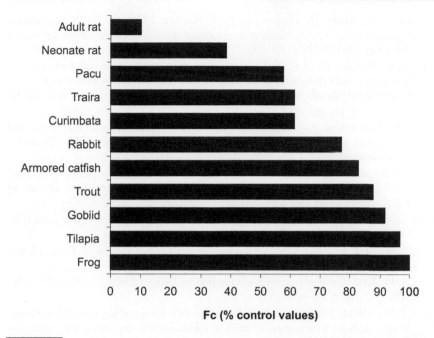

Fig. 4.12 The effects of ryanodine on Fc (% of values before the rest period) of ventricular strips of several species already studied. Modified from Tibbits *et al.* (1991).

variable O_2-availability of water relative to atmospheric air. In this context, the impact of temperature is important since metabolism and O_2-hemoglobin affinity are temperature dependent. In contrast to ectothermic tetrapods vertebrates, teleost fish tend to maintain a constant O_2-extraction of the gas exchange organ, when faced with changes of ambient temperature. Once again, this reflects that respiratory control is linked to ambient O_2, while adjustments of acid-base status are largely accomplished by modulation of bicarbonate levels.

References

Aguiar, L.H., A.L. Kalinin and F.T. Rantin. 2002. The effects of temperature on oxygen uptake, gill ventilation and ECG waveforms in serrasalmid fish, *Piaractus mesopotamicus*. *Journal of Thermal Biology* 27: 299-308.

Aho, E. and M. Vornanen. 1998. Ca^{2+}-ATPase activity and Ca^{2+} uptake by sarcoplasmic reticulum in fish heart: Effects of thermal acclimation. *Journal of Experimental Biology* 201: 252-232.

Aho, E. and M. Vornanen. 1999. Contractile properties of atrial and ventricular myocardium of the heart of rainbow trout *Oncorhynchus mykiss*: Effects of thermal acclimation. *Journal of Experimental Biology* 202: 2663-2677.

Albers, C., R. Manz, D. Muster and G.M. Hughes. 1983. Effects of acclimation temperature on oxygen transport in the blood of carp, *Cyprinus carpio. Respiration Physiology* 52: 165-179.

Andersen, N.A.A., J.S. Laursen and G. Lykkeboe. 1985. Seasonal variations in hematocrit, red cell hemoglobin and nucleoside triphosphate concentration in the European eel *Anguilla anguilla. Comparative Biochemistry and Physiology* A81: 87-92.

Anelli-Jr, L.C., C.D. Olle, M.J. Costa, F.T. Rantin and A.L. Kalinin. 2004. Effects of temperature and calcium availability on ventricular myocardium from the neotropical teleost *Piaractus mesopotamicus* (Holmberg 1887—Teleostei, Serrasalmidae). *Journal of Thermal Biology* 29: 103-113.

Bailey, J.R. and W.R. Driedzic. 1990. Enhanced maximum frequency and force development of fish hearts following temperature acclimation. *Journal of Experimental Biology* 149: 239-254.

Bers, D.M. 2000. Calcium fluxes involved in control of cardiac myocyte contraction. *Circulation Research* 87: 275-281.

Bers, D.M. 2001. *Excitation-Contraction Coupling and Cardiac Contractile Force*. Kluwer Academic Publishers, Dordrecht.

Bers, D.M. 2002. Cardiac excitation-contraction coupling. *Nature* (London) 415: 198-205.

Bers, D.M., J.W.M. Bassani and R.A. Bassani. 1993. Competition and redistribution among calcium transporting systems in rabbit cardiac myocytes. *Cardiovascular Research* 27: 1772-1777.

Cameron, J.N. 1971. Oxygen dissociation characteristics of the blood of the rainbow trout, *Salmo gairdneri. Comparative Biochemistry and Physiology* A38: 699-704.

Cameron, J.N. and D.J. Randall. 1972. The effect of increased ambient CO_2 on arterial CO_2 tension, CO_2 content and pH in rainbow trout. *Journal of Experimental Biology* 57: 673-680.

Cech, J.J., D.W. Bridges, D.M. Rowell and P.J. Balzer. 1976. Cardiovascular responses of winter flounder, *Pseudopleuronectes americanus* (Walbaum), to acute temperature increase. *Canadian Journal of Zoology* 54: 1383-1388.

Chugun, A., T. Oyamada, K. Temma, Y. Hara and H. Kondo. 1999. Intracellular Ca^{2+} storage sites in the carp heart: Comparison with the rat heart. *Comparative Biochemistry and Physiology* A123: 61-71.

Costa, M.J., L. Rivaroli, F.T. Rantin and A.L. Kalinin. 2000. Cardiac tissue function of the teleost fish *Oreochromis niloticus* under different thermal conditions. *Journal of Thermal Biology* 25: 373-379.

Coyne, M.D., C.S. Kim, J.S. Cameron and J.K. Gwathmey. 2000. Effects of temperature and calcium availability on ventricular myocardium from rainbow trout. *American Journal of Physiology* 278: R1535-R1544.

Dejours, P. 1981. *Principles of Comparative Respiratory Physiology*. 2nd Edition. Elsevier/ North Holland, Amsterdam.

Desforges, P.R., S.S. Harman, K.M. Gilmour and S.F. Perry. 2002. Sensitivity of CO_2 excretion to blood flow changes in trout determined by carbonic anhydrase availability. *American Journal of Physiology* 282: R501-R508.

Driedzic, W.R. and H. Gesser. 1985. Ca^{2+} protection from the negative inotropic effect of contraction frequency on teleost hearts. *Journal of Comparative Physiology* B156: 135-142.

Driedzic, W.R. and H. Gesser. 1988. Differences in force-frequency relationships and calcium dependency between elasmobranch and teleost hearts. *Journal of Experimental Biology* 140: 227-241.

Driedzic, W.R. and H. Gesser. 1994. Energy metabolism and contractility in ectothermic vertebrate hearts: hypoxia, acidosis, and low temperature. *Physiological Reviews* 74: 221-258.

Fabiato, A. 1983. Calcium-induced release of calcium from the cardiac sarcoplasmic reticulum. *American Journal of Physiology* 245: C1-C14.

Farrell, A.P. 2002. Cardiorespiratory performance in salmonids during exercise at high temperature: Insights into cardiovascular design limitations in fishes. *Comparative Biochemistry and Physiology* A132: 797-810.

Farrell, A.P. and D.R. Jones. 1992. The heart. In: *Fish Physiology*, W.S. Hoar, D.J. Randall and A.P. Farrell (eds.). Academic Press, San Diego, Vol. 12A, pp. 1-88.

Farrell, A.P. and S.M. Clutterham. 2003. On-line venous oxygen tensions in rainbow trout during graded exercise at two acclimation temperatures. *Journal of Experimental Biology* 206: 487-496.

Farrell, A.P., A.K. Gamperl, J.M.T. Hicks, H.R. Shiels and K.I. Jain. 1996. Maximum cardiac performance of rainbow trout (*Oncorhynchus mykiss*) at temperatures approaching the upper lethal limit. *Journal of Experimental Biology* 199: 663-672.

Fernandes, M.N. and F.T. Rantin. 1989. Respiratory responses of *Oreochromis niloticus* (Pisces, Cichlidae) to environmental hypoxia under different thermal conditions. *Journal of Fish Biology* 35: 509-519.

Gesser, H. 1996. Cardiac force-interval relationship, adrenaline and sarcoplasmic reticulum in rainbow trout. *Journal of Comparative Physiology* B166: 278-285.

Gilmour, K.M. and S.F. Perry. 1994. The effects of hypoxia, hyperoxia of hypercapnia on the acid-base disequilibrium in arterial blood of the rainbow trout. *Journal of Experimental Biology* 192: 269-284.

Glass, M.L. and R. Soncini. 1999. Physiological shunts in gills of teleost fish: assessment of the evidence. In: *Biology of Tropical Fishes*, A.L. Val and V.M.F. Almeida-Val (eds.). INPA, Manaus, pp. 333-341.

Glass, M.L. and S.C. Wood. 1981. Gas exchange and control of breathing in reptiles. *Physiological Reviews* 63: 232-260.

Glass, M.L., N.A. Andersen, M. Kruhøffer, E.M. Williams and N. Heisler. 1990. Combined effects of environmental PO_2 and temperature on ventilation and blood gases in the carp *Cyprinus carpio* L. *Journal of Experimental Biology* 148: 1-17.

Graham, M.S. and A.P. Farrell. 1990. Myocardial oxygen consumption in trout acclimated to 5°C and 15°C. *Physiological Zoology* 63: 536-554.

Graham, M.S. and A.P. Farrell. 1985. The seasonal intrinsic cardiac performance of a marine teleost. *Journal of Experimental Biology* 118: 173-183.

Hajdu, S. 1969. Mechanism of the Woodworth staircase phenomenon in heart and skeletal muscle. *American Journal of Physiology* 216: H206-H214.

Harder, V., R.H.S. Souza, W. Severi, F.T. Rantin and C.R. Bridges. 1999. The South American lungfish — Adaptation to an extreme habitat. In: *Biology of Tropical Fishes*, A.L. Val and V.M.F. Almeida-Val (eds.), Institute Nacional de Pesquisas da Amazônia (INPA), Manaus, pp. 87-98.

Harrison, S.M. and D.M. Bers. 1990. Temperature dependence of myofilament Ca sensitivity of rat, guinea pig and frog ventricular muscle. *American Journal of Physiology* 258: C274-C281.

Hartmund, T. and H. Gesser. 1991. ATP, creatine phosphate and mechanical activity in rainbow trout myocardium under inhibition of glycolysis and cell respiration. *Journal of Comparative Physiology* B160: 691-697.

Harwood, C.L., F.C. Howarth, J.D. Altringham and E. White. 2000. Rate-dependent changes in cell shortening, intracellular Ca^{2+} levels and membrane potential in single, isolated rainbow trout (*Oncorhynchus mykiss*) ventricular myocytes. *Journal of Experimental Biology* 203: 493-504.

Hazel, J.R.W., W.S. Garlick and P.A. Sellner. 1978. The effects of assay temperature on upon the pH optima of enzymes from poikilothermic animals: A test of the imidazole alphastat hypothesis. *Journal of Comparative Physiology* 123: 97-104.

Heisler, N. 1984. Acid-base regulation in fishes. In: *Fish Physiology*, W.S. Hoar and D.J. Randall (eds.). Academic Press, New York, Vol. 10A, pp. 315-401.

Hove-Madsen, L. 1992. The influence of temperature on ryanodine sensitivity and the force-frequency relationship in the myocardium of rainbow trout. *Journal of Experimental Biology* 167: 47-60.

Hove-Madsen, L. and H. Gesser. 1989. Force-frequency relation in the myocardium of rainbow trout: Effects of K^+ and adrenaline. *Journal of Comparative Physiology* B159: 61-69.

Hove-Madsen, L. and L. Tort. 1998. L-type Ca^{2+} current in the excitation-contraction coupling in single atrial myocytes from rainbow trout. *American Journal of Physiology* 275: R2061-R2069.

Hove-Madsen, L., A. Llach and L. Tort. 1998. Quantification of Ca^{2+} uptake in the sarcoplasmic reticulum of trout ventricular myocytes. *American Journal of Physiology* 275: R2070-R2080.

Hove-Madsen, L., A. Llach and L. Tort. 2000. Na^+/Ca^{2+}-exchange activity regulates contraction and SR Ca^{2+} content in rainbow trout atrial myocytes. *American Journal of Physiology* 279: R1856-R1864.

Jackson, D.C. 1973. Ventilatory responses to hypoxia in turtles at various temperatures. *Respiration Physiology* 18: 178-187.

Jackson, D.C. 1989. Control of breathing: Effects of temperature. In: *Comparative Pulmonary Physiology. Current Concepts*, S.C. Wood (ed.). Marcel Dekker, New York, pp. 621-645.

Jobling, M. 1981. Temperature tolerance and the final preference — rapid methods for assessment of optimum growth temperature. *Journal of Fish Biology* 19: 439-455.

Keen, J.E., A.P. Farrell, G.F. Tibbits and R.W. Brill. 1992. Cardiac physiology in tunas. II. Effect of ryanodine, calcium, and adrenaline on force-frequency relationship in atrial strips from skipjack tuna, *Katsuwonus pelamis*. *Canadian Journal of Zoology* 70: 1211-1217.

Keen, J.E., D.M. Vianzon, A.P. Farrell and G.F. Tibbits. 1994. Effect of temperature and temperature acclimation on the ryanodine sensitivity of the trout myocardium. *Journal of Comparative Physiology* B164: 438-443.

Kelsch, S.W. and W.H. Neil. 1990. Temperature preference versus acclimation in fishes: selection for changing metabolic optima. *Transactions of the American Fisheries Society* 119: 601-610.

Koch-Weser, J. and J.R. Blinks. 1963. The influence of the interval between beats on myocardial contractility. *Pharmacological Reviews* 15: 601-652.

Korsmeyer, K.E., N.C. Lai, R.E. Shadwick and J.B. Graham. 1997. Heart rate and stroke volume contribution to cardiac output in swimming yellowfin tuna: Response to exercise and temperature. *Journal of Experimental Biology* 200: 1975-1986.

Krogh, A. 1914. The quantitative relation between temperature and standard metabolism in animals. *Internationale Zeitschrift für Physiologische und Chemische Biologie* 1: 491-508.

Krogh, A. 1968. *The Comparative Physiology of Respiratory Organisms.* Dover, New York.

Kruhøffer, M., M.L. Glass, A.S. Abe and K. Johansen. 1987. Control of breathing in an amphibian *Bufo paracnemis*: Effects of temperature and hypoxia. *Respiration Physiology* 69: 267-275.

Kuhn, H.J., K. Guth, B. Drexler, W. Berberich and J.C. Rüegg. 1979. Investigation of the temperature dependence of the crossbridges parameters for attachment, force generation and detachment as deduced from mechano-chemical studies in glycerinated single fibers from the dorsal longitudinal muscles of *Lethocerus maximus*. *Biophysics of Structure and Mechanism* 6: 1-29.

Laurent, P., S. Holmgren and S. Nilsson. 1983. Nervous and humoral control of the fish heart: structure and function. *Comparative Biochemistry and Physiology* A76: 525-542.

Lennard, R. and H. Huddart. 1989. Purinergic modulation of cardiac activity in the flounder during hypoxic stress. *Journal of Comparative Physiology* B159: 105-113.

Lennard, R. and H. Huddart. 1991. The effect of thermal stress on electrical and mechanical responses and associated calcium movements of flounder heart and gut. *Comparative Biochemistry and Physiology* A98: 221-228.

Lennard, R. and H. Huddart. 1992. Hypoxia-induced changes in electrophysiological responses and associated calcium movements of flounder (*Platichthys flesus*) heart and gut. *Comparative Biochemistry and Physiology* A101: 717-721.

Lomholt, J.P. and K. Johansen. 1979. Hypoxia acclimation in carp—how it affects O_2 uptake, ventilation and O_2 extraction from water. *Physiological Zoology* 52: 38-49.

Maricondi-Massari, M., A.L. Kalinin, M.L. Glass and F.T. Rantin. 1998. The effects of temperature on oxygen uptake, gill ventilation and ECG waveforms in the Nile tilapia *Oreochromis niloticus*. *Journal of Thermal Biology* 23: 283-290.

Matikainen, N. and M. Vornanen. 1992. Effect of season and temperature acclimation on the function of crucian carp (*Carassius carassius*) heart. *Journal of Experimental Biology* 167: 203-220.

Maylie, J.M., M.G. Nunzi and M. Morad. 1979 Excitation-contraction coupling in ventricular muscle of dogfish (*Squalus acanthias*). *Bulletin of Mount Desert Island Laboratory* 19: 84-87.

Mill, J.G., D.V. Vassallo and C.M. Leite. 1992. Mechanisms underlying the genesis of post-rest contractions in cardiac muscle. *Brazilian Journal of Medical and Biological Research* 25: 399-408.

Milsom, W.K. and R.W. Brill. 1986. Oxygen sensitive afferent information arising from the first gill arch of yellow-finned tuna. *Respiration Physiology* 66: 193-203.

Møller-Nielsen, T. and H. Gesser. 1992. Sarcoplasmic reticulum and excitation-contraction coupling at 20 and 10°C in rainbow trout myocardium. *Journal of Comparative Physiology* B162: 526-534.

Morita, A. and H. Tsukuda. 1995. The effect of thermal acclimation on the eletrocardiogram of goldfish. *Journal of Thermal Biology* 19: 343-348.

Nikinmaa, M. 1990. *Vertebrate Red Blood Cells: Adatations of Function to Respiratory Requirements*. Springer-Verlag, Berlin.

Olle, C.D. 2003. Função cardíaca do teleósteo *Hoplias malabaricus* (Teleostei, Erythrinidae) diferentes condições térmicas. Master's Thesis, Federal University of São Carlos, SP, Brazil.

Piiper, J. and P. Scheid. 1984. Model analysis of gas transfer in fish gills. In: *Fish Physiology*, W.S. Hoar and D.J. Randall (eds.). Academic Press, New York, Vol. 10A, pp. 228-262.

Purup-Hansen, S. and H. Gesser. 1987. Extracellular Ca^{2+}, force, and energy state in cardiac tissue of rainbow trout. *American Journal of Physiology* 253: R838-R847.

Randall, D.J. and J.N. Cameron. 1973. Respiratory control of arterial pH as temperature changes in rainbow trout (*Salmo gairdneri*). *American Journal of Physiology* 225: 997-1002.

Rantin, F.T., H. Gesser, A.L. Kalinin, C.D.R. Guerra, J.C. Freitas and W.R. Driedzic. 1998. Heart performance, Ca^{2+} regulation and energy metabolism at high temperatures in *Bathygobius soporator*, a tropical marine teleost. *Journal of Thermal Biology* 23: 31-39.

Reeves, R.B. 1972. An imidazole alphastat hypothesis for vertebrate acid-base regulation: tissue carbon dioxide content and body temperature in bullfrogs. *Respiration Physiology* 14: 218-236.

Reeves, R.B. 1976. Temperature-induced changes in blood acid-base status, pH and PCO_2 in a binary buffer. *Journal of Applied Physiology* 40: 752-761.

Reynolds, W.W. and M.E. Casterlin. 1979. Behavioral thermoregulation and the "final preferendum" paradigm. *American Zoologist* 19: 211-224.

Roberts, J.L. 1979. Seasonal modulation of thermal acclimation and behavioral thermoregulation in aquatic animals. In: *Marine Pollution: Functional Responses*, W.B. Vernberg, A. Calebrese, F.P. Thurberg and F.J. Vernberg (eds.). Academic Press, New York, pp. 365-388.

Rousseau, E., J.S. Smith and G. Meissner. 1987. Ryanodine modifies conductance and gating behaviour of single Ca^{2+} release channels. *American Journal of Physiology* 253: C364-C368.

Saunders, R.L. 1962. The irrigation of the gills in fishes. II. Efficiency of oxygen uptake in relation to respiratory flow activity and concentrations of oxygen and carbon dioxide. *Canadian Journal of Zoology* 40: 817-862.

Schmidt-Nielsen, K. 1975. *Animal Physiology—Adaptations and Environment*. Cambridge University Press, London.

Schurmann, H. and J.S. Schou-Christiansen. 1994. Behavioral thermoregulation and swimming activity in the cod (*Boreogadus saida*) and the navaga (*Eleginus navaga*). *Journal of Thermal Biology* 19: 207-212.

Shiels, H.A. and A.P. Farrell. 1997. The effect of temperature and adrenaline on the relative importance of the sarcoplasmic reticulum in contributing Ca^{2+} to force development in isolated ventricular trabeculae from rainbow trout. *Journal of Experimental Biology* 200: 1607-1621.

Shiels, H.A. and A.P. Farrell. 2000. The effect of ryanodine on isometric tension development in isolated ventricular trabeculae from Pacific mackerel (*Scomber japonicus*). *Comparative Biochemistry and Physiology* A125: 331-341.

Shiels, H.A., E.V. Freund, A.P. Farrell and B.A. Block. 1999. The sarcoplasmic reticulum plays major role in atrial muscle of yellowfin tuna. *Journal of Experimental Biology* 202: 881-890.

Shiels, H.A., E.D. Stevens and A.P. Farrell. 1998. Effects of temperature, adrenaline and ryanodine on power production in rainbow trout *Oncorhynchus mykiss* ventricular trabeculae. *Journal of Experimental Biology* 201: 2701-2710.

Shiels, H.A., M. Vornanen and A.P. Farrell. 2002. The force-frequency relationship in fish hearts — A review. *Comparative Biochemistry and Physiology* 132: 811-826.

Soncini, R. and M.L. Glass. 1997. The effects of temperature and hyperoxia on arterial PO_2 and acid-base status in *Piaractus mesopotamicus. Journal of Fish Biology* 51: 225-233.

Soncini, R. and M.L. Glass. 2000. Oxygen and acid-base status related drives to gill ventilation in carp. *Journal of Fish Biology* 56: 528-541.

Souza, R.H.S., R. Soncini, M.L. Glass, J.R. Sanches and F.T. Rantin. 2001. Ventilation, gill perfusion and blood gases in dourado, *Salminus maxillosus* Valenciennes (Teleostei, Characidae), exposed to graded hypoxia. *Journal of Comparative Physiology* B171: 483-489.

Stecyk, J.A. and A.P. Farrell. 2002. Cardiorespiratory responses of the common carp (*Cyprinus carpio*) to severe hypoxia at three acclimation temperatures. *Journal of Experimental Biology* 205: 759-768.

Stevens, E.D. and W.D. Neill. 1978. Body temperature relations of tuna, specially skipjack. In: *Fish Physiology*, W.S. Hoar and D.J. Randall (eds.), Academic Press, New York, Vol. 7, pp. 315-424.

Stevens, E.D. and A.M. Sutterlin. 1976. Heat transfer between the fish and the ambient water. *Journal of Experimental Biology* 65: 131-145.

Stevens, E.D., H.M. Lan and J. Kendall. 1974. Vascular anatomy of the countercurrent heat exchanger of skipjack tuna. *Journal of Experimental Biology* 61: 145-153.

Sundin, L., S.G. Reid, A.L. Kalinin, F.T. Rantin and W.K. Milsom. 1999. Cardiovascular and respiratory reflexes: the tropical fish, traíra (*Hoplias malabaricus*) O_2 chemoresponses. *Respiration Physiology* 116: 181-199.

Sundin, L., S.G. Reid, F.T. Rantin and W.K. Milsom. 2000. Branchial receptors and cardiorespiratory reflexes in neotropical fish, the tambaqui (*Colossomma macropomum*). *Journal of Experimental Biology* 203: 1225-1239.

Sutko, J.L., D.M. Bers and J.P. Reeves. 1986. Postrest inotropy in rabbit ventricle: Na^+-Ca^{2+} exchange determines sarcoplasmic reticulum Ca^{2+} content. *American Journal of Physiology* 250: H654-H661.

Tibbits, G.F. 1996. Towards a molecular explanation of the high performance of the tuna heart. *Comparative Biochemistry and Physiology* A113: 77-82.

Tibbits, G.F., L. Hove-Madsen and D.M. Bers. 1991. Calcium transport and the regulation of cardiac contractility in teleosts: a comparison with higher vertebrates. *Canadian Journal of Zoology* 69: 2014-2019.

Tibbits, G.F., C.D. Moyes and L. Hove-Madsen. 1992. Excitation-contraction coupling in the teleost heart. In: *Fish Physiology*, W.S. Hoar, D.J. Randall and A.P. Farrell (eds.). Academic Press, New York, Vol. 12A, pp. 267-303.

Tiitu, V. and M. Vornanen. 2002. Regulation of cardiac contractility in a cold stenothermal fish, the burbot *Lota lota* L. *Journal of Experimental Biology* 205: 1597-1606.

Tompe, P.P. and H.E.D.J. Keurs. 1990. Force and velocity of sarcomere shortening in trabeculae from rat heart. Effects of temperature. *Circulation Research* 66: 1239-1254.

Vornanen, M. 1989. Regulation of contractility of the fish (*Carassius carassius* L.) heart ventricle. *Comparative Biochemistry and Physiology* C94: 477-483.

Vornanen, M. 1996. Effect of extracellular calcium on the contractility of warm- and cold-acclimated crucian carp heart. *Journal of Comparative Physiology* B165: 507-517.

Vornanen, M. 1997. Sarcolemmal Ca influx through L-type Ca channels in ventricular myocytes of a teleost fish. *American Journal of Physiology* 272: R1432-R1440.

Vornanen, M. 1998. L-type Ca^{2+} current in fish cardiac myocytes: Effects of thermal acclimation and β-adrenergic stimulation. *Journal of Experimental Biology* 201: 533-547.

Vornanen, M., H. Shiels and A.P. Farrell. 2002. Plasticity of excitation-contraction coupling in fish cardiac myocytes. *Comparative Biochemistry and Physiology* 132: 827-846.

Wood, S.C. 1982. The effect of oxygen affinity on arterial PO_2 in animals with central vascular shunts. *Journal of Applied Physiology* 53: R1360-R1364.

Yamamitsu, S. and Y. Itazawa. 1990. Effects of preload, afterload and temperature on the cardiac function and ECG in the isolated perfused heart of carp. *Nippon Suisan Gakkaishi* 56: 229–237.

Zummo, G., R. Acierno, C. Agnisola and B. Tota. 1995. The heart of the icefish: Bioconstruction and adaptation. *Brazilian Journal of Medical and Biological Research* 28: 1265-1276.

Oxygen Consumption during Embryonic Development of the Mudskipper (*Periophthalmus modestus*): Implication for the Aerial Development in Burrows

Aya Etou[1], Tatsusuke Takeda[1], Yu Yoshida[2] and Atsushi Ishimatsu[2],*

INTRODUCTION

Mudskippers are a highly specialized group of goby, belonging to the subfamily Oxudercinae. They inhabit intertidal mudflats of tropical and temporal zones with the highest species diversity in Asia. Mudskippers have developed physiological abilities for respiration, mobility and vision in air, and these abilities enable them to lead amphibious mode of life, although the extent of terrestriality ranges widely between species. During low tide when mudflat surface is exposed, mudskippers volitionally emerge

Authors' addresses: [1]Department of Animal and Marine Bioresource Science, Faculty of Agriculture, Kyushu University, Fukuoka 812-8581, Japan.
[2]Institute for East China Sea Research, Nagasaki University, Tairamachi, Nagasaki 851-2213, Japan.
Corresponding author: E-mail: a-ishima@nagasaki-u.ac.jp

from burrows and spend most of their time on the mudflat surface for feeding, territory defence, and courtship (Clayton, 1993). During high tide when the mudflat surface is inundated with water, most species seem to reside in their burrows, but some may spend hours perching on exposed stones or stalks of halophytic vegetation (Ikebe and Oishi, 1996)

Mudskippers are also exceptional among fishes in their reproductive strategy. The most remarkable feature of mudskipper reproduction is the deposition of air of their burrows probably to ensure oxygen supply to developing embryos. This seems to be an environmental adaptation that is of vital importance since water filling mudskipper burrows are so hypoxic that embryos would not be able to develop (Atkinson and Taylor, 1991; Ishimatsu et al., 1998, 2000).

Periophthalmus modestus is distributed on the coasts of southern Japan, Korea and China (Murdy, 1989). Spawning season of *P. modestus* in southern Japan is from mid-May to the end of July. *P. modestus* builds J-shaped burrows in mudflats, and the eggs are deposited in the terminal upturned end. The average number of the eggs in one burrow is ca. 6,000, and eggs are laid in a monolayer on the wall of the egg chamber (Kobayashi et al., 1971). *P. modestus* eggs are ellipsoidal in shape, the size of fertilized eggs being 0.90 mm (longer axis) × 0.64 mm (shorter axis). The eggs have filamentous attachment threads on the animal pole (Kobayashi et al., 1972). Like many other gobies, the male guards burrows during egg incubation period, which lasts for ca. one week. At the end of the egg incubation period, eggs hatch and larvae escape from the burrow for dispersion. It is currently unknown whether the larvae swim by themselves through the main shaft of the burrow to escape or the paternal fish help them in the process.

The purposes of the present study are to investigate developmental changes in oxygen consumption rates of *P. modestus* eggs in air and water, effects of aerial hypoxia on oxygen consumption rate, and survival of the eggs and newly hatched larvae in hypoxic water.

MATERIALS AND METHODS

Preparation of the Eggs

Eggs of *P. modestus* were collected in June and July, 2003, at an estuarine mudflat along a tributary of Rokkaku River, Saga, Japan. After excavating egg chambers of burrows, the chambers complete with eggs were extracted and brought back to the laboratory. Eggs were separated from the mud and repeatedly rinsed in 20‰ sea water to remove mud particles covering the eggs. The cleaned eggs were placed on a piece of wetted filter paper, which drooped into 20‰ sea water in a plastic box. The eggs were incubated in moist air in the box, and not immersed in sea water. Eggs were daily cleaned by rinsing, and any dead or damaged eggs were removed. Egg incubation and all measurements were conducted at 24°C. Developmental stages were determined from the drawings by Kobayashi *et al.* (1972) who reported embryonic development of artificially fertilized *P. modestus* eggs in normoxic water (Fig. 5.1). The eggs, from a total of 30 burrows, were used for measurements within a few days after collection.

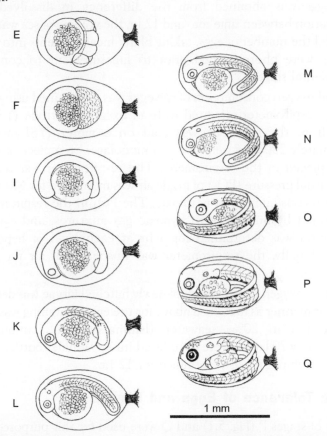

Fig. 5.1 Developmental stages of the *Periophthalmus modestus* eggs used in this chapter. E: 8-cell stage, 125 min after fertilization, F: morula stage, 290 min, I: formation of embryo, Kuppfer's vesicle appeared, 29 hr, J: eye-vesicle appeared, 30 hr, K: 9-somite stage, 35 hr, L: 17-somite stage, otocysts formed, 39 hr, M: heart beat began, 52 hr, N: 18-somite stage, 63 hr, O: xanthophores appeared, 99 hr, P: pectoral fins and swimbladder developed, 117 hr, Q: competent of hatching, 170 hr. Eggs were obtained by hormone treatment, and artificially fertilized. Fertilized eggs were incubated in aerated 50% seawater at 19-20°C. Reproduced after Kobayashi *et al.* (1972) with permission.

Developmental Changes in Oxygen Consumption

Oxygen consumption of the eggs was determined in both air and water. Aquatic oxygen consumption of the eggs was determined with the Winkler method (Cech, 1990). Approximately 60 eggs were transferred into a 50-ml Winkler bottle filled with well-aerated 20‰ sea water. Three bottles were prepared for each clutch to determine egg oxygen consumption in 12 hr at 24°C; another three bottles served as blank. Initial concentration of dissolved oxygen was determined by immediately adding reagents into the other set of three bottles. Oxygen consumption of the eggs was obtained from the difference in dissolved oxygen concentration between time zero and 12 hr, the volume of sea water in the bottle, and the number of eggs, taking blank measurements into account. The eggs were photographed prior to measurement to confirm the developmental stage.

Aerial oxygen consumption of the eggs was determined using a 1.5 ml plexiglass respirometer mounted with an oxygen electrode (YSI 5750). Output from the electrode was read on a meter (YSI model 58). Approximately 250 eggs were laid in a monolayer on a piece of plankton net and placed in the respirometer. The respirometer was sealed and oxygen partial pressure (PO_2) of inside air was monitored for 6 hr to obtain egg oxygen consumption in normoxia. The PO_2 in the respirometer was then lowered by flushing with hypoxic gas mixtures, and egg oxygen consumption was determined for 3 hr at each level of hypoxia (see Fig. 5.3). Finally, the respirometer was flushed with air for recovery measurement.

Aquatic oxygen consumption of newly hatched larvae was determined in a similar manner as described above for the eggs. Hatching was induced by immersion in 20‰ seawater (Ishimatsu et al., unpublished). Approximately 20 larvae were introduced into a Winkler bottle, and their oxygen consumption was determined for 12 hr.

Hypoxic Tolerance of Eggs and Larvae

The eggs of stages P (Fig. 5.1) and Q were used for this purpose. Six eggs were placed on a piece of plankton net in triplicate, and placed in a plastic chamber supplied with different levels of hypoxic gas mixture. The oxygen level of the chamber air was lowered from 100% air to 50%, 20% and 10%. Heart rate, and eye and body movements were counted at 0, 3, and 8 hr.

The PO_2 of the chamber air was continuously monitored with the oxygen meter.

Twenty individuals of newly hatched larvae were introduced into a 500 ml Erlenmeyer flask filled with 20‰ sea water, of which the dissolved oxygen levels ranged from 26 to 11%. Larvae were observed every hour. When a larva did not respond to shaking the flank, it was judged to be dead. Dissolved oxygen concentration was determined by the Winkler titration immediately before and after the test.

Data are expressed in mean ± SD, wherever possible.

RESULTS

Developmental Changes in Oxygen Consumption Rate

Figure 5.2 shows developmental changes in aquatic and aerial oxygen consumption rates in normoxia. The aquatic oxygen consumption rate increased 4-fold during development from 15.11 nl hr^{-1} (stage E-F; 2 to 5 hr after fertilization N = 1) to 59.01 ± 3.50 (stage P-Q; 117-170 hr, N = 2). Aerial oxygen consumption rate increased from 18.5 ± 5.3 nl hr^{-1} (N = 4) to 76.5 ± 12.0 (N = 5). There was no statistically significant difference between oxygen consumption rates in air and water where data are available at comparable developmental stages (P > 0.05, t-test).

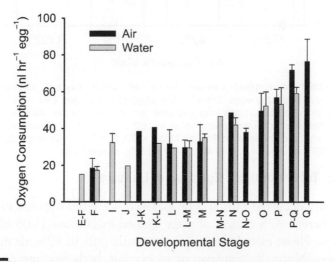

Fig. 5.2 Developmental changes in oxygen consumption rate (Mean + SD) of *Periophthalmus modestus* eggs in air (black bars) and normoxic water (gray bars). The number of measurements ranged from one (without an error bar) to seven. There was no significant difference between those groups where pairwise comparison was possible (P > 0.05, t-test).

Oxygen consumption rate of newly hatched larvae was 138.7 ± 32.9 nl hr^{-1} (N = 29).

Effect of Aerial Hypoxia on Egg Oxygen Consumption

Oxygen consumption rate of the eggs significantly decreased in the lowest hypoxic condition (21% air (=4.3 kPa PO$_2$), P < 0.05, Friedman Repeated Measures Analysis of Variance on Ranks). Oxygen consumption significantly increased during recovery (P < 0.05, Fig. 5.3). There was no obvious difference between developmental stages in this regard.

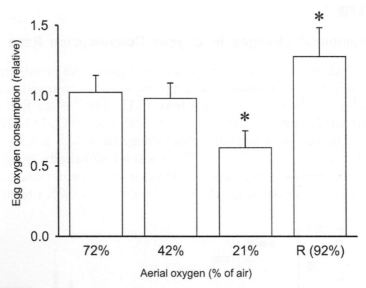

Fig. 5.3 Effect of ambient oxygen levels on aerial oxygen consumption of *Periophthalmus modestus* eggs. Data for the eggs of the stages L-Q were pooled for comparison since no difference was detected among the stages. Asterisks indicate significant differences from the control value at 100% air (P < 0.05, Friedman Repeated Measures Analysis of Variance on Ranks). Mean + SD, N = 9. R: recovery. PO$_2$ of the 72, 42, 21, and 92% air are 14.8, 8.7, 4.3, and 19.0 kPa, respectively.

Hypoxic Tolerance of Eggs and Larvae

No mortality occurred during 8 hr exposure to aerial hypoxia in eggs of stage P, whereas 90% of eggs of stage Q died in 10% air (2.06 kPa PO$_2$) within 8 hr. Heart rate decreased significantly only in 10% air in eggs of both stages. Normally transparent embryonic body became opaque in those individuals that developed slowed heart rate. Body movement appeared to be suppressed in hypoxic conditions (10% air for stage P and 20 and 10% air for stage Q).

Larval survival decreased with severity of aquatic hypoxia (Fig. 5.4). At dissolved oxygen levels of 10% saturation, all larvae died within 1 hr.

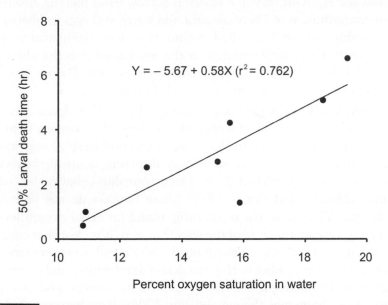

Fig. 5.4 The relationship between levels of aquatic hypoxia and 50% death time (hr) of *Periophthalmus modestus* larvae. The 50% death time was obtained from the relationship between survival rate and exposure time at each hypoxic level.

DISCUSSION

Air has several advantages over water regarding oxygen supply. First, oxygen concentration in air is approximately 30 times higher than that in water at the same PO_2. Second, diffusion coefficient of oxygen is 8000 times higher in air than in water (Dejours, 1981). These characteristics of air facilitate oxygen diffusion through the egg surface and reduce thickness of the boundary layer, which may constitute a significant diffusion barrier in aquatic environment. In spite of these differences, the oxygen consumption rates of *P. modestus* eggs were indistinguishable when measured in air and in normoxic water (Fig. 5.2). Thus, if *P. modestus* eggs were incubated in normoxic water in the mudflat burrows, they would develop normally.

However, *P. modestus* eggs in fact cannot develop in the burrows. Burrows in intertidal sediments usually contain hypoxic water (Atkinson and Taylor, 1991), and in particular those burrows excavated in mudflats create extremely hypoxic conditions due to decomposition of organic matters by microorganisms and poor exchange of interstitial water (Raffaelli and Hawkins, 1996). *P. modestus* burrow water had the dissolved oxygen concentration of 7% of air-saturated water, and eggs incubated in water of this oxygen level died within two days (Ishimatsu *et al.*, unpublished). Thus, air deposition in the egg chamber is the absolute necessity for the eggs to develop in mudflat burrows, and *P. modestus* eggs develops normally in air as long as enough humidity is given.

Oxygen levels of the gas samples obtained from *P. modestus* burrows generally ranged from 48 to 93% of outside air level (PO_2 ranging from 10 to 19.2 kPa), although gaseous PO_2 were often considerably lower when collected from burrows with guarding male displaying courtship behavior (Ishimatsu *et al.*, unpublished data). Since courtship behavior precedes spawning (Matoba and Dotsu, 1977), these burrows do not probably contain eggs. Thus, from the relationship found for aerial oxygen levels and oxygen consumption rate of the eggs (Fig. 5.3), it is clear that oxygen levels in the egg chamber are maintained high enough to ensure oxygen uptake of developing embryos. Hypoxia delays development and growth of fish embryos (Rombough, 1988), has teratogenic effects, and disturbs balance of sex hormones (Shang and Wu, 2004). How burrow-guarding males maintain oxygen environment in the egg chamber remains obscure.

Once hatched within the egg chamber, the larvae must quickly escape from the burrow. Sensitivity to hypoxia generally increases as development proceeds, and newly hatched larvae are significantly more sensitive than embryos before hatch (Rombough, 1988). The present results demonstrated that the hatched larvae die within 1 hr if they remain in the burrow. Therefore, there must be some mechanisms for the larvae to come out to ensure their dispersion. We do not currently know if burrow-guarding male helps them escape. Further observation is needed to clarify this point.

References

Atkinson, R.J.A. and A.C. Taylor. 1991. Burrows and burrowing behaviour of fish. *Symposium of the Zoological Society* (London) 63: 133-155.

Cech, J.J. Jr. 1990. Respirometry. In: *Methods for Fish Biology*, C.B. Schreck and P.B. Moyle (eds.). American Fisheries Society, Bethesda, pp. 335-362.

Clayton, D.A. 1993. Mudskippers. *Oceanography Marine Biology Annual Reviews* 31: 507-577.

Dejours, P. 1981. *Principles of Comparative Respiratory Physiology.* 2nd edition. Elsevier/ North Holland, Amsterdam.

Ikebe, Y. and T. Oishi. 1996. Correlation between environmental parameters and behaviour during high tides in *Periophthalmus modestus. Journal of Fish Biology* 49: 139-147.

Ishimatsu, A., T. Takeda, T. Kanda, S. Oikawa and K.H. Khoo. 2000. Burrow environment of mudskippers in Malaysia. *Journal of Bioscience* 11: 17-28.

Ishimatsu, A., Y. Hishida, T. Takita, T. Kanda, S. Oikawa, T. Takeda and K.H. Khoo. 1998. Mudskippers store air in their burrows. *Nature* (London) 391: 237-238.

Kobayashi, T., Y. Dotsu and N. Miura. 1972. Egg development and rearing experiments of the larvae of the mud skipper, *Periophthalmus cantonensis. Bulletin of Faculty of Fisheries, Nagasaki University* 33: 49-62.

Kobayashi, T., Y. Dotsu and T. Takita. 1971. Nest and nesting behavior of the mud skipper, *Periophthalmus cantonensis* in Ariake Sound. *Bulletin of Faculty of Fisheries, Nagasaki University* 32: 27-40.

Matoba, M. and Y. Dotsu. 1977. Prespawning behavior of the mud skipper, *Periophthalmus cantonensis* in Ariake Sound. *Bulletin of Faculty of Fisheries, Nagasaki University* 43: 23-33.

Murdy, E.O. 1989. A taxonomic revision and cladistic analysis of the Oxudercine gobies (Gobiidae: Oxudercinae). *Records of Australian Museum* 11 (Supplement): 1-93.

Raffaelli, D. and S. Hawkins. 1996. *Intertidal Ecology.* Chapman and Hall, London.

Rombough, P.J. 1984. Respiratory gas exchange, aerobic metabolism, and effects of hypoxia during early life. In: *Fish Physiology*, W.S. Hoar and D.J. Randall (eds.). Academic Press, San Diego, Vol. 11A, pp. 59-161.

Shang, E.H.H. and R.S.S. Wu. 2004. Aquatic hypoxia is a teratogen and affects fish embryonic development. *Environment Science and Technology* 38: 4763-4767.

Gill Morphological Adjustments to Environment and the Gas Exchange Function

Marisa Narciso Fernandes*, Sandro Estevan Moron and Marise Margareth Sakuragui

INTRODUCTION

The gills are the main respiratory organs of exclusively water breathing fish species. Facultative or obligatory air-breathing species have the skin and accessory organs for breathing air. Like other respiratory surfaces, the gills have large surface area contained in a small space, and a small diffusion distance between water and blood for oxygen and carbon dioxide exchange. During the respiratory cycle, the double pumping mechanism generates an almost continuous water flow across the gills in a counter-current direction to the blood flowing through the blood spaces in the secondary lamellae. Striated and smooth muscles precisely control the movements of gill arches and filaments ensuring their position in the water

Authors' address: Laboratory of Zoophysiology and Comparative Biochemistry, Department of Physiological Sciences, Federal University of São Carlos, via Washington Luis, km 235, 13565-905–São Carlos, SP, Brazil.
*Corresponding author: E-mail: dmnf@power.ufscar.br

flow. External and internal chemoreceptors in the gills contribute to reflexes that ensure an efficient breathing pattern and proper ventilation-perfusion matching at the respiratory surface (Sundin *et al.*, 1999; Reid *et al.*, 2000)

In addition to the respiratory function the gills are the site of other important functions related to animal homeostasis such as water and ion regulation, acid-base balance and excretion of waste products that are essential for life (Evans *et al.*, 2005). Although these functions correspond to specialized epithelia with distinct blood vascular pathways they are superimposed. The large respiratory surface of gills facilitates passive fluxes of water and ions across gill epithelia unfavorable for homeostasis and has to counteract by active ion transport.

Freshwater environments are unstable and characterized by daily or/ and seasonal fluctuations of temperature, oxygen, carbon dioxide, pH, dissolved ions, and organic and inorganic substances. The gills have to cope with all the functions to allow fish to survive in such environment. Physiological changes in response to changes in the environment may result in morphological adjustments of gill epithelia that, as pointed out by Laurent (1989), may improve one function in a given environmental situation in detriment to another.

This chapter focuses the gill morphology in relation to the respiratory function, the adaptations to ionic disturbances and the interference in the gas transfer.

GENERAL MORPHOLOGY OF TELEOST GILLS

The anatomy and general structure of fish gills depend on the group in question and have been described in extensive reviews (Hughes and Morgan, 1973; Hughes, 1984a; Laurent, 1884; Wood 2001). Briefly, the gills of teleost fish consist of four gill arches on each side of pharynx, each supporting two rows of filaments or primary lamellae (holobranches) that varies in number and length depending on species and fish development (Fig. 6.1A). The upper and lower surfaces of each filament bear regularly spaced lamellae or secondary lamellae, which are the units for gas exchange, forms a "curtain" through which the water flows (Fig. 6.1B). The gill components, filaments and lamellae, are adjusted to the plastic form features of other functional head components such as the shape of the sternohyoid muscle, ceratobranquial length and head streamlining to meet the functional demand of each species. The available space for the

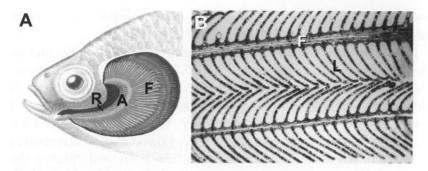

Fig. 6.1 **A** Localization of the four gill arches in the fish head. **B.** Detail showing two gill filaments and the position of lamellae. A, gill arch; F, filament; L, lamella; R, rakers.

gill apparatus and the gill shape are related to the water velocity over the gills and the feeding habits (Adapted after Galis and Barel, 1980).

Internally, a cartilaginous rod provides support for the gill filament and the filament striated abductor and adductor muscles connected to the gill arch bones (Fig. 6.2), the epibranchial and ceratobranchial, adjust the position of filaments during ventilation (Laurent, 1984; Fernandes and Perna, 1995; Fernandes *et al.*, 1995). Smooth muscle bundles scattered among collagen fibers connecting successive filaments attest to the fine adjustments of filament position during the respiratory cycle ensuring a

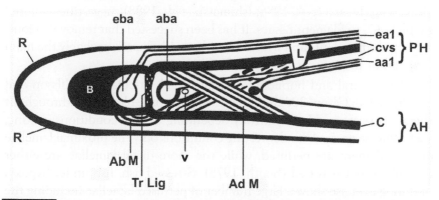

Fig. 6.2 Schematic cross-section of a gill arch showing the gill structure. For clarity, the vascular components, afferent (aba) and efferent (eba) branchial arteries, afferente (aa1) and efferent (ea1) filamental or primary arteries, central venous sinus (cvs) and branchial vein (v) are represented only in the upper side of the figure. One lamella associated with the filamental vasculature is represented. The sustaining components, the gill arch bone (B) and filament cartilage rod (C), the gill muscles: abductor (Ab M) and adductor (Ad M) muscles, and the transverse ligament (Tr Lig) are also shown. AH, anterior hemibranch; PH, posterior hemibranch; R, rakers.

flow of water on the lamellae, although water flow bypass may occur at high ventilatory volumes (Fernandes and Perna, 1995).

The blood supply to gill tissue consists of an arterio-arterial and an arterio-venous pathway (Laurent, 1984; Olson, 2002a). In the arterio-arterial pathway, the deoxygenated blood from the ventral aorta is distributed to branchial arteries into each gill arch, to the filamental afferent artery at each gill filament and then, via short afferent arterioles, to the lamella, where the gas exchange takes place (Fig. 6.2). Oxygenated blood is collected in very short efferent arterioles and reaches the filamental efferent artery towards the branchial efferent artery, which coalesces in the dorsal aorta. The arterio-venous vasculature consists of two systems. One of these systems is called nutritive, and consists of short arteries originating from the efferent branchial and filamental efferent arteries. Blood is drained into filamental veins and then to branchial vein in the gill arch. The second system consists of a network of blood vessels, the central venous sinus or venolymphatic system in the filament and arterio-venous anastomoses from the arterio-arterial blood pathway supplying the central venous sinus, which is drained into filamental veins and from there into the branchial vein. The central venous sinus occupies the full length of filament and varies in complexity among species. The functional importance of these blood systems has been discussed as, *in vivo*, the arterio-venous flow being between 7-10% of the arterio-arterial blood flow (Iwama *et al.*, 1986; Ishimatsu *et al.*, 1988) suggesting a limited supply of O_2 to filament tissues. It has been suggested that removal of ions, acid-base equivalents and wastes from the epithelium is probably the main function of this system (Laurent, 1989).

Both neural and hormonal control of gill vasculature (Nilsson and Sundin, 1998; Olson, 2002b) provides an adequate blood flow throughout the arterio-arterial blood pathway depending on fish conditions: resting, exercising or stressed. Under resting conditions only the proximal lamellae on the filament are perfused, while the more distal lamellae are either partially or not perfused (Booth, 1978). Stressed fish, fish under hypoxia or during exercise show a large number of perfused lamellae including the distal ones (Booth, 1979; Sundin, 1995). Increasing blood pressure into branchial and filamental vessels provides the perfusion of lamellar blood spaces by surpassing the critical opening pressures of afferent and efferent arterioles or pillar channels. Circulating catecholamines tend to increase these effects by dilating the afferent arterioles, via β-adrenergic receptors, and constricting the efferent arterioles via α-adrenergic receptors, thereby favoring the blood flow throughout the lamellae including the more distal ones (Nilsson and Sundin, 1998). Furthermore, catecholamines tend to constrict the arterio-venous pathway favoring the arterio-arterial blood pathway and causing lamellar distention, which increases the lamellar height and consequently the lamellar area and may also reduce the water blood distance (Laurent and Hebibi, 1989; Sollid *et al.*, 2003). Icefish gills have large blood spaces in the lamella and a prominent and continuous marginal channel; noradrenaline increases the diameter of blood spaces and lamellar height, and the pillar cells control lamellar perfusion (Rankin and Tuurala, 1998). Conversely, cholinergic stimulation tends to reduce the number of perfused lamellae, causing an overall vasoconstriction effect on the arterio-arterial blood pathway.

The lamellae consist of a series of pillar cells covered with the basement membrane and two epithelial cell layers continuous with the filament epithelium. Pillar cells contain large central nuclei that may be indented in the region of the collagen columns connecting the two surfaces of basement membrane. These cells are distributed have polygonally or linearly throughout lamella (Fig. 6.3). The flanges of pillar cells form spaces in which the blood circulates in the opposite direction of the water flow. The proximal channel of the lamella is buried in the filament and is undifferentiated from the rest of the lamellar network. The marginal channel around the free edge of the lamella comprises endothelial cells and the flange of pillar cells, and is usually larger than the blood channels comprising by the flanges of pillar cells (Fig. 6.3). The contact region between the pillar cell flanges may be simple abutment. The basement membrane of the lamella is continuous with that of the filament and contains collagen strands connecting the two sides of the basement membrane in the lamella (Hughes, 1984a; Olson, 2002). The inner epithelial cell layer is undifferentiated and is separated from the outer epithelial cell layer by the lymphoid space, which may contain leukocytes. Adhesion structures are not usually found between the cells in the two cell layers. However, tight junctions and numerous desmosomes in the tortuous junction region characterize the junctions between neighboring epithelial cells.

Gill Epithelia

Two types of epithelia surround the gills entirely. The epithelium surrounding the filament is multilayered, has several types of cells, is

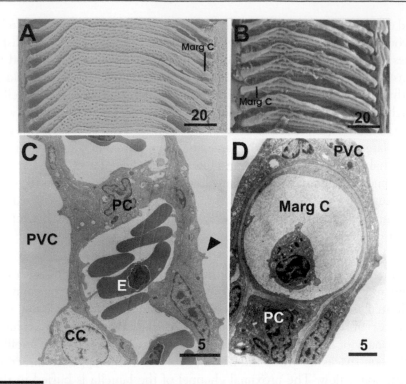

Fig. 6.3 A and B. Vasculature of lamella of two erythrinid fish, the water-breathing *Hoplias malabaricus* (**A**) and facultative air-breathing fish *Hoplerythrinus unitaeniatus* (**B**). The holes in the lamellar casts represent the localization of pillar cells, which were digested. Note the almost polygonal arrangement of pillar cells in the water-breathing erythrinid and the parallel arrays arrangement of these cells in the facultative air-breathing erythrinid. The lamellar marginal channel (Marg C) is larger than the blood channels formed by the pillar cell flanges. Scale bars in µm. **C and D.** Cross section of lamella showing the blood spaces formed by the pillar cell flanges and pillar cell body (PC) of base of lamella in **C** and the marginal channel (Marg C) in **D** of *H. malabaricus*. Note the smooth surface membrane of the pavement cells (PVC) in the outermost lamellar epithelium cell layer and microridges (arrowhead). CC, chloride cell; E, erythrocyte. Scale bars in µm.

closely associated with the arterio-venous vasculature and is called non-respiratory epithelium. The epithelium that covers the lamella is thin in most fish, consisting of two layers of flat epithelial cells; it is associated to arterio-arterial vasculature and is called respiratory epithelium.

Filament Epithelium

The majority of cells found in the filament epithelium from the inner to the outmost layer are nondifferentiated, neuroepithelial, chloride, mucous and pavement cells (Fig. 6.4). Other cell types that may be found in the

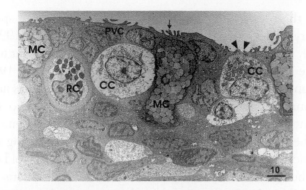

Fig. 6.4 TEM micrography of filament epithelium of *H. malabaricus*. Note the pavement cells with microridges (arrow), chloride cell (CC) showing several mitochondria and its apical surface in contact with the surrounding water (arrowheads), mucous cells (MC) with mucus vesicles and a rodlet cell (RC). Scale bars in μm.

filament epithelium, but are not common to most fish species, are the rodlet cells (Fig. 6.4) which are secretory cells restricted to the filament epithelium, whose function is still unknown, for they are present in several other fish organs (see review: Manera and Dezfuli, 2004; Mazon *et al.*, 2007). Macrophages and mast cells are also, in some cases, found among the filament epithelium cells.

Nondifferentiated cells are abundant in the innermost epithelial layer. The neuroepithelial cells described by Dunel-Erb *et al.* (1982) are located on the serosal side of the epithelium, resting on the basement membrane, along the leading edge of filament. These cells have a spheroid or bilobed nucleus, densely packed microfilaments, and mitochondria in the perinuclear and apical region. The cells are mainly characterized by dense-cored vesicles of 80 to 100 nm scattered within the cytosol without any apparent pattern of distribution. Green fluorescence induced by formaldehyde treatment of these cells indicates the presence of biogenic amine serotonin (Baily *et al.*, 1992). The close proximity of nerve endings and the degranulation of dense-cored vesicles caused by environmental hypoxia suggest a possible chemoreceptor role.

Mucous cells are distributed mainly on the leading and trailing edges of filaments and are scattered in the interlamellar region (Laurent, 1989). These cells have the nucleus located in the basal region of the cell close to mitochondria, and possess a Golgi apparatus. Mucous vesicles are concentrated in the apical region of the cell (Fig. 6.4). They are almost completely covered by the pavement cells and using SEM, it is difficult to distinguish these cells from the chloride cells in some fish species (Fig. 6.5A) (Perry and Laurent, 1993; Moron and Fernandes, 1996, Moron et al., 2003). However, they are easily recognized under a light microscope due to the properties of the mucosubstances (Gona, 1979). The mucosubstances (glycoproteins with sulfated or acid residues and neutral glycoproteins) produced by theses cells are released into the water environment by fusing the mucous vesicles into apical cell membranes and releasing their contents in the form of droplets or releasing the entire mucous vesicles into water. Unstressed fish appear to be devoid of mucous substances but may show an extensive mucus layer in situations of stressed (Handy and Eddy, 1991). Stressed animals appear to release large amounts of mucus vesicles at once (Fig. 6.5B) (Cabral, unpublished). Mucosubstances may be involved in fish's defense against mechanical injuries, toxic substances and pathogens in water (Shepheard, 1994). The charged radicals of mucus attract ions close to cell surface resulting in a ion gradient from gill membrane to water flow during respiration (Handy, 1989) which may help ion regulation. Furthermore, mucus complex with metals washes out toxic metals from the gill surface and then reduces the diffusion rate into fish (Handy et al., 1989; Tao et al., 2000).

Chloride cells are usually distributed mainly in the interlamellar region and at the trailing edge of filament (Laurent, 1984; Wood, 2001). In freshwater fish they may also be found distributed in the leading edge of filament close to lamellae. These cells are round or elongated and have

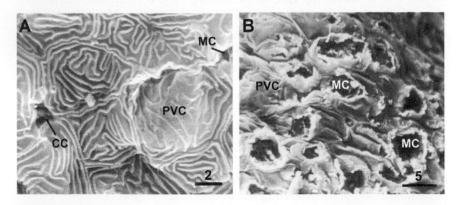

Fig. 6.5 A. SEM micrography of filament epithelium of *H. malabaricus* showing pavement cells (PVC) with microridges overlapping the chloride (CC) and mucous (MC) cells. Scale bars in μm. Modified from Moron and Fernandes (1996.) **B.** Mucous cells from stressed fish (*Rhinelepis strigosa*). Scale bars in μm.

large nuclei. They are characterized by a large number of mitochondria, small vesicles and a network system of tubular membranes contacting the interstitial medium (Fig. 6.4). Immunogold labeling of Na^+-K^+-ATPase is largely distributed in the tubular system and is higher in mature chloride cells (Dang *et al.*, 2000). Two chloride cell types, α- and β-chloride cells, was described in freshwater fish based in their morphological features and location in the filament epithelium (Pisam *et al.*, 1987, 1993; Pisam and Rambourg, 1991; Powell *et al.*, 1994). α-chloride cells are electron-opaque, elongated in shape, with a smooth apical membrane and are found at the base of lamellae close to the basal membrane. β-chloride cells are electron-dense, round or ovoid in shape, located in the interlamellar region. However, these differences may be related to the stage of development or degeneration of chloride cells as discussed by Wendelaar Bonga and van der Meij (1989) and Wendelaar Bonga *et al.* (1990).

In seawater fish chloride cells are single and characterized by an apical pit or are found in clusters sharing invagination, in the apical crypt. An accessory cell indented in the lateral surface of adjacent chloride cells and inserted between chloride cell and pavement cells constitutes the main difference between the filament epithelium of seawater and freshwater fish (for further details see: Laurent, 1984; Wood, 2001). The junctional complexes between chloride and pavement cells are characterized by tight junctions and desmosomes while a type of leaky junction is present between chloride and accessory cells (Laurent, 1984)

In freshwater fish the apical surface of chloride cells varies among species and environmental conditions ranging from an apical surface with microvilli to an indented, smooth or sponge-like surface (Fig. 6.6). In some species the chloride cells are buried in the epithelium and the apical surface is almost entirely covered by the pavement cells (Fig. 6.5A). Extensive tight junctions with several desmosomes characterize the intercellular junctions between chloride and pavement cells (Moron and Fernandes, 1996). Depending on external or internal environmental conditions the apical chloride cell surface exposed to water changes in a few hours by dynamic changes in pavement cell coverage (Goss *et al.*, 1995; Laurent *et al.*, 1995b; Perry, 1997; Fernandes *et al.*, 1998; Fernandes and Perna-Martins, 2002; Moron *et al.*, 2003). The chloride cell function in seawater fish is well established. They are the main sites for active trans-cellular excretion of Cl^- which is coupled with passive paracellular excretion of Na^+ (Wood and Marshall, 1994; McCormick, 1995; Wood, 2001). In freshwater fish the function of chloride cells is related to the

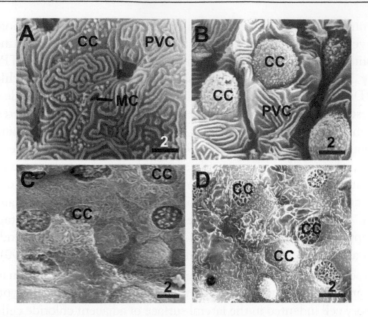

Fig. 6.6 SEM micrography of filament epithelium showing the apical surface of chloride cells of four tropical fish: erythrinid *Hoplerythrinus unitaeniatus* (**A**), cichlid *Geophagus brasiliensis* (**B**), loricariids *Hypostomus tietensis* (**C**) and *Rhinelepis strigosa* (**D**). Scale bars in μm.

active uptake of Cl^-, Na^+, Ca^{2+} and probably other divalent ions (Flik *et al.*, 1993). The active uptake of Na^+ by chloride cells remains uncertain in particular environmental situations such as hypercapnia and acid waters (Goss *et al.*, 1994).

Pavement cells are flat, polygonal, with few mitochondria and an abundant endoplasmic reticulum and Golgi complex. The junctional complex between pavement cells includes numerous desmosomes and long tight junction at the apical surface. The cells are overlaid with glicocalyx and their apical surface membrane may be smooth with long microridges limiting cells, or present short or long convoluted microridges like 'fingerprint' (Figs. 6.5, 6.6). These features increase the cell surface in contact with the external medium, and may trap mucus and/or create an unstirred boundary layer close to the cell membrane.

Lamellar Epithelium

The lamellar epithelium consists two layers of epithelial cells (Fig. 6.3C, D). The innermost cell layer is undifferentiated, while the outermost cell layer consists of pavement cells like those described for the filament epithelium. Mucous cells rarely appear in the lamellar epithelium (Laurent, 1989; Fergunson *et al.*, 1992) but may often be found in the lamellae of the air-breathing fish, *Hoplerythrinus unitaeniatus* (Moron *et al.*, 2003). In freshwater fish chloride cells are present in the lamella of air-breathing fish and fish living in soft and/or polluted waters (Fernandes and Perna-Martins, 2001; Mazon *et al.*, 2002; Moron *et al.*, 2003). Leukocytes can be found between the two epithelial cell layers in the lamellae. The apical surface membrane of pavement cells may be similar to filamental pavement cells or presenting short microvilli or smooth surfaces (Fig. 6.7) (Moron and Fernandes, 1996; Fernandes and Perna-Martins, 2001; Moron *et al.*, 2003). The two flat epithelial cell layers of lamellae make the diffusion distance between water and blood very short, ranging from 3 µm in fast swimming fish up to 10-14 µm in benthic sluggish fish (Hughes, 1972) or in air-breathing fish (Hakin *et al.*, 1978; Costa *et al.*, 2007). However, this is not a rule, as evidenced by some tropical fish living in hypoxic waters, e.g., *Hoplias malabaricus* which, although not very active, presents a very small water-blood distance (Sakuragui *et al.*, 2003).

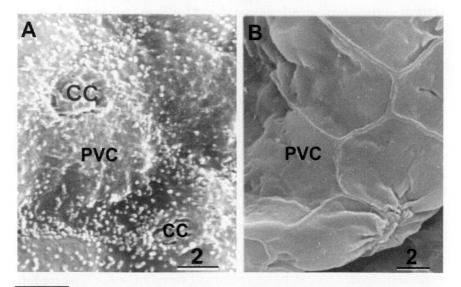

Fig. 6.7 SEM micrography of lamellar epithelium of loricariid *Hypostomus tietensis* showing pavement cells with microvilli (**A**) and erythrinid *Hoplias malabaricus* showing a smooth pavement cell surface characterized by microridges defined within in the cell limits (**B**). Scale bars in µm.

GILL MORPHOMETRY AND RESPIRATORY FUNCTION

Knowledge about the gill respiratory surface area and the area of accessory air-breathing organs in fish is essential for quantitative studies to compare the effectiveness of these respiratory organs for gas exchange. Hughes's method (1966) has been intensively used to date. The method consists of three main measurements: the total length of the filaments (L), number of secondary lamellae/mm on both sides of filament (n) and the estimated bilateral area of the lamella: total gill area $= L \cdot n \cdot bl$. Each measurement can be taken from a sample or samples of whole gills. As pointed out by Hughes (1972, 1984c), depending on the number of filaments in each gill arch measurements are made using every 5th, 10th or even more filaments in large gills and the lamellae area is estimated using the measured filaments of the second gill arch at their base, medium and apical region, since the lamellae are different in size along filaments and between filaments. Sampling methods inevitably contain intrinsic errors and, although most fish gills are geometrically organized in filaments and lamellae, some air-breathing species have highly modified gills making Hughes's method difficult to apply. A new stereological method using vertical sections was recently proposed and has been applied to obligatory and facultative air-breathing species as well to obligatory water-breathing species (Moraes et al., 2005; Costa et al., 2007). The new method is less time consuming, and is applicable to all types of gills. In this chapter, all gill morphometric discussion is based on data obtained by Hughes's method (1966), since most of the available data from the gill area of fish have been obtained by this method.

Fish respiration is limited by the physical and chemical properties of water (Dejours, 1981). The high viscosity and low O_2 solubility of water implies in a high metabolic cost of respiration. Respiration may be aggravated by numerous environmental factors such as temperature, organic matter and pollution. The efficiency of the respiratory function of fish gills depends on the gill's diffusing capacity. Fick's law of diffusion capacity is given as Dphysiol $= \dot{V}O_2/\Delta PO_2$, where $\dot{V}O_2$ is the quantity of O_2 transferred and ΔPO_2 is the O_2 partial pressure difference between water and blood. This can be expressed as Dmorphol $= KA/t$, where K is the tissue specific Krogh diffusion constant for respiratory gases, A is the respiratory surface area and t is the water-blood diffusion distance. This means that the diffusion capacity is proportional to the respiratory gill area and inversely proportional to the diffusion distance.

The respiratory surface area and the thickness of the gas diffusion distance between species, as already emphasized above, contribute to shed light on fish physiology and are of crucial importance in studies on fish growth, taxonomy and the effects of environmental changes on the respiratory function (Hughes 1980, 1984b; Hughes and Al-Kadhomiy, 1988; Santos *et al.*, 1994; Perna and Fernandes, 1996; Mazon *et al.*, 1998, 2002; Fernandes and Perna-Martins, 2002). For instance, the respiratory surface area (lamellae area) of the erythrinid fish *Hoplias malabaricus* (obligatory water-breather) and *Hoplerythrinus unitaeniatus* (facultative air-breather) illustrates the differences in the gill components and respiratory surface area (lamellae area) as fish grow (Table 6.1). *H. malabaricus* has greater number of long filaments and of small lamellae. The allometric relationship between the gill components and/or the respiratory gill surface area (Y) and the body mass (M_B) given by the equation $Y=aM_B^b$, where a is the value of 1 g of fish and b is the allometric or scaling constant, shows similar increases in the number of filaments and a lesser increase in length in the air-breathing fish. However, the major difference between species is the fewer lamellae/mm of filament (b = − 0.09 in *H. malabaricus* and b = −0.16 in *H. unitaeniatus*) and the larger bilateral area of lamellae in *H. malabaricus* as the fish grows (b = 0.69 in *H. malabaricus* and b = 0.43 in *H. unitaeniatus*). These features, combined with the size-specific growth in filament length, results in a larger respiratory surface area in *H. malabaricus* (b = 1.14 for total lamellae surface area) than in *H. unitaeniatus* (b = 0.66 for total lamellae surface area) (Fernandes *et al.*, 1994). The water-blood thickness diffusion distance of lamellae is lower (harmonic mean = 3.16 μm) in

Table 6.1 Values for the gill components and respiratory surface area from erythrinids *Hoplias malabaricus* and *Hoplerythrinus unitaeniatus* of 10, 100 and 1000 g based in the alometric relationship between the gill components and area (Y) and body mass (M_B) $Y = aM_B^b$ (see text for details). Calculated from Fernandes *et al.* (1994).

Gill components	H. malabaricus			H. unitaeniatus		
	10 g	100 g	1000 g	10 g	100 g	1000 g
Total filament number	2134	2687	3383	1766	2275	2931
Total filament length (mm)	2959	10260	35576	2638	6042	13842
Lamellae number/mm filament	28	23	19	33	23	16
Lamellar bilateral area (mm²)	0.010	0.048	0.235	0.016	0.043	0.116
Total lamellae area (mm²)	1739	24005	331361	2739	12521	57231
Respiratory area/M_B (mm²/g)	173.91	240.07	331.39	269.86	123.35	56.38

H. malabaricus (Sakuragui *et al.*, 2003) than in *H. unitaeniatus* (harmonic mean = 5.44 µm) (Moron *et al.*, unpublished). Considering that under resting conditions or in normoxic water the entire gill surface area is not utilized (Booth, 1978; Hughes 1980), the large number of lamellae of *H. malabaricus* gills, combined with the high O_2 affinity of its hemoglobin (Johansen *et al.*, 1978; Perry *et al.*, 2004) and the changes in blood pressure and flow during hypoxia due to catecholamines releases (Nilsson and Sundin, 1998; Perry *et al.*, 2004), favor the increase of blood distribution into lamellae (Soivio and Tuurala, 1981), resulting in additional functional gill surface area (Randall and Daxboeck, 1984). These factors represent an advantage for O_2 transfer from water to blood during hypoxia in this species, mainly in the larger specimens, and enable *H. malabaricus* to live in the same ponds as *H. unitaeniatus*, whose respiratory process is more flexible, since it takes O_2 from air.

   H. malabaricus has a low O_2 demand when at rest; lower than that reported for *Hoplias lacerdae*, a more active water-breathing erythrinid species that inhabits well-oxygenated waters and is less tolerant to hypoxia (Rantin *et al.*, 1992; Kalinin *et al.*, 1993), *H. unitaeniatus* (Mattias *et al.*, 1996; Oliveira *et al.*, 2004) and other teleosts (Fernandes and Rantin, 1989, 1991; Fernandes *et al.*, 1995; Mattias *et al.*, 1998; Takasusuki *et al.*, 1998) even at high temperatures (Cameron and Wood, 1978; Rantin *et al.*, 1985). Small specimens ($\sim$ 40 g) of *H. malabaricus* were less tolerant to hypoxia than large specimens (350 g) (Kalinin *et al.*, 1993). Under gradual hypoxia, small *H. malabaricus* and *H. unitaeniatus* (without access to air) maintain an almost constant O_2 uptake up to PO_2 = 27-30 mmHg and 80 mmHg, respectively (Kalinin *et al.*, 1993; Mattias *et al.*, 1996), and the large ones up to 18 and 40 mmHg (Kalinin *et al.*, 1993; Oliveira *et al.*, 2004). Below these PO_2s they behave as oxygen conformers. To meet their needs during hypoxia the ventilation increase is higher in *H. unitaeniatus* but oxygen extraction is lower (Mattias *et al.*, 1996). *H. unitaeniatus*, with free access to air, starts to breath air at water PO_2 around 60-40 mmHg (Mattias *et al.*, 1996; Oliveira *et al.*, 2004; Perry *et al.*, 2004), and under acute hypoxia (water PO_2 = 5 mmHg) the oxygen consumption from air is 40 $mlO_2 \cdot kg^{-1} \cdot h^{-1}$ (Jucá-Chagas, 2004). Differences in respiratory gill area and the water-blood barrier may be the morphometric base for hypoxia tolerance, and the blood properties and other adjustments are physiological components that explain the high tolerance to low oxygen level in the environment.

Gill Chloride Cells and Respiration

Seawater or freshwater teleost fish have to maintain osmolalities of around 250-350 mosmol relative to the external medium (1050 mosmol in 100% seawater and <10 mosmol in freshwater). Seawater, with the exception of estuarine regions, is a more stable environment while the freshwater may varies from <10 mosmol to almost distilled water. Consequently, passive influx of ions (NaCl) and efflux of water occurs in seawater fish whereas a passive loss of ions (NaCl) and gain of water occurs in freshwater fish. The gills, with their large surface area and reduced water-blood distance, are the main fish structure where the diffusive processes take place. Conversely, as already mentioned, the gills are also the site for osmotic and ionic regulation and acid-base equilibrium. The gill chloride cells are known to be responsible for Cl^- secretion in seawater fish concomitant with Na^+ excretion, via paracellular pathway (Fosket and Scheffey, 1982; Marshall, 1995; Wood, 2001). In freshwater fish the role of these cells is more complex. Several studies have shown that the chloride cells are the site of active uptake of Na^+ and Cl^- (Avella *et al.*, 1987; Perry *et al.*, 1992; Laurent *et al.*, 1995a; Perry, 1997), and Ca^{2+} uptake (Flik *et al.*, 1996). However, at least in situations of acute hypercapnia, there is evidence that chloride cells play a role in acid-base regulation (Goss *et al.*, 1992, 1994, 1995, 1998). In this case, trans-epithelial Cl^- uptake occurs in the chloride cells (Cl^-/HCO_3^- exchange) whereas Na^+ uptake occurs in the pavement cells, via Na^+ channel on the apical membrane linked to an electrogenic proton pump (H^+-ATPase). The current models for ion secretion in seawater and ion uptake in freshwater of gills are given by Evans *et al.* (2005), Wood (2001) and a molecular approach is found in Hirose *et al.* (2003).

The number and distribution of chloride cells in freshwater fish vary among species, even in those living in the same environment. In general they are restricted to the leading and trailing edges of filament close to and at the base of lamellae and in the filament's interlamellar region. Chloride cells are occasionally found in the lamellae. The chloride cells density in a given freshwater species is not directly related with higher whole body ion uptake as evidenced in a study on ion transport in different fish species (Perry *et al.*, 1992) and in fish exposed to metals (Mazon *et al.*, 2002). However, higher ion absorption is direct related to chloride cell fractional area (CCFA) (Perry *et al.*, 1992). Chloride cell proliferation, induced by cortisol in response to changes in the internal or external medium, concomitantly increases Ca^{2+} (Flik and Perry, 1989) and NaCl (Bindon et al., 1994a) uptake in freshwater fish. However, the increase in CC number and ion uptake is not proportional. Cortisol increases the chloride cell turnover resulting in numerous immature or senescent chloride cells (Dang et al., 2000; Mazon et al., 2002), which do not have high Na^+/K^+-ATPase units in the basolateral membrane, and hence, they have lower transport activity than mature chloride cells (Dang et al., 2000).

Most fish living in soft waters or chronically exposed to soft water have numerous chloride cells on the filament and throughout the lamellae, suggesting an adaptation to maintain ion regulation. However, some of them rarely present chloride cells in the lamellae. For instance, the erythrinids, H. malabaricus and H. unitaeniatus living in the same soft water (ion concentration in mM: $Ca^{2+} = 2.13 \pm 0.09$, $Mg^{2+} = 0.073 \pm 0.001$, $Na^+ = 0.50 \pm 0.01$, $Cl^- = 0.678 \pm 0.015$) ponds shows distinct differences in chloride cell density (Fig. 6.8). H. malabaricus rarely have

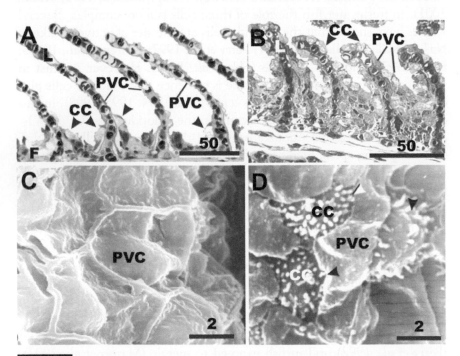

Fig. 6.8 Light and scanning electron micrograph showing the lamellae of the erythrinids, *Hoplias malabaricus* (**A and C**) and *Hoplerythrinus unitaeniatus* (**B and D**) from the same environment. Note the presence of numerous chloride cells in the *H. unitaeniatus* lamellae and the differences in the apical membrane of chloride and pavement cells. Scale bars in μm. modified from Moron *et al.* (2003).

chloride cells in the lamella, while *H. unitaeniatus* have around 20 and 9 chloride cell/mm^2·10^2 in the lamellae (Fernandes and Perna-Martins, 2002; Moron *et al.*, 2003; Sakuragui *et al.*, 2003). Neither species responds to exposure to a 3-part dilution of soft water, but *H. malabaricus* and *H. tietensis* exhibit high chloride cell proliferation in the filament and lamella after 24 h of exposure to distilled water (Fig. 6.9), whereas *H. unitaeniatus* shows chloride cell proliferation in the lamellae only after 48 h of exposure. Nevertheless, the total chloride cell density in contact with the external medium does not increase in either species. The apical surface area goes from 10 to 55 μm in *H. malabaricus* and from 15 to 35 μm in *H. unitaeniatus*, resulting in a significant increase of CCFA that is higher in *H. malabaricus*. However, *H. tietensis*, a loricariid fish, from the same environment as *H. malabaricus* and *H. unitaeniatus* has numerous chloride

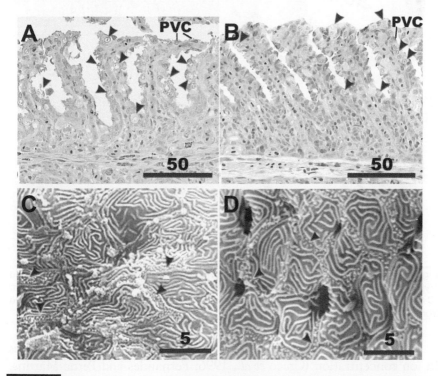

Fig. 6.9 Light and scanning electron micrograph showing the lamellae and filament epithelium of the erythrinids, *Hoplias malabaricus* (**A and C**) and *Hoplerythrinus unitaeniatus* (**B and D**) exposed to distilled water. Note the thickness of lamellae and the apical surface of chloride cells. Scale bars in μm. Modified from Moron *et al.* (2003).

cells with large apical surface area in the lamellae (Fig. 6.10A). The chloride cells of *H. tietensis* exposed to distilled water show high proliferation in lamellae and 66% from them exhibit sharp reduction on their apical surface in direct contact with the external medium by forming a sponge-like structure recessed under the adjacent pavement cells resulting in a 50% reduction of the CCFA (Fig. 6.10B) (Fernandes and Perna-Martins, 2002). In all cases, the morphological adjustments of gill chloride cell efficiently maintain ion homeostasis.

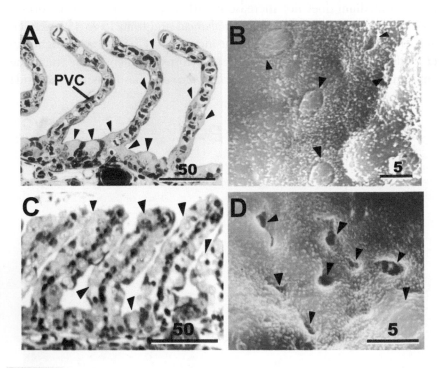

Fig. 6.10   Light and scanning electron micrograph showing the lamellae of the loricariid *Hypostomus tietensis* from natural habitat (**A and B**) and after exposure to distilled water (**C and D**). Note the changes in the apical surface of chloride cells. Scale bars in μm. Modified from Fernandes and Perna-Martins (2002).

The presence of chloride cells on the lamellae in response to changes in ion concentration (Greco *et al.*, 1996; Fernandes and Perna-Martins, 2002; Moron *et al.*, 2003), or contaminants (Dang *et al.*, 2000; Mazon *et al.*, 2002, 2004) in water or circulating hormones, cortisol and growth hormone (Bindon *et al.*, 1994a, b), potentially affect the respiratory function of gills by increasing the water-blood diffusion distance for gas exchange. The decrease in the thickness of the water-blood diffusion barrier reported by Laurent and Hebibi (1989) and by Sakuragui *et al.* (2003) is related to stretch lamellar epithelium due to enlargement of blood spaces in the lamella caused by gill hyperemia. The significant reduction in arterial PO_2 observed by Sakuragui *et al.* (2003) during the exposure of *H. malabaricus* to deionized water (1-7 days) in normoxia evidenced that the decrease of water-blood diffusion distance by epithelial stretching and lamellar hyperemia or the increase of water-blood diffusion distance by chloride cell proliferation both lead to impairment of the gill respiratory function. According to Fick's diffusion law, although the rate of gas diffusion across the lamellar epithelium is inversely related to the thickness of water-blood distance, lamellar hyperemia favors gas transfer only to erythrocytes close to epithelium. On the other hand, chloride cell proliferation on lamellae limits the gas transfer from water to blood.

The degree of oxygen transfer limitation may depend on the intensity of chloride cell proliferation in the lamellae and on whether there adjustments occur to compensate for it. Studies in *Oncorhynchus mykiss* experiencing chloride cell proliferation in the lamellae showed reduction in the oxygen transfer (lowering arterial PO_2) in normoxia and graded hypoxia (Thomas and Motais, 1990), in moderate and severe hypoxia (Bindon *et al.*, 1994b; Greco *et al.*, 1995) and only under normoxia and moderate hypoxia but not in severe hypoxia (Perry *et al.*, 1996, Perry, 1998). *H. malabaricus* kept in deionized water for 1, 2, and 7 days and exhibiting chloride cell proliferation in the lamellae show arterial PO_2 reduction in normoxia but did not show arterial PO_2 reduction compared with fish kept in control water under the same degree of environmental hypoxia (Fig. 6.11A) (Sakuragui *et al.*, 2003). The activation of mechanisms to improve O_2 transfer from water to blood should be involved in the effectiveness of O_2 transfer in the fish kept in deionized water under severe hypoxia (see below).

Recent data (Perry *et al.*, 2004) on arterial PO_2 of *H. unitaeniatus* (fish with a large number of chloride cell in the lamellae in their natural habitat) revealed low arterial PO_2 in normoxia with free access to air (50 mmHg) and lower if the access to air was denied (30 mmHg). During moderate to severe hypoxia, the arterial PO_2 dropped to almost 20 to 15 mmHg in fish allowed breathing air and decreased to 10 mmHg in fish kept under water.

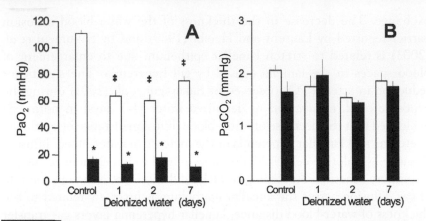

Fig. 6.11 Arterial oxygen partial pressure (PaO$_2$) (**A**) and arterial carbon dioxide partial pressure (PaCO$_2$) (**B**) in normoxia (PO$_2$ = 135 mmHg) (open bars) and hypoxia (PO$_2$ = 21 mmHg) (black bars) of *Hoplias malabaricus* kept in deionized water. The bars indicate the mean values and error bars indicate + SEM. A double-dagger symbol indicates a significant difference from controls. The asterisk indicates a significant difference from normoxia values. Modified from Sakuragui *et al.* (2003).

CO$_2$ excretion may be expected to be affected by chloride cell proliferation in the lamellae. Bindon *et al.* (1994b) reported an elevation of arterial PCO$_2$ during normoxia and hypoxia in *O. mykiss* with increasing thickness of water-blood diffusion distance caused by the presence of chloride cell in the lamellae. However, Greco *et al.* (1995) and Perry *et al.* (1996) failed to demonstrate such response in the same species in normoxia. No change in the arterial PCO$_2$ was reported by Sakuragui *et al.* (2003) in *H. malabaricus* experiencing chloride cell proliferation (kept in deionized) in normoxia and hypoxia (Fig 6.11B).

Hyperventilation favoring CO$_2$ excretion and O$_2$ uptake compensates, at least in part, the detrimental effect of chloride cell in the lamellae (Greco *et al.*, 1995; Perry *et al.*, 1996; Sakuragui *et al.*, 2003). In the case of *O. mykiss* (Greco *et al.*, 1995) breathing frequency was maximal in normoxia, which may limit other important compensatory responses during hypoxia exposure, such as respiratory alkalosis, which is associated with the elevation of red blood cell pH and the increase in blood O$_2$ transfer. In the tropical fish, *H. malabaricus* (Sakuragui *et al.*, 2003), the overall gill ventilation did not differ in normoxia although breathing frequency was higher in fish kept for one day in deionized water, the ventilation volume was higher in fish kept 2 and 7 days, and breathing frequency was higher in moderate and severe hypoxia.

Perry *et al.* (1996) showed that the increase in hemoglobin-oxygen binding affinity and the release of catecholamines reduce the effects of increased water-blood thickness diffusion barrier by the lamellar proliferation of chloride cell during hypoxia in *O. mykiss*. In their study of the adrenergic response in tropical fish exposed to acute hypoxia, Perry *et al.* (2004) demonstrated that the release of catecholamine in *H. malabaricus* and *H. unitaeniatus* occur when blood O_2 concentration is reduced to approximately 50-60% of the normoxic value, which occurs at water PO_2 below 20 and 40 mmHg, respectively the critical PO_2 at which there is an abrupt transition from O_2-uptake regulation to a reduction of metabolic rate of both species. Furthermore, the authors also demonstrated a clearly positive relationship between the threshold arterial PO_2 for catecholamine release and the affinity of hemoglobin-oxygen binding (P_{50}) as already found in other teleosts.

CONCLUSIONS AND PERSPECTIVES

The gills of most teleost fish are similar in their general structure but interspecific variability of the gill components, the blood pathway inside the lamellae and the thickness of water-blood diffusion distance influence the effectiveness of gas transfer. Chloride cell proliferation is a common response of fish to increased ion uptake. However, it depends on previous chloride cell density in the gill epithelium. Short-term exposure to soft water induces to intense proliferation of chloride cells, particularly in the lamellar epithelium. Nevertheless, fish that live in soft and ion-poor waters exhibit considerable variations in the number of chloride cells in the lamellar epithelium and in the chloride cell fractional area (CCFA). Further researches on the mechanisms to avoid ion loss and favor ion uptake in species living in soft waters will undoubtedly help us gain a better understanding of this variability. The presence of chloride cells in the lamellar epithelium increases the water-blood diffusion distance causing respiratory impairment. Compensatory physiological adjustments relating to increased gill ventilation differ among species in normoxia and hypoxia. Researches addressing other physiological factors that may be involved will also contribute to shed light on this variability.

Acknowledgments

FAPESP, CNPq and CAPES grants financially supported original work by M.N. Fernandes cited in this chapter.

References

Avella, M., A. Masoni, M. Bornancin and N. Mayer-Gostan. 1987. Gill morphology and sodium influx in the rainbow trout (*Salmo gairdneri*) acclimated to artificial freshwater environments. *Journal of Experimental Biology* 241: 159-169.

Baily, Y., S. Dunel-Erb and P. Laurent. 1992. The neuroepithelial cells of the fish gill filament: indolamine-immunocytochemistry and innervation. *Anatomical Record* 233: 143-161.

Bindon, S.D., J. C. Fenwick and S.F. Perry. 1994a. Branchial chloride cells proliferation in the rainbow trout, *Oncorhynchus mykiss*: implications for gas exchange. *Canadian Journal of Zoology* 72: 1395-1402.

Bindon, S.D., K.M. Gilmour, J.C. Fenwick and S.F. Perry. 1994b. The effects of branchial chloride cell proliferation on respiratory function in the rainbow trout, *Oncorhynchus mykiss*. *Journal of Experimental Biology* 187: 47-63.

Booth, J.H. 1978. The distribution of blood flow in the gills of fish: application of a new technique to rainbow trout (*Salmo gairdneri*). *Journal of Experimental Biology* 73: 119-129.

Booth, J.H. 1979. The effects of oxygen supply, epinephrine, and acetylcholine on the distribution of blood flow in trout gills. *Journal of Experimental Biology* 83: 31-39.

Cameron, J.N. and C.M. Wood. 1978. Renal function and acid-base regulation in two Amazonian erythrinid fishes: *Hoplias malabaricus*, a water breather, and *Hoploerythrinus unitaeniatus*, a facultative air breather. *Canadian Journal of Zoology* 56: 917-930.

Costa, O.T.F., A.C.E. Pedretti, A. Schmitz, S.F. Perry and M.N. Fernandes. 2007. Stereological estimation of surface area and barrier thickness of fish gills in vertical sections. *Journal of Microscopy* 225: 1-9.

Dang, Z., R.A.C. Lock, G. Flik and S.E. Wendelaar Bonga. 2000. Na^+/K^+-ATPase immunoreactivity in branchial chloride cells of *Oreochromis mossambicus* exposed to copper. *Journal of Experimental Biology* 203: 379-387.

Dejours, P. 1981. *Principles of Comparative Respiratory Physiology*. 2nd edition. Elsevier/North Holland, Amsterdam.

Dunel-Erb, S., Y. Baily and P. Laurent. 1982. Neuroepithelial cells in fish gill primary lamellae. *Journal of Applied Physiology* 53: 1342-1353.

Evans, D.H., P.M. Piermarini and K.P. Chae. 2005. The multifunctional fish gill: Dominant site of gas exchange, osmoregulation, acid-base regulation, and excretion of nitrogenous waste. *Physiological Review* 85: 97-177.

Fergunson, H.W., D. Morrison, V.E. Ostland, J. Lumsden and P. Byrne. 1992. Responses of mucus-producing cells in gill disease of rainbow trout (*Oncorhynchus mykiss*). *Journal of Comparative Pathology* 106: 255-265.

Fernandes, M.N. and S.A. Perna Martino. 1995. Internal morphology of the gill of a loricariid fish, *Hypostomus plecostomus*: Arterio-arterial vasculature and muscle organization. *Canadian Journal of Zoology* 73: 2259-2265.

Fernandes, M.N. and S.A. Perna-Martins. 2001. Epithelial gill cells in the armoured catfish, *Hypostomus* cf. *tietensis* (Loricariidae). *Brazilian Journal of Biology* 61: 69-78.

Fernandes, M.N. and S.A. Perna-Martins. 2002. Chloride cells responses to long-term exposure to distilled and hard water in the gill of the armoured catfish, *Hypostomus tietensis* (Loricariidae). *Acta Zoologica* (Stockholm) 83: 321-328.

Fernandes, M.N. and F.T. Rantin. 1989. Respiratory responses of *Oreochromis niloticus* (Pisces, Cichlidae) to environmental hypoxia, under different thermal conditions. *Journal of Fish Biology* 35: 509-519.

Fernandes, M.N. and F.T. Rantin. 1994. Relationships between oxygen availability and metabolic cost of breathing in Nile tilapia (*Oreochromis niloticus*): Aquaculture consequences. *Aquaculture* 127: 339–346.

Fernandes, M.N., F.T. Rantin, A.L. Kalinin and S.E. Moron. 1994. Comparative study of gill dimensions of three erythrinid species in relation to their respiratory function. *Canadian Journal of Zoology* 72: 160-165.

Fernandes, M.N., W.R. Barrionuevo and F.T. Rantin. 1995. Effect of thermal stress on the respiratory responses to hypoxia of a South America Prochilodontidae fish, *Prochilodus scrofa*. *Journal of Fish Biology* 46: 123-133.

Fernandes, M.N., S.A. Perna, C.T.C. Santos and W. Severi. 1995. The gill filament muscles in two loricariid fish (genera *Hypostomus* and *Rhinelepis*). *Journal of Fish Biology* 46: 1082-1085.

Fernandes, M.N., S.A. Perna and S.E. Moron. 1998. Chloride cells apical surface changes in gill epithelia of the armoured catfish, *Hypostomus plecostomus* L. during exposure to distilled water. *Journal of Fish Biology* 52: 844-849.

Flik, G. and S.F. Perry. 1989. Cortisol stimulates whole body calcium uptake and the branchial calcium pump in freshwater rainbow trout. *Journal of Endocrinology* 120: 75-82.

Flik, G., J.A. van der Velden, K.J. Dechering, P.M. Verbost, T.J.M. Schoenmakers, Z.L. Kolar and S. Wendelaar Bonga. 1993. Ca^{2+} and Mg^{2+} transport in gills and gut of tilapia, *Oreochromis mossambicus*: a review. *Journal of Experimental Zoology* 265: 356-365.

Flik, G., P.H.M. Klaren, T.J.M. Schoenmakers, M.J.C. Bijvelds, P.M. Verbost and S.E. Wendelaar Bonga. 1995. Cellular calcium transport in fish: Unique and universal mechanisms. *Physiological Zoology* 69: 403-417.

Foskett, J.K. and C. Scheffey. 1982. The chloride cell: definitive identification as the salt-secreting cells in teleosts. *Science* 215: 164-166.

Galis, F. and C.P.N. Barel. 1980. Comparative functional morphology of the gills of African lacustrine cichlidae (Pisces, Teleostei). An eco-morphological approach. *Netherlands Journal of Zoology* 30: 392-430.

Gona, O. 1979. Mucous glycoproteins in teleostean fish: a comparative histochemical study. *Histochemical Journal* 11: 709-718.

Goss, G.G., S.F. Perry, C.M. Wood and P. Laurent. 1992. Mechanisms of ion and acid-base regulation at the gills of freshwater fish. *Journal of Experimental Zoology* 263: 143-159.

Goss, G.G., P. Laurent and S.F. Perry. 1994. Gill morphology during hypercapnia in brown bulhead (*Ictalurus nebulosus*): Role of chloride cells and pavement cells in acid-base regulation. *Journal of Fish Biology* 45: 705-718.

Goss, G.G., S.F. Perry and P. Laurent. 1995. Ultrastructural and morphometrics studies on ion and acid-base transport processes in freshwater fish. In: *Fish Physiology,* C.M. Wood and T.J. Shuttleworth (eds.), Academic Press, San Diego, Vol. 14: *Cellular and Molecular Approaches to Fish Ionic Regulation,* pp. 257-284.

Goss, G.G., S. Perry, J. Fryer and P. Laurent. 1998. Gill morphology and acid-base regulation in freshwater fishes. *Comparative Biochemistry and Physiology* A119: 107-116.

Greco, A.M., K.M. Gilmour, J.C. Fenwick and S.F. Perry. 1995. The effect of soft-water acclimation on respiratory gas transfer in the rainbow trout, *Oncorhynchus mykiss. Journal of Experimental Zoology* 198: 2557-2567.

Greco, A.M., J.C. Fenwick and S.F. Perry. 1996. The effect of softwater acclimation on the gill structure in the rainbow trout *Oncorhynchus mykiss. Cell and Tissue Research* 285: 75-82.

Hakin, A., J.S. Datta Munshi and G.M. Hughes. 1978. Morphometrics of the respiratory organs of the Indian green snake-headed fish, *Channa punctata. Journal of Zoology* (London) 184: 519-543.

Hand, R.D. 1989. The ionic composition of rainbow trout body mucus. *Comparative Biochemistry and Physiology* A9: 571-575.

Hand, R.D. and F.B. Eddy. 1991. The absence of mucus on the secondary lamellae of unstressed rainbow trout, *Oncorhynchus mykiss* (Walbaum). *Journal of Fish Biology* 38: 153-155.

Hand, R.D., F.B. Eddy and G. Roman. 1989. *In vitro* evidence for the ionoregulatory role of rainbow trout mucus in acid, acid/aluminum and zinc toxicity. *Journal of Fish Biology* 35: 737-747.

Hirose, S., T. Kaneko, N. Naito and Y. Takei. 2003. Molecular biology of major components of chloride cells. *Comparative Biochemistry and Physiology* B136: 593-620.

Hughes, G.M. 1966. The dimensions of gills in relation to their function. *Journal of Fish Biology* 45: 177-195.

Hughes, G.M. 1972. Morphometrics of fish gills. *Respiration Physiology* 14: 1-25.

Hughes, G.M. 1980. Morphometric of fish gas exchange organ in relation to their respiratory function. In: *Environmental Physiology of Fishes,* M.A. Ali (ed.), Plenum Press, New York, pp. 33-56.

Hughes, G.M. 1984a. General anatomy of the gills. In: *Fish Physiology,* W.S. Hoar and D.J. Randall (eds.), Academic Press, Orlando, Vol. 10 A, pp. 1-72.

Hughes, G.M. 1984b. Scaling of respiratory areas in relation to oxygen consumption of vertebrates. *Experientia* 40: 519-524.

Hughes, G.M. 1984c. Measurements of gill area in fishes: practices and problems. *Journal of the Marine Biological Association of the United Kingdom* 64: 637-655.

Hughes, G.M. and N.K. Al-Kadhomiy. 1988. Changes in scaling of respiratory systems during the development of fishes. *Journal of the Marine Biological Association of the United Kingdom* 68: 489-498.

Hughes, G.M. and M. Morgan. 1973. The structure of fish gills in relation to their respiratory function. *Biological Reviews* 48: 419-475.

Ishimatsu, A., G.K. Iwama and N. Heisler. 1988. *In vivo* analysis of partitioning of cardiac output between systemic and central venous sinus circuits in rainbow trout: A new approach using chronic cannulation of the branchial vein. *Journal of Fish Biology* 137: 75-88.

Iwama, G.K., A. Ishimatsu and N. Heisler. 1986. *In vivo* characterization of the lamellar and CVS compartments in the gill of rainbow trout. *Physiologist* 28: 282.

Johansen, K., C.P. Mangum and G. Lykkeboe. 1978. Respiratory properties of the blood of Amazon fishes. *Canadian Journal of Zoology* 56: 898-906.

Jucá-Chagas, R. 2004. Air-breathing of the neotropical fishes *Lepidosiren paradoxa*, *Hoploerythrinus unitaeniaus* and *Hoplosternum littorale* during aquatic hypoxia. *Comparative Biochemistry and Physiology* A139: 49-53.

Kalinin, A.L., F.T. Rantin and M.L. Glass. 1993. Dependence on body size of respiratory function in *Hoplias malabaricus* (Teleostei, Erythrinidae) during graded hypoxia. *Fish Physiology and Biochemistry* 12: 47-51.

Laurent, P. 1984. Gill internal morphology. In: *Fish Physiology*, W.S. Hoar and D.J. Randall (eds.), Academic Press, Orlando, Vol. 10A, pp.73-183.

Laurent, P. 1989. Gill structure and function. In: *Comparative Pulmonary Physiology*, S.C. Wood (ed.), Marcel Dekker, New York, pp. 69-120.

Laurent, P. and N. Hebibi. 1989. Gill morphometry and fish osmoregulation. *Canadian Journal of Zoology* 67: 3055-3063.

Laurent, P., S. Dunel-Erb, C. Chevalier and J. Lignon. 1995a. Gill epithelial cell kinetics in a freshwater teleost, *Oncorhynchus mykiss*, during adaptation to ion-poor water and hormonal treatments. *Fish Physiology and Biochemistry* 13: 353- 370.

Laurent, P., J.N. Maina, H.L. Bergman, A. Narahara, P.J. Walsh and C.M. Wood. 1995b. Gill structure of a fish from an alkaline lake. Effect of exposure to pH 7.0. *Canadian Journal of Zoology* 73: 1170-1181.

Manera, M. and B.S. Dezfuli. 2004. Rodlet cells in teleosts: a new insight into their nature and functions. *Journal of Fish Biology* 65: 567-619.

Marshall, W.S. 1995. Transport processes in isolated Teleost epithelia: opercular epithelium and urinary bladder. In: *Fish Physiology*, C.M. Wood and T.J. Shuttleworth (eds.), Academic Press, New York, Vol. 14: *Cellular and Molecular Approaches to Fish Ionic Regulation* pp. 1-23.

Mattias, A.T., S.E. Moron and M.N. Fernandes. 1996. Aquatic respiration during hypoxia of the facultative air-breathing fish, *Hoploerythrinus unitaeniatus*. A comparison with the water-breathing, *Hoplias malabaricus*. In: *Physiology and Biochemistry of the Fishes of the Amazon*, A.L. Val, V.M.F. Almeida-Val and D.J. Randall (eds.), INPA, Manaus, pp. 203-211.

Mattias, A.T., F.T. Rantin and M.N. Fernandes. 1998. Changes in gill respiratory parameters during progressive hypoxia in the facultative air-breathing fish, *Hypostomus regani* (Loricariidae). *Comparative Physiology and Biochemistry* A120: 311-315.

Mazon, A.F., C.C.C. Cerqueira and M.N. Fernandes. 2002. Gill cellular changes induced by copper exposure in the South American tropical freshwater fish *Prochilodus scrofa*. *Environmental Research* A 88: 52-63.

Mazon, A.F., M.N. Fernandes, M.A. Nolasco and W. Severi. 1998. Functional morphology of gills and respiratory area of two active rheophilic fish species, *Plagioscion squamosissimus* and *Prochilodus scrofa*. *Journal of Fish Biology* 52: 50-61.

Mazon, A.F., M.O. Huising, A.J. Taverne-Thiele, J. Bastiaans, B.M.L. Verlurg-van Kemenade. 2007. The first appearance of rodlet cells in carp (*Cyprinus carpio* L) ontogeny and their possible role during stress and parasite infection. *Fish & Shellfish Immunology* 22: 27-37.

Mazon, A.F., D.T. Nolan, R.A.C. Lock, M.N. Fernandes and S.E. Wendelaar-Bonga. 2004. A short-term in vitro gill culture system to study the effects of toxic (copper) and non-toxic (cortisol) stressors in the rainbow trout, *Oncorhynchus mykiss* (Walbaum). *Toxicology in Vitro* 18: 691-701.

McCormick, S.D. 1995. Hormonal control of gill Na^+, K^+-ATPase and chloride cell function. In: *Fish Physiology*, C.M. Wood and T.J. Shuttleworth (eds.), Academic Press, San Diego, Vol. 14: *Cellular and Molecular Approaches to Fish Ionic Regulation*. pp. 285-315.

Moraes, M.F.P.G., S. Höller, O.T.F. Costa, M.L. Glass, M.N. Fernandes and S.F. Perry. 2005. Morphometric comparison of the respiratory organs of the South American lungfish, *Lepidosiren paradoxa* (Dipnoi). *Physiological and Biochemical Zoology* 78: 546-559.

Moron, S.E. and M.N. Fernandes. 1996. Pavement cell structural differences on *Hoplias malabaricus* gill epithelia. *Journal of Fish Biology* 49: 357-362.

Moron, S.E., E.T. Oba, C.A. Andrade and M.N. Fernandes. 2003. Chloride cell responses to ion challenge in two tropical freshwater fish, the erythrinids *Hoplias malabaricus* and *Hoploerythrinus unitaeniatus*. *Journal of Experimental Biology* A 298: 93-104.

Nilsson, S. and L. Sundin. 1998. Gill blood flow control. *Comparative Biochemistry and Physiology* A 119: 137-147.

Oliveira, R.D., J.M. Lopes, J.R. Sanches, A.L. Kalinin, M.L. Glass and F.T. Rantin. 2004. Cardiorespiratory responses of the facultative air-breathing fish jeju, *Hoploerythrinus unitaeniatus* (Teleostei, Erythrinidae), exposed to graded ambient hypoxia. *Comparative Biochemistry and Physiology* A 139: 479-485.

Olson, K.R. 2002a. Vascular anatomy of the fish gill. *Journal of Experimental Biology* 293: 214-231.

Olson, K.R. 2002b. Gill circulation: Regulation of perfusion distribution and metabolism of regulatory molecules. *Journal of Experimental Biology* 293: 320-335.

Perna, S.A. and M.N. Fernandes. 1996. Gill morphometry of the facultative air-breathing loricariid fish, *Hypostomus plecostomus* (Walbaum) with special emphasis to aquatic respiration. *Fish Physiology and Biochemistry* 15: 213-220.

Perry, S.F. 1997. The chloride cell: structure and function in the gill of freshwater fishes. *Annual Review of Fish Physiology* 59: 325-347.

Perry, S.F. 1998. Relationships between branchial chloride cells and gas transfer in freshwater fish. *Comparative Biochemistry and Physiology* A 119: 9-16.

Perry, S.F., G.G. Goss and P. Laurent. 1992. The interrelationships between gill chloride cell morphology and ionic uptake in four freshwater teleosts. *Canadian Journal of Zoology* 70: 1775-1786.

Perry, S.F., S.G. Reid, E. Wankiewicz, V. Iyer and K.G. Gilmour. 1996. Physiological responses of rainbow trout (*Oncorhynchus mykiss*) to prolonged exposure to softwater. *Physiological Zoology* 69: 1419-1441.

Perry, S.F., S.G. Reid, K.M. Gilmour, C.L. Boijink, J.M. Lopes, W.K. Milsom and F.T. Rantin. 2004. A comparison of adrenergic stress responses in three tropical teleosts exposed to acute hypoxia. *American Journal of Physiology* 287: R188-R197.

Pisam, M. and A. Rambourg. 1991. Mitochondria-rich cells in the gill epithelium of teleost fishes: an ultrastructural approach. *International Review of Cytology* 130: 191-232.

Pisam, M., A. Caroff and A. Rambourg. 1987. Two types of chloride cells in the gill epithelium of a freshwater-adapted euryhaline fish: *Lebistes reticulates*; their modifications during adaptation to seawater. *American Journal of Anatomy* 179: 40-50.

Pisam, M., B. Auperin, P. Prunet, F. Rentier-Delrue, J. Martial and A. Rambourg. 1993. Effects of prolactin on α and β chloride cells in the gill epithelium of the saltwater adapted tilapia *Oreochromis niloticus*. *Anatomical Record* 235: 275-284.

Powell, M.D., D.J. Speare and G.M. Wright. 1994. Comparative structural morphology of lamellar epithelial chloride and mucous cell glycocalyx of the rainbow trout (*Oncorhynchus mykiss*) gill. *Journal of Fish Biology* 44: 725-730.

Randall, D. and C. Daxboeck. 1984. Oxygen and carbon dioxide transfer across fish gills. In: *Fish Physiology*, W.S. Hoar and D.J. Randall (eds.), Academic Press, Orlando, Vol. 10A, pp. 263-314.

Rankin, J.C. and H. Tuurala. 1998. Gills of Antarctic fish. *Comparative Biochemistry and Physiology* A119: 149-163.

Rantin, F.T., A.L. Kalinin and M.N. Fernandes. 1992. Respiratory responses to hypoxia in relation to mode of life of two erythrinid species (*Hoplias malabaricus* and *Hoplias lacerdae*). *Journal of Fish Biology* 41: 805-812.

Rantin, F.T., M.N. Fernandes, M.C.H. Furegato and J.R. Sanches. 1985. Thermal acclimation in the teleost, *Hoplias malabaricus* (Pisces, Erythrinidae). *Boletin de Fisiologia animal*, Univ. São Paulo, 9: 103-109.

Reid, S.G., L. Sundin, A.L. Kalinin, F.T. Rantin and W.K. Milsom. 2000. Cardiorespiratory and respiratory reflexes in the tropical fish, traira (*Hoplias malabaricus*): CO_2/pH chemoreceptors. *Respiration Physiology* 120: 47-59.

Sakuragui, M.M., J.R. Sanches and M.N. Fernandes. 2003. Gill chloride cell proliferation and respiratory responses to hypoxia of the neotropical erythrinid fish *Hoplias malabaricus*. *Journal of Comparative Physiology* B173: 309-317.

Santos, C.T.C., M.N. Fernandes and W. Severi. 1994. Respiratory gill surface area of a facultative air-breathing loricariid fish, *Rhinelepis strigosa*. *Canadian Journal of Physiology* 72: 2009-2013.

Shepheard, K.L. 1994. Functions for fish mucus. *Review in Fish Biology and Fisheries* 4: 401-429.

Soivio, A. and H. Tuurala. 1981. Structural and circulatory responses to hypoxia in the secondary lamellae of *Salmo gairdneri* gills at two temperatures. *Journal of Comparative Physiology* B145: 37-43.

Sollid, J., P. De Angellis, K. Gundersen and G.E. Nilsson. 2003. Hypoxia induces adaptive and reversible gross morphological changes in crucian carp gills. *Journal of Experimental Biology* 206: 3667-3673.

Sundin, L.I. 1995. Responses of the branchial circulation to hypoxia in the Atlantic cod, *Gadus morhua. American Journal of Physiology* 268: R771-R778.

Sundin, L.I., S.G. Reid, A.L. Kalinin, F.T. Rantin and W.K. Milsom. 1999. Cardiovascular and respiratory reflexes: The tropical fish, traira (*Hoplias malabaricus*) O_2 chemoresponses. *Respiratory Physiology* 116:181-199.

Takasusuki, J., M.N. Fernandes and W. Severi. 1998. The occurrence of aerial respiration in the armoured catfish, *Rhinelepis strigosa* (Loricariidae) during progressive hypoxia. *Journal of Fish Biology* 52: 369-379.

Tao, S., H. Li and C.F. Liu. 2000. Fish uptake of inorganic and mucus complexes of lead. *Ecotoxicology and Environmemtal Safety* 46:174-180.

Thomas, S. and R. Motais. 1990. Acid-base balance and oxygen transport during hypoxia in fish. *Journal of Comparative Physiology* 6: 76-91.

Wendelaar Bonga, S.E. and J.C.A. van der Meij. 1989. Degeneration and death, by apoptosis and necrosis, of the pavement and chloride cells in the gills of the teleost, *Oreochromis mossambicus. Cell Tissue Research* 255: 235-243.

Wendelaar Bonga, S.E., G. Flik, P.H.M. Balm and J.C.A. van der Meij. 1990. The ultrastructure of chloride cells in the gills of the teleost *Oreochromis mossambicus* during exposure to fresh water. *Cell Tissue Research* 259: 575-585.

Wood, C.M. 2001. Toxic responses of the gill. In: *Target Organ Toxicity in Marine and Freshwater Teleosts*, D. Schlenk and W.H. Benson (eds.). Taylor & Francis, London, Vol 1: Organs, pp. 1-89.

Wood, C.M. and W.S. Marshall. 1994. Ion balance, acid-base regulation and chloride cell function on the common killifish, *Fundulus heteroclitus*, a euryhaline estuarine teleost. *Estuaries* 17: 34-52.

Behavior and Adaptation of Air-breathing Fishes

Marisa Fernandes-Castilho[1,2], Eliane Gonçalves-de-Freitas[1,3],
Percilia Cardoso Giaquinto[1,4], Christiane Patricia Feitosa de
Oliveira[5], Vera Maria de Almeida-Val[6] and Adalberto Luís Val[6,*]

INTRODUCTION

Air-breathing fishes are those who can uptake oxygen from the aerial
environment, using lungs, mouth, gut, gills, skin and other kinds of thin
and abundantly vascularized tissue. According to Graham (1997), 'air
breathing' is used to describe a fish respiratory mechanism, in opposition
to 'non-air breathing', that describes the fishes who do not use air to
breathe. 'Water-breathing', usually named to describe fishes that do not
breathe air is a non apropos term, since most of air-breather species also
respire in water at some extent (Graham, 1997).

Authors' addresses: [1]Research Center on Animal Welfare - RECAW.
[2]Laboratory of Studies on Animal Stress, Dept. of Physiology, Universidade Federal do
Paraná, Curitiba, Brazil.
[3]Laboratory of Animal Behavior, Dept. of Zoology and Botany, Universidade Estadual
Paulista, São José do Rio Preto, Brazil.
[4]Laboratory of Physiology and Animal Behavior, Dept. of Physiology, Faculty of Medicine,
Universidade de São Paulo, Ribeirão Preto, Brazil.
[5]Ph.D. student in Biotechnology, Universidade Federal do Amazonas, Manaus, Brazil.
[6]Laboratory of Ecophysiology and Molecular Evolution, National Institute for Amazon
Research, Manaus, Brazil.
Corresponding author: E-mail: dalval@inpa.gov.br

Air-breathing fishes are classified in two major types: amphibious and aquatic air-breathers. According to Graham (1997), amphibious fishes breathe air mainly during periods they are out of the water and are either active on land, volitional exposure as *Periophthalmus* and *Andamia*, or inactive on land, enduring brief exposures to the air or during estivation, as *Blennius* and *Protopterus*, respectively. Otherwise, aquatic air-breathers remain in water and surface periodically to gulp air. Graham (1997) divides this group in two subgroups: (1) facultative air-breathers are fishes that do not breathe air in normoxic water, but breathe air when facing hypoxia or hypercapnia, or in response to increased O_2 requirements, as during increased water temperature or increased organic activity. (2) continuous air-breathers are fishes that breathe air at regular intervals in any aquatic conditions (from hypoxia to hyperoxia). Continuous obligatory air-breathers always need aerial oxygen, even in normoxic water conditions. On the other hand, continuous non-obligatory air-breathers do not require aerial O_2 to survive in normoxic conditions. Therefore, continuous non-obligatory fishes have the same requirements of facultative ones, since both can breathe in normoxic water. The difference is based only at the intervals (or manner) of O_2 uptake. However, in hypoxic or hypercapnic conditions, both facultative and continuous non-obligatory became obligatory air breathing. In fact, as Graham (1997) had already pointed, such classification is far from an exact science, because environmental conditions, and even age and size of the individual, can affect air-breathing activity.

Complete understanding of air-breathing in fishes depends on several kinds of information that include morphology, physiology, behavior and environment. In this chapter we summarize the main points related to behavior and adaptation of air-breathing fishes. Moreover, we will also take into account environmental changes that can affect such fish group. This issue is gaining interest because of environmental changes caused by man activities, such as river damming, mercury contamination and oil pollution, which are potentially dangerous for aquatic ecosystem and aquatic organisms, including fish (Val, 1999).

DISTRIBUTION OF AIR-BREATHING FISHES

Air-breathing fish species are distributed among 49 families belonging to the Subclasses Sarcopterygii (smaller group) and Actinopterygii. They live in brackish water, freshwater and marine water, widespread over all continents in the world. The greatest diversity at the family level, however, is in Africa, Asia, Central and South America (Table 7.1).

Table 7.1 Air-breathing fish families around the world (based on Nelson, 1994; Graham, 1997).

Continent	Fish family		
	Subclass Sarcopterygii[1]		
Africa	Protopteridae	Latimeeiidae	
Asia	Lepidosirenidae		
Australia	Ceratodontidae		
South America	Lepidosirenidae		
	Subclass Actinopterygii[2]		
Africa	Anabantidae	Eleotridae	Notopteridae
	Aplocheilidae	Galaxiidae	Osteoglossidae
	Channidae	Gobiidae	Pantodontidae
	Clariidae	Gymnarchidae	Phractolemidae
	Cobitididae	Mastacembelidae	Polipteridae
	Cyprinodontidae		
Asia	Anabantidae	Cyprinodontidae	Notopteridae
	Aplocheilidae	Gobiidae	Osphronemidae
	Belontiidae	Helostomatidae	Osteoglossidae
	Channidae	Heteropneustidae	Pangasiidae
	Clariidae	Luciocephalidae	Synbranchidae
	Cobitididae	Mastacembelidae	
Central America	Aplocheilidae	Gobiesocidae	Lepisosteidae
	Callichthyidae	Gymnotidae	Loricariidae
	Cyprinodontidae	Hypopomidae	Megalopidae
	Electrophoridae	Labrisomidae	Synbranchidae
	Erythrinidae	Lebiasinidae	Trichomycteridae
Europe	Anguillidae	Umbridae	
North America	Amiidae	Eleotridae	Megalopidae
	Aplocheilidae	Lepisosteidae	Umbridae
	Cyprinodontidae		
Oceania[3]	Cottidae	Gobiesocidae	Synbranchidae
	Eleotridae	Lepdogalaxiidae	Tripterygiidae
	Galaxiidae	Osteoglossidae	
South America	Aplocheilidae	Erythrinidae	Loricariidae
	Aspredinidae	Galaxiidae	Osteoglossidae
	Callichthyidae	Gymnotidae	Synbranchidae
	Cyprinodontidae	Hypopomidae	Trichomycteridae
	Electrophoridae	Lebiasinidae	

[1]Subclass Sarcopterygii includes 3 genera of lungfishes and *Latimeria* (coelacanth)
[2]Subclass that includes the subdivision Teleostei.
[3]Australia, New Zealand and small islands

Additionally to the species occurring endemically to these places, there are species belonging to some families of wide distribution. It is the case of marine families, as Stichaidae, Pholididae and Bleniidae, that are found in North Atlantic, North Pacific and Indian oceans (Nelson, 1994). Gobiidae is another large family widely distributed in shallow waters of tropical and subtropical regions (Graham, 1997).

EVOLUTION OF AIR-BREATHING ACTIVITY IN FISH

Fossils

Among the fossils of air-breathing fishes, the most ancient are the Actinopterygian fossils from Paleozoic (570-245 million years before present) (Graham, 1997). Fifty percent of Dipnoan fossils (lungfishes) are from Devonian period (400-370 mybp) and the remainder 50% lived during the Carboniferous (360-330 mybp), the Permian (286-248 mybp) and post-Paleozoic (Marshall, 1986). These fossils are testimonies that air-breathing ability in vertebrates existed before the emergence of animals on the land, and that such event occurred well before the evolution of amphibians, as suggested by Johansen (1968).

Environment

The evolution of air-breathing fishes is close-related to changes in the composition of atmosphere and water. According to Johansen (1968), during the Devonian Period, a respiration crisis developed for vertebrates inhabiting freshwater because high temperatures and dead organic matter caused a drop of dissolved O_2 to marginal levels. In such a scenario, fishes equipped with organs to uptake O_2 from air were selected.

In fact, further studies (which were summarized by Gans et al., 1999 and commented below) revealed substantial changes in O_2 and CO_2 contents in both water and atmosphere over the geological periods. From Silurian to end-Devonian atmospheric O_2 was between 14-18% and in freshwater between 4.6-5.9 ml·L^{-1} (less than current levels, that are 20.9% and 7ml·L^{-1} in air and freshwater, respectively). Atmospheric CO_2 content, on the other hand, was 0.4-0.3% from Silurian to end-Devonian and water CO_2 was 3.8-3.1 ml·L^{-1} during this period (also different from nowadays values). Thus, such marginal levels of O_2 and increased CO_2 in the water during this geological period should have led some fishes to the water surface to uptake O_2 directly from the air.

At the end-Carboniferous, O_2 rose to 35% and CO_2 declined to 0.03% in the atmosphere, as an effect of terrestrialization of plants, deposition of carbonates and subsequent terrestrial weathering (Berner, 1997). Such a change in gas content must have improved gill breathing to pulse-pumping and then, aspiration breathing of aerial gases (Gans *et al.*, 1999). A rapid diffusion of both O_2 (from air to blood circulation) and CO_2 (from blood to air) because of high gradient of the gas would be advantageous for fishes. Between end-Carboniferous and end-Permian, atmospheric O_2 decay again (whenever followed by an opposite behavior in CO_2 content) and back to rise at Cretaceous reaching today's levels (Gans *et al.*, 1999). During such changes in aerial and aquatic environment, evolution of air-breathing fishes occurred in several ways, giving rise to many the species that we can observe today.

Air-breathing Organs

Air-breathing organs used by extant species (gills, skin, pharyngeal and mouth pouches, stomach, intestine, gas bladder and lung) were obviously selected in the course of cited events. All organs have a general design for respiration, i.e., a thin layer and well-vascularized tissue plus a wide surface that facilitates rapid gas diffusion in both directions, from and to external and internal environment. Among those organs, evolutionary transitions between gas bladder and lung are the central thread (Graham, 1999). According to this author, the lung and respiratory gas bladder are functional air-breathing organs of the primitive fishes and the loss of gas bladder capacity to serve as a respiratory function led to the acquisition of novel air-breathing organs in higher teleosts.

Another adaptation involves a convergent evolution of surfactants in lung and gas bladder to avoid surface collapse during breathing. It is present in several fish species (including non air-breathing species) and its composition indicates that gas bladder surfactants can represent the vertebrate proto-surfactant (Daniels *et al.*, 2004).

The presence of air-breathing organs (ABO) is just the first step toward effective air breathing. Changes in fishes' blood circulation had to be selected together to provide an effective transportation of oxygenated blood from the ABO to the body tissues (Johansen, 1968). Thus, fish that use mouth as ABO (e.g., *Synbranchus marmoratus*) have extensive vascularization in the mouth that carries oxygenated blood to heart before entering body circulation. Similarly, vessels coming from the lungs (e.g., in lungfishes) carrying oxygenated blood enter the heart before entering body circulation (Johansen, 1968; Graham, 1997).

Air-breathing Organs and Cardiovascular Implications

Independently of the strategies used by the different fish species, all of them depend on accessory air-breathing organs, which can vary widely (Table 7.2). Such organs are essentially hollow spaces with richly vascularized walls that can be ventilated periodically.

Air-breathing activity, as a supplement to water breathing or as the main pathway for gas exchange, has independently evolved several times in fishes. Consequently, a diversity of characteristics in air-breathing organs is found, including its importance relative to gill breathing and its arterial supply and venous drainage. Burggren and Johansen (1986) masterly showed the various cardiovascular arrangements and pathways associated with air-breathing organs in fishes (Fig. 7.1). According to Burggren *et al.* (1997), an important characteristic of all these cardiovascular arrangements is that oxygen-rich blood from the air-breathing organ is returned to the central veins, where it mixes with relatively oxygen-poor blood draining from the systemic tissues. The direct effect is an increase in venous blood oxygen content and partial pressure of oxygen (PO_2). This may be advantageous in supplying oxygen to the heart, especially since these fishes often experience aquatic hypoxia. Obviously, oxygen added to the blood in the capillaries of the air-breathing organ likely reaches the systemic tissues after first transiting the branchial circulation. However, oxygen acquired in the air-breathing organs and transferred to central venous blood has the potential to be lost again to severely hypoxic water surrounding the gills if the venous PO_2 in the gill lamellae exceeds the water PO_2. An adjustment showed in such animals, in contrast to exclusively water breathing fishes that increase gill ventilation with hypoxia, is that many air-breathing fishes show a reduced rate of gill ventilation in severely hypoxic water, shifting the burden of gas exchange to the air-breathing organ. Burggren *et al.* (1997) consider, inclusive, that this characteristic of complete mixing of oxygen-rich blood from the air-breathing organs with oxygen-poor blood from the systemic tissues before reaching the heart represent, in most air-breathing fishes, a little selection for specializations leading to separate blood streams within atrium and ventricle. The exception is the central circulation of the Dipnoi (lungfishes), where systemic and pulmonary veins returns reach the heart separately. With oxygen-rich and oxygen-poor blood entering into separate regions of the atria, there now existed a strong selection pressure for the evolution of anatomical and physiological mechanisms to

Table 7.2 Classification of fish air-breathing organs and some representative species of each group.

A—Lungs and respiratory gas bladders of the more primitive bony fishes	LUNGS		*Lepidosiren paradoxa* *Protopterus* sp. *Neoceratodus forsteri*
	RESPIRATORY GAS BLADDERS		*Arapaima gigas* *Heterotis niloticus* *Pantodon buchholzi* *Erythrinus erythrinus* *Hoplerythrinus unitaeniatus* *Pangasius hypophthalmus* *Gymnotus carapo* *Lebiasiana bimaculata*
	ORGANS IN THE HEAD REGION	Buccal and pharyngeal epithelial surfaces	*Ophisternon aenigmaticum* *Hypopomus brevirostris* *H. occidentalis* *Electrophorus electricus* *Synbranchus marmoratus*
B—Air-breathing organs of the more advanced teleosts		Branchial and opercular epithelial surfaces Pouches formed adjacent to the pharynx	*Boleophthalmus chinensis* *Periophthalmus vulgaris* *Chuma punctatus* *C. striatus* *C. gachua* *C. marulius* *Monopterus* (= *Amphipnous*) *cuchia*

(Table 7.2 Contd.)

(Table 7.2 Contd.)

		M. fossarius *M. indicus*
	Branchial diverticulae	*Clarias mossambicus* *C. batrachus* *C. lazera* *Heteropneustes fossilis*
	Gills	*Hypopomus* sp. *Synbranchus* sp.
ORGANS LOCATED ALONG THE DIGESTIVE TUBE	Pneumatic duct	*Anguilla anguilla*
	Esophagus	*Dallia pectoralis*
	Stomach	*Bunocephalus amaurus* *Pygidium striatum* *Eromophilus mutisii* *Ancistrus chagresi* *Liposarcus pardalis* *Pterygoplichthys multiradiatus*
	Intestine	*Corydoras aeneus* *Brochis splendens* *Hoplosternum littorale*
C- Skin		*Neochanna burrowsius* *Anguilla vulgaris* *Heteropneues fossilis* *Electrophorus electricus* *Dormitator latifrons*

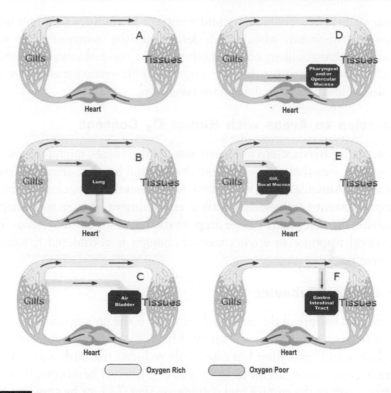

Fig. 7.1 Schematic representation of circulatory organization of various air-breathing fishes showing the direction of the flow of oxygenated and deoxygenated blood related to the air-breathing organs. Modified from Burggren and Johansen (1986).

preserve the identity of these separate streams of blood as they passed through the chamber of the heart, the ventral aorta, and branchial circulation. Indeed, the atrium, ventricle and bulbous cordis have partial septa divisions in all three extant genera of lungfishes, with the South American lungfish *Lepidosiren* showing the greatest degree of separation.

BEHAVIORAL AND PHYSIOLOGICAL ADAPTATIONS

Behavioral and physiological strategies related to air-breathing adaptation are close related events, sometimes difficult to separate between them. This is because behavioral changes depend on physiological changes, and many times, behavioral alterations mean changes in internal environment. Here we describe some types of behavior as well as physiological events that provide air breathing in fishes. Types of behavior that we can immediately observe are the ways used to uptake air from surface and the search for water under better O_2 conditions (or how fishes behave). Important, also, is to determine the moment that such behavioral changes likely occur. In other words, these observations lead to speculations on how fish behave under specific environmental oxygen conditions and what regulate those behaviors.

Migration to Areas with Higher O_2 Content

When facing hypoxic environment many fish species migrate to higher oxygenated area that can be far from the original hypoxic area. According to Val and Almeida-Val (1995) many water-breathing species are able to detect a current O_2 depletion in their environment or even to anticipate it, leaving the place before large drops in the water O_2 content occur. This behavioral response to environmental changes is considered the fastest adaptive response.

Air Gulping Behavior

The central feature of the air-breathing cycle is the air gulp by the aquatic air breathers. In many aquatic air breathers, the mechanical coupling of exhalation and inhalation has essentially reduced the air-breath cycle to a single component, an air-holding phase. This can be accomplished by swimming up to the surface and gulping air (Fig. 7.2), or by crawling onto land and gulping air or by passively exchange gases with the atmosphere across several respiratory surfaces.

According to the type of air-breather, amphibious and aquatic, different patterns of air acquisition can be found. The aquatic air-breathers remain in the water and gulp air at the surface. Facultative air-breathers usually turn to air-breathing when the water becomes hypoxic

A   **B**

Fig. 7.2   Generalized air-breathing cycles. **(A)** An amphibious blenny while emergent take air gulps, and **(B)** aquatic air-breathers remain in water and gulp air at the surface. Extracted and modified from Graham (1997).

or when the oxygen demand increases. Continuous air-breathers take breaths more regularly regardless the oxygen content of the water. The frequency of air-breathing is determined by oxygen levels in the water, water temperature, and activity level.

The ventilatory patterns of air-breathing fish are commonly described as 'arrhythmic' or 'irregular' because the variable periods of breath-holding are punctuated by seemingly unpredictable air-breathing events (Shelton *et al.*, 1986). Despite that, some studies have show rhythmic cycles of air-breathing patterns. Hedrick *et al.* (1994) demonstrated that the aerial ventilation in *Amia calva* is indeed periodic. Maheshwari *et al.* (1999) tested air-gulping behavior of two species (*Heteropneustes fossilis* and *Clarias batrachus*) to observe rhythms of air-breathing activity. Observations were collected 4 times a day for 15 months, in fish kept at a light-dark period of 12:12. These Indian siluroids increased breathing frequency during the summer season (low O_2 in water) and during the rainy season (high O_2), when reproduction demands high energetic input.

Regarding circadian variation and air-gulping activity, Yadu and Shepdure (2002) found that air-gulping activity exhibits a significant 24 hours rhythm in *Clarias batrachus*. The peak of the air-gulping activity rhythm was in the dark phase of the day cycle. Despite these examples, much of the variability seen for the duration of fish air-breath appears to be real. Even under controlled laboratory conditions, rhythmic or periodic aerial respiration patterns and uniform inter-breath intervals are seldom seen.

Under some circumstances an aquatic air breather may hyperventilate (a series of rapid inhalations and exhalations) before submerging, or small amounts of gas may be released for air-breathing organ at various times during the breath-hold phase (Hedrick and Jones, 1993). Also, air-gulping behavior depends on fish size. The smallest known air-breathing fish is 10 mm in length. Such size limitation might result from the requirement of the fish to overcome both the surface tension of the water when gulping air and to counter-force the increased buoyancy resulting from such ingested air.

High density and size hierarchy also elevates the surfacing activity and consequently affects food conversion in air-breathing fishes (Sampath and Pandian, 1985). Generally, surfacing frequency increases in grouped individuals: the increase is two fold in small individuals and about 4 times in medium and large size fishes (Pandian and Sampath, 1984). Swimming activity associated with surfacing behavior costs considerable quantity of energy. The energy drained on the surface activity decreases the proportion of the energy allocated for growth. In addition, the temporal cost in aquatic air-breathing animals affects vital activities like food searching and reproduction (Halliday and Sweatman, 1976).

Environmental pressures can be also critical on air-breathing patterns. Randle and Chapman (2004) recognize two main pressures on air-breathing of the African anabantid *Ctenopoma muriei*: surface travel and aerial predation. They found that the fish was predominantly found in shallow waters of 15-30 cm with ample vegetation, an optimum between lowest travel distance to the surface and best protection from predator detection. Therefore, habitats with minimal costs of air-breathing would be favored, that is, low predation and minimal surface travel distance.

Skimming the Water Surface

Skimming the O_2-rich surface layers of water is a specific adaptation to uptake O_2 from the water-air interface. It does not involve gulping air (Fritsche and Nilsson, 1994; Val and Almeida-Val, 1995; Watters and Cech Jr, 2003). Although appearing with different intensities and at different thresholds, skimming is observed in many different non-related fish species, an indication of the evolutionary convergence of this behavioral character (Val and Almeida-Val, 1995). These animals do not have any accessory air-breathing organ; they just pump the O_2-rich layer of water from air-water interface across the surface of the gills where gas exchange takes place. Some species that use this strategy have a morphological adaptation to improve oxygen uptake even more, as is the swelling of the lower lips found in tambaqui (*Colossoma macropomum*), matrinchã (*Brycon* cf *melanopterus*) and other fish species of the Amazon (Val and Almeida-Val, 1995; Val, 1996). Such expanded lip does not contain any blood vascularization and the function seems to be exclusively mechanical, serving to improve skimming and consequently the uptake of O_2 by the gills.

Air-deposition Behavior

Mudskippers constitute a group of 25 air-breathing species belonging to four genera and are the most derived and amphibious of the teleost subfamily Oxudercinae. These fish spend extensive periods of time out of the water and have numerous physiological, morphological and behavioral specializations for amphibious life (Gordon *et al.*, 1969; Graham, 1997; Lee and Graham, 2002). Mudskippers store air in their 'J'-shaped, intertidal burrows, resulting in an air phase pattern (Ishimatsu *et al.*, 1998) (see Fig. 7.3).

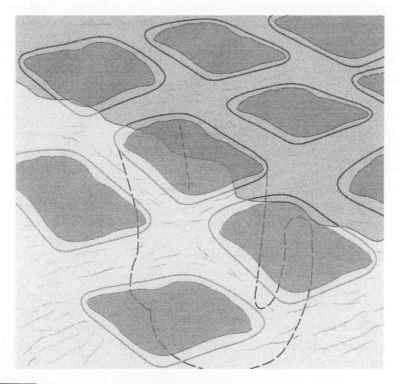

Fig. 7.3 Mudskippers store air in their intertidal burrows, resulting in an air phase pattern. Burrows are built in the littoral zone and most are located in anoxic mud. The burrow is shaped like a **j** or like a **y**, with two entrances.

In order to establish an air phase, mudskipper must perform the air-deposition behavior. This consists of a rapidly repeated series of actions that includes inflating the buccal chamber with air on the mudflat surface, transporting it into the burrow, releasing it, and returning to the surface with a deflated buccal chamber. Because the burrows have an upturned portion that is not connected to the surface (Matoba and Dotsu, 1977; Clayton and Vaughn, 1986), deposited air stays under ground and forms an air phase rather then floating back to the surface. Mudskippers build their burrows in the littoral zone and most are located in anoxic mud (Scholander *et al.*, 1955; Takita *et al.*, 1999). During high tide, the burrows are covered by water and the fish remains confined there (for up to 10 h) for protection from predators.

The air phase in the mudskipper burrow has been hypothesized to be a source of oxygen for respiration or a medium for embryonic development (oxygenation of developing eggs), or both (Ishimatsu *et al.*, 1998). Lee *et al.* (2005) showed that the deposition rates do not differ between males and females, giving further support to the idea that air phase is also important for adult intra-burrow respiration.

According to Gans (1970) and Val and Almeida-Val (1995), although the utilization of air as a source of O_2 include many advantages for fish, as for example, independence from the fluctuations in dissolved O_2, reduced work in pumping an O_2-rich fluid, and decreased size of the pump needed, there are also some disadvantages, like disturbance of the hydrostatic balance mechanisms, exposure to aerial predation and exposure to frequent temperature changes. In fact, air-breathing behavior is only one of the many strategies used by fish to uptake enough O_2 under unfavorable environmental conditions, though that even in such a case it does not mean complete independence of water. Air-breathing fishes are still dependent on water to excrete carbon dioxide, nitrogenous waste and regulate ion levels, among other process using gills and skin (Randall *et al.*, 1978; Val, 1999). Thus, in addition to behavioral features, other strategies are used to uptake oxygen by fish under hypoxic environment and include physiological and biochemical adjustments to optimize O_2 uptake and/or to decrease the O_2 needs (Val and Almeida-Val, 1995; Almeida-Val, 1999), in spite of the O_2-capacitance coefficient in air be about 30 times that of water (Dejours, 1981). The capacity of fish to retrieve O_2 from air can be influenced by many factors such as the amount of O_2 and CO_2 in the aquatic and aerial environment, blood oxygen affinity, organic phosphates in the red blood cells, temperature, body mass, acclimation, growth period activity and behavior, among others (Geiger *et al.*, 2000; Val, 2000; Amin-Naves *et al.*, 2004; Oliveira *et al.*, 2004; Seymour *et al.*, 2004). Indeed, different fish species use different mechanisms directed towards optimizing gas transfer that include, among others, regulation of the proportions of different Hb fraction, adjustment of intraerythrocytic levels of organic phosphates, changes in hematocrit, Hb concentration and metabolic depression (Randall, 1990; Val *et al.*, 1992; Almeida-Val *et al.*, 1993).

O$_2$/CO$_2$ Implications on Air-breathing Behavior

Theoretically, air-breathing behavior in tropical fish should allow the organism to have a higher oxidative metabolism, both due to the increase in the amount of O$_2$ intake and to the effect of the higher temperature. However, obligate air-breathing fishes depend almost exclusively on their aquatic environment for nitrogen and CO$_2$ excretion (Brauner and Val, 1996) and so respiring oxygen from air have not been coupled with increasing energy fluxes (ATP turnover rates); instead, these fish live at the lower limits of energy turnover (Val and Almeida-Val, 1995). According to Almeida-Val (1999), such apparently unexpected responses are correlated with the lifestyle of these animals; for example, lungfish burrow and spend part of their lives aestivating, pirarucu (*Arapaima*) and electric eel (*Electrophorus*) are sluggish and must keep themselves submerged most of the time in such a manner that they have only short time intervals for oxygen uptake. In general the oxygen uptake rate is variable and usually correlated with the general behavior and ecophysiological characteristic of the animal, and not only with the respiratory apparatus or behavior. Almeida-Val and Farias (1996) found that the tissue oxygen consumption rates of *Hoplosternum littorale*, a facultative air-breather, are as high as those of the migratory water-breathing fish, *Semaprochilodus insignis*. On the other hand, *Liposarcus pardalis*, another facultative air-breather, shows the lowest rate of tissue oxygen uptake among the analyzed species. *Hoplosternum littorale* is an active fish and continuously breathes air while *L. pardalis* is a sluggish fish, remaining on the bottom of lakes, breathing air only when the environmental oxygen is depleted (Almeida-Val, 1999).

THE IMPACT OF ENVIRONMENTAL POLLUTION: CHANGES IN AIR-BREATHING BEHAVIOR

Aquatic pollution means any alteration of physical, chemical and biological properties of the aquatic environment that causes physiological disturbance and or damage to tissues leading to an unhealthy state of the organism living in such an environment (Heath, 1995; Costa and Costa, 2004). Aquatic changes are not often caused by increment of chemical or toxic substances in the environment. Abrupt changes of temperature and availability of oxygen can occur as an effect of devastation of gallery forest, being considered in this case an example of habitat degradation. The effects of temperature and water hypoxia on the physiology, biochemistry and behavior of fish have been intensively analyzed (Almeida-Val and Val, 1995; Val and Almeida-Val, 1995; Sanchez *et al.*, 2001; Amin-Neves *et al.*, 2004; Jucá-Chagas, 2004).

However, in the last decades, deposition of dangerous chemical substances increased dramatically in many water bodies. Pesticides, transition metals, organophosphorus, petroleum compounds, among others, have reached aquatic habitats and many studies have been conducted to estimate their effects on fish and fish food web (Heath, 1995; Nemcsók and Bnedeczky, 1995; Val, 1997; Brauner *et al.*, 1999; Val and Almeida-Val, 1999; Scott *et al.*, 2003; Scott and Sloman, 2004). The major sources of these substances are domestic discharges and industrial effluents.

Aquatic Pollution and the General Effects on Behavior of Fish

Many studies have been carried out to evaluate the effect of the pollutants on the physiology and fish survival. The majority of the methods is traditionally based on the tests of acute lethality such as the LC_{50} 96 h test. However, the impact of pollution on the development, growth, reproduction and behavior must be considered, especially because acute lethality tests ignore "ecological death" that may occur at concentrations well below those causing significant mortality (Scott and Sloman, 2004). These authors suggest that the lowest observable effect concentration (LOEC) can be a good common measure for comparison among different studies and species.

Analysis of the effect of pollutants on behavior of fish has been intensively studied, as behavioral change is one of the first responses of fish, both aquatic and air-breathing fish, to changes in environmental quality (Agarwal, 1991; Santhakumar *et al.*, 2000a, b; Scott and Sloman, 2004). Although environmental pollution can affect several kinds of behavioral responses, a particular emphasis on the effect of the environmental pollution on the respiratory behavior of the air-breathing fishes is presented below.

Hypoxia and Temperature Effect on the Respiratory Behavior

Low levels of dissolved oxygen associated with high temperatures are one of the main environmental constraints in the tropics. A classic example occurs in water bodies of the Amazon region, where the oxygen levels can fall to zero in the floodplain waters at night and reach over-saturated levels at noon in the next day (reviewed by Val, 1995). Simultaneously, high temperature directly affects oxygen solubility and promotes the increase of the metabolic activity of fishes, increasing thus the oxygen requirement of them.

Changes in water temperature and changes in dissolved oxygen elicit different behavioral responses of air-breathing fish (Val, 1995; Val and Almeida-Val, 1995; Mattias *et al.*, 1998; Sanchez *et al.*, 2001; Fernandes-Castilho *et al.*, 2004; Jucá-Chagas, 2004; Oliveira, 2004). When exposed to extreme aquatic hypoxia, facultative air-breathers, for example, are forced to survive exclusively on the oxygen uptake from air and use, as mentioned above, different physiological mechanisms. Strictly air-breathers, however, adjust several physiological parameters if exposed to hypoxic environment.

The South American lungfish, *Lepidosiren paradoxa*, an obligate air breather, increased significantly the frequency of air breathes (fr), from 10.4, to 17.2 and 23.3 breaths·h^{-1}, when exposed to PO_2 of 150, 72, and 38 mmHg, respectively. Thus, the ventilation volume (V_I) increased three-fold at a PO_2 of 38 mmHg compared to the normoxic control value (Sanchez *et al.*, 2001). According to these authors, aquatic hypoxia, with access to normoxic air, had no significant effect on the mean rate or amplitude of the air breaths for *L. paradoxa*.

Jucá-Chagas (2004) described a set of behavioral events in three air-breathing fish species, *Lepidosiren paradoxa*, *Hoplerythrinus unitaeniatus* and *Hoplosternum littorale* exposed to hypoxia (PO_2 < 5 mmHg). The author observed that *L. paradoxa* props itself slowly on the side wall of a home tank, remaining for a few seconds in contact with the surface and then returning to the bottom where it stays until the next respiratory episode. *Hoplerythrinus unitaeniatus*, moves to the surface to gulp air through the mouth and eliminates it through the opercular openings. *Hoplosternum littorale* ascends vertically to reach the air/water interface, expands the oral cavity to gulp air and, afterwards, turns its body 180° to the bottom. In this interval it eliminates the respiratory gas through the anal opening. This research showed that in aquatic hypoxia, the fr of *H. unitaeniatus* (5.66 breaths·h^{-1}) was higher than that of *L. paradoxa* (3.31 breaths·h^{-1}) and *H. littorale* (2.66 breaths·h^{-1}). The steep increase in fr of *H. unitaeniatus* means that this species is less tolerant to hypoxia than *L. paradoxa* and *H. littorale*, respectively.

Oliveira *et al.* (2004), studying specimens of *H. unitaeniatus* exposed to graded aquatic hypoxia, PO_2 varying from 140 to 20 mmHg, showed that air-breathing frequency and air-breathing duration increased gradually, reaching maximum values at a PO_2 of 17 mmHg, thought that under normoxia, i.e, PO_2 above 64 mmHg, aerial respiration in this species was absent. In *Hypostomus regani*, a facultative air-breathing fish, air-breathing occurs only under severe hypoxia. Mattias *et al.* (1998) observed that air-breathing frequency increased as PO_2 was progressively reduced and that below 20 mmHg, some fish exhibited uncoordinated swimming movements, followed by a loss of equilibrium and subsequent recovering.

The effects of temperature on air breathing of *L. paradoxa* under experimental conditions were evaluated by Amin-Naves *et al.* (2004). The authors exposed specimens of *L. paradoxa* at 15, 25 and 35°C and observed a positive relationship of respiratory frequency and pulmonary ventilation with temperature.

These experiments clearly demonstrate that changes in oxygen levels and changes in temperature drive changes in respiratory behavior and that different species respond differently to these environmental variables.

Effects of Petroleum on Respiratory Behavior

Pollution of aquatic environment with crude oil has generated great concern as the accidents involving oil spill in aquatic water bodies are, in general, of great amplitude. In many cases, they result in disappearance of aquatic habitats, being extremely limiting for their recovery (Val, 1997).

According to Val and Almeida-Val (2004), fish exposed to crude oil face two sets of problem. The crude oil reduces the oxygen diffusion from the air interface and the presence of oil on the top of the water column reduces the penetration of solar light, reducing the photosynthesis, promoting further reduction of dissolved oxygen. Thus, the film of petroleum formed on the top of the water column represents a challenge for both obligatory and facultative air-breathing fishes.

Specimens of the obligatory air-breathing *Arapaima gigas* exposed to crude oil reduces in about 3 times the respiratory frequency after 20 minutes of exposition and maintain this frequency 24 hours later, what results in an increased oxygen debt and in a almost 10 times reduction in locomotion activity (Val and Almeida-Val, 2004; Fernandes-Castilho *et al.*, 2004). Val *et al.* (2003) film recorded air uptaking by *A. gigas* exposed to crude oil and observed that the animal ingests oil when approaching water surface that is expelled through the opercular opening.

Colossoma macropomum, the tambaqui, when exposed to hypoxia expands the lower lip to capture the oxygen-rich water layer on surface of the water column. This adaptation also appears in other species of the Amazon. Fully expanded lip occurs in about two hours after this species is exposed to hypoxia (Almeida-Val and Val, 1995; Val and Almeida-Val, 1995). Thus, the water-air interface is vital for the fish species using this adaptation. In the case of an oil spill, the oil deposited on the top of the water column, i.e., at the water-air interface, poses an additional constraint to the animals. *Colossoma macropomum*, for example, pumps oil and water across the gills what make this adaptation irrelevant as the oiled gills are inadequate for gas exchange (Val and Almeida-Val, 1999).

Another problem related to the presence of crude oil in aquatic environment is the water-soluble fraction (WSF) which contains mono and poly aromatic hydrocarbons. These compounds mixed with the water severely affect fish respiration, causing a severe impairment of gas exchange. Brauner *et al.* (1999) submitted *Hoplosternum littorale*, a facultative air breather fish, to different concentrations of WSF (12.5, 25, 37.5, 50%) and observed that even the exposure to the lower concentration (12.5% WSF of crude oil) for 45 min resulted in a significant increase in air-breathing frequency relative to control values. Exposure to higher concentrations resulted in a sustained elevated air-breathing frequency rate, almost three times higher than the control values. According to these authors, the increase in air-breathing frequency during exposure to the WSF may be a generalized response to contaminated water.

Other Pollutants and Respiratory Behavior

Environmental pollutants such as heavy metals, organophosphates, hydrogen sulphide and other chemicals represent serious constraints to many aquatic organisms. Many studies have been conducted to estimate the effects of this kind of pollution on fishes. In a polluted environment, as fishes are in close contact with toxicants, these are absorbed causing many biochemical and physiological adjustments in order to reduce pollutant toxicity. A classical example of these effects is the stimulation of mucus production (Lichtenfels *et al.*, 1996). The hypersecretion of mucus is an important mechanism to reduce absorption rate of the pollutants.

However, increased mucus secretion on gills and skin in fishes may be maladaptive from a respiratory standpoint. This is because it adds an unstirred layer next to the lamellar surface and therefore increases oxygen diffusion distance, leading internal hypoxia (reviewed by Heath, 1995).

Oliveira (2003) observed an increased mucus production and a discrete expansion of lower lips in *Colossoma macropomum* exposed to copper, in normoxic water, what suggest that this Amazonian fish species responds to copper as similar as temperate fish species to reduce copper absorption. The appearance of the lips clearly suggests that the animal faces some hypoxia as a consequence of the mucus on the gills.

In addition to these effects, some toxicants, as metals, can cause gill damage, as edema, epithelial lifting, separation of epithelial layers, necrosis, aneurysm, epithelial rupture and fusion of the secondary lamellae (Mazon and Fernandes, 1999; Mazon et al., 2002). Fishes exposed to pollutants showed several gill histopatological lesions that compromise gill respiratory function (Rajan and Banerjee, 1993; Mazon et al., 2002). Such lesions promote alterations of respiratory behavior. In the case of air-breathing fishes, a significant increase of respiratory frequency is described.

Behavioral changes of air breathing of *Anabas testudineus* was observed when individuals of this species were exposed to acute concentration of the insecticide Monocrotophos® (organophosphate). The changes included decreased opecular movement, increased surfacing behavior, irregular swimming activity, loss of equilibrium and increased mucus secretion all over the body (Santhakumar and Balaji, 2000b). Brauner et al. (1995) studying the effects of acid and hydrogen sulphide exposure on the facultative air breather *H. littorale*, also described an increase in the air-breathing frequency of this Amazonian fish species.

CONCLUSION

The animal behavior reflects its well being state, considering it is an expression of a link among physiological, biochemical and ecological process. Therefore, behavioral indicators are important tools for assessing the effects of aquatic pollutants or any other environmental disturbance on fish.

Air-breathing fishes could be considered the most representative example of surviving in extreme water environment conditions. Changes in water quality undoubtedly affect air-breathing fishes. Thus, integrative studies are required to understand the impact of environmental changes on such organisms, mainly in the tropics, where several fish species are still unknown.

References

Agarwal, S.K. 1991. Bioassay evaluation of acute toxicity levels of mercuric-chloride to an air-breathing fish *Channa punctatus* (Bloch)—Mortality and behavior study. *Journal of Environmental Biology* 12: 99-106.

Almeida-Val, V.M.F. 1999. Phenotypic plasticity in Amazon fishes: Water versus air-breathing. In: *Water/Air Transition in Biology*, A.K. Mittal, F.B. Eddy and J.S. Datta Munshi (eds.). Science Publishers, Inc. Enfield (NH), USA, pp. 101-115.

Almeida-Val, V.M.F. and I.P. Farias. 1996. Respiration in fish of the Amazon: Metabolic adjustments to chronic hypoxia. In: *Physiology and Biochemistry of the Fishes of the Amazon*, A.L. Val, V.M.F. Almeida-Val and D.J. Randall (eds.). Instituto Nacional de Pesquisas da Amazonia (INPA), Manaus, Brazil, pp. 257-271.

Almeida-Val, V.M.F. and A.L. Val. 1995. Adaptação de peixes aos ambientes de criação. In: *Criando Peixes na Amazônia*, A.L. Val, and A. Honczarynk (eds.). Instituto Nacional de Pesquisas da Amazônia (INPA), Manaus, pp. 45-59.

Almeida-Val, V.M.F., A.L. Val and P.W. Hochachka. 1993. Hypoxia tolerance in Amazon fishes: status of an under-explored biological "goldmine". In: *Surviving Hypoxia: Mechanisms of Control versus Adaptation*, P.W. Hochachka, P.W. Lutz, T. Sick, M. Rosenthal and G. Van den Thillard (eds.). CRC Press, Boca Raton, pp. 435-445.

Amin-Neves, J., H. Giusti and M.L. Glass. 2004. Effects of acute temperature changes on aerial and aquatic gas exchange, pulmonary ventilation, and blood gas in the South American lungfish, *Lepidosiren paradoxa*. *Comparative Biochemistry and Physiology* A 138: 133-139.

Berner, R.A. 1997. The rise of plants and their effect on weathering and atmospheric CO_2. *Science* 291: 339-376.

Beumont, M.W., P.J. Butler and E.W. Taylor. 1995. Exposure of brown trout, *Salmo trutta*, to sub-lethal copper concentrations in soft acidic water and its effect upon sustained swimming performance. *Aquatic Toxicology* 33: 45-63.

Brauner, C.J., C.L. Ballantyne, D.J. Randall and A.L. Val. 1995. Air breathing in the armoured catfish (*Hoplosternum littorale*) as an adaptation to hypoxic, acidic and hydrogen sulphide rich waters. *Canadian Journal of Zoology* 73: 739-744.

Brauner, C.J. and A.L. Val. 1996. The interaction between O_2 and CO_2 exchange in the obligate air-breathing, *Arapaima gigas*, and the facultative air-breather *Liposarcus pardalis*. In: *Physiological and Biochemistry of the Fishes of the Amazon*, A.L. Val, V.M.F. Almeida-Val and D.J. Randall (eds.). Instituto Nacional de Pesquisas da Amazônia (INPA), Manaus, Brazil, pp. 101-110.

Brauner, C.J., C.L. Ballantyne, M.M. Vijayan and A.L. Val. 1999. Crude oil exposure affects air-breathing frequency, blood phosphate levels and ion regulation in an air-breathing teleost fish, *Hoplosternum littorale*. *Comparative Biochemistry and Physiology* C 123: 127-134.

Burggren, W.W. and K. Johansen. 1986. Circulation and respiration in lungfishes (Dipnoi). *Journal of Morphology* 1 (Supplement): 217-236.

Burggren, W.W., A. Farrel and H. Lillywhite. 1997. Vertebrate cardiovascular systems. In: *Comparative Physiology (Handbook of Physiology; section* 13), W.H. Dantzler (ed.). New York, pp. 215-308.

Clayton, D.A. and T.C. Vaughan. 1986. Territorial acquisition in the mudskipper *Boleophthalmus boddarti* (Pisces:Gobiidae) on the mudflats of Kuwait. *Journal of Zoology* 209: 501-519.

Costa, M.A.G. and E.C. Costa. 2004. *Poluição Ambiental: Herança para Gerações Futuras*. Orium, Santa Maria.

Daniels, C.B., S. Orgeig, L.C. Sullivan, N. Ling, M.B. Bennet, S. Schurch, A.L. Val and C.J. Brauner. 2004. The origin and evolution of the surfactant system in fish: insights into the evolution of lungs and swim bladders. *Physiological and Biochemical Zoology* 77: 732-749.

Fernandes-Castilho, M., A.L. Val and E.M. Silva. 2004. Effect of crude oil on respiratory and locomotion behavior of Amazon fish pirarucu (*Arapaima gigas*). In: *Proceeding of the VI International Congress on the Biology of Fish. Advances in Fish Biology*, A. Val and D. Mackinlay (eds.). Manaus, Amazonas, pp. 279-282.

Fritsche, R. and S. Nilsson. 1994. Cardiovascular and ventilatory control during hypoxia. In: *Fish Ecophysiology*, J.C. Rankin and F.B. Jensen (eds.). Chapman and Hall, London, pp. 180-206.

Gans, C. 1970. Strategy and sequence in the evolution of the external gas exchangers of ectothermal vertebrates. *Forma et Function* 3: 61-104.

Gans, C., R. Dudley, N.M. Aguilar and J.B. Graham. 1999. The Pre-Devonian Carbon Dioxide Crash, the Late-Paleozoic Oxygen Pulse and Associated Shifts in Ventilatory Mechanisms. In: *Water/Air Transition in Biology*, A.K. Mittal, F.B. Eddy and J.S. Datta Munshi (eds.). Science Publishers, Inc., Enfield (NH), USA, pp. 31-43.

Geiger, S.P., J.J. Torres and R.E. Crabtree. 2000. Air breathing and gill ventilation frequencies in juvenile tarpon, *Megalops atlanticus*: Responses to changes in dissolved oxygen, temperature, hydrogen sulfide, and pH. *Environmental Biology of Fishes* 59: 181-190.

Gordon, M.S., I. Boetius, D.H. Evans, R. McCarthy and L.C. Oglesby. 1969. Aspects of the physiology of terrestrial life in amphibious fishes l: the mudskipper, *Periophthalmus sobrinus*. *Journal of Experimental Biology* 50: 141-149.

Graham, J.B. 1997. *Air-Breathing Fishes: Evolution, Diversity and Adaptation*. Academic Press, San Diego.

Graham, J.B. 1999. Comparative aspects of air-breathing fish biology: An agenda for some Neotropical species. In: *Biology of Tropical Fishes*, A.L. Val and V.M.F. Almeida-Val (eds.). Instituto Nacional de Pesquisas da Amazônia (INPA), Manaus. pp. 317-331.

Halliday, T.R. and H.P.A. Sweatman. 1976. To breathe or not to breathe? The Newt's problem. *Animal Behavior* 24: 551-561.

Heath, A.G. 1995. *Water Pollution and Fish Physiology*. CRC Press, Boca Raton, Lewis Publishers. 2nd edition.

Hedrick, M.S. and D.R. Jones. 1993. The effects of altered aquatic and aerial respiratory gas concentrations on air-breathing patterns in a primitive fish (*Amia calva*). *Journal of Experimental Biology* 181: 81-94.

Hedrick, M.S., S.L. Katz and D.R. Jones. 1994. Periodic air-breathing behavior in a primitive fish revealed by spectral analysis. *Journal of Experimental Biology* 197: 429-436.

Ishimatsu, A., Y. Hishida, T. Takita, T. Kanda, S. Oikawa, T. Takeda and K.H. Khoo. 1998. Mudskippers store air in their burrows. *Nature* (London) 391: 237-238.

Johansen, K. 1968. Air-breathing fishes. *Scientific American* 219: 102-111.

Jucá-Chagas, R. 2004. Air breathing of the neotropical fishes *Lepidosiren paradoxa*, *Hoploerythrinus unitaeniatus* and *Hoplosternum littorale* during aquatic hypoxia. *Comparative Biochemistry and Physiology* A 139: 49-53.

Lee, H.J. and J.B. Graham. 2002. Their game is mud. *Natural History* 9: 42-47.

Lee, H.J., C.A. Martinez, K.J. Hertzberg, A.L. Hamilton and J.B. Graham. 2005. Burrow air phase maintenance and respiration by the mudskipper *Scartelaos histophorus* (Gobiidae:Oxudercinae). *Journal of Experimental Biology* 208: 169-177.

Lichtenfels, A.J.F.C., G. Lorenzi-Filho, E.T. Guimarães, M. Macchione and P.H.N. Saldiva. 1996. Effects of water pollution on the gill apparatus of fish. *Journal of Comparative Pathology* 115: 47-60.

Maheshwari, R., A.K. Pati and S. Gupta. 1999. Annual variation in air-gulping behavior in two Indian siluroids, *Heteropneustes fossilis* and *Clarias batrachus*. *Indian Journal of Animal Science* 69: 66-72.

Marshal, C.R. 1986. A List of Fossils and Extant Diponoan. *Journal of Morphology* 1 (Supplement): 15-23.

Matoba, M. and Y. Dotsu. 1977. Prespawning behavior of the mudskipper *Periophthalmus cantonensis* in Ariake Sound. *Bulletin of the Faculty of Fisheries*. Nagasaki University, 43: 23-33.

Mattias, A.T., F.T. Rantin and M.N. Fernandes. 1998. Gill respiratory parameters during hypoxia in the facultative air-breathing fish, *Hypostomus regani* (Loricariidae). *Comparative Biochemistry and Physiology* A120: 311-315.

Mazon, A.F. and M.N. Fernandes. 1999. Toxicity and differential tissue accumulation of copper in the tropical freshwater fish, *Prochilodus scrofa* (Prochilodontidae). *Bulletin of Environmental Contamination and Toxicology* 63: 797-804.

Mazon, A.F., C.C.C. Cerqueira and M.N. Fernandes. 2002. Gill cellular changes induced by copper exposure in South American tropical freshwater fish *Prochilodus scrofa*. *Environmental Research* A 88: 52-63.

Nelson, J.S. 1994. *Fishes of the World*. 3rd edition. John Wiley & Sons, Inc., New York.

Nemcsók, J. and I. Benedeczky. 1995. Pesticide metabolism and the adverse effects of metabolites on fishes. In: *Biochemistry and Molecular Biology of Fishes. Environmental and Ecological Biochemistry*, P.W. Hochacka and T.P. Mommsen (eds.). Elsevier, Amsterdam, Vol. 5, pp. 313-348.

Oliveira, C.P.F. 2003. Efeito de cobre e chumbo, metais pesados presentes na água de formação derivada da extração do petróleo da província petroleira do Urucu – Am, sobre o tambaqui, *Colossoma macropomum* (Curvier, 1818). M.Sc. Thesis. Instituto Nacional de Pesquisas da Amazônia (INPA), Manaus, Brazil.

Oliveira, R.D., J.R. Lopes, J.R. Sanches, A.L. Kalinin, M.L. Glass and F.T. Rantin. 2004. Cardiorespiratory responses of the facultative air-breathing fish jeju, *Hoploerythrinus unitaeniatus* (Teleostei, Erythrinidae), exposed to graded hypoxia. *Comparative Biochemistry and Physiology* A 139: 479-485.

Pandian, T.J. and K. Sampath. 1981. Air-breathing fishes. *Science Reporter* 18: 646-649.

Rajan, M.T. and T.K. Banerjee. 1993. Histopathological changes in the epidermis of the air-breathing catfish *Heteropneustes fossilis* exposed to sublethal concentration of mercuric chloride. *Biomedical and Environmental Science* 6: 405-412.

Randall, D.J. 1990. Control and co-ordination of gas exchange in water breathers. In: *Advances in Comparative and Environmental Physiology*, R.G. Boutilier (ed.). Springer-Verlag, Berlin, Vol. 6, pp. 253-278.

Randall, D.J., A.P. Farrell and M.S. Haswell. 1978. Carbon dioxide excretion in the pirarucu (*Arapaima gigas*), an obligate air-breathing fish. *Canadian Journal of Zoology* 56: 977-982.

Randle, A.M. and L.J. Chapman. 2004. Habitat use by the African anabantid fish *Ctenopoma muriei*: Implications for costs of air breathing. *Ecology of Freshwater Fish* 13: 37-45.

Sampath, K. and T.J. Pandian. 1985. Effects of size hierarchy on surfacing behavior and conversion rate in an air-breathing fish *Channa striatus*. *Physiology and Behavior* 34: 51-55.

Sanchez, A., R. Soncini, T. Wang, P. Koldjaer, E.W. Taylor and M.L. Glass. 2001. The differential cardio-respiratory responses to ambient hypoxia and systemic hypoxaemia in the South American lungfish, *Lepidosiren paradoxa*. *Comparative Biochemistry and Physiology* A130: 677-687.

Santhakumar, M., M. Balaji, K.R. Saravanan, D. Soumady and K. Ramudu. 2000a. Effect of monocrotophos on optomotor behaviour of an air-breathing fish *Anabas testudineus* (Bloch). *Journal of Environmental Biology* 21: 65-68.

Santhakumar, M. and M. Balaji. 2000b. Acute toxicity of an organophosphorus insecticide monocrotophos and its effects on behaviour of an air-breathing fish *Anabas testudineus* (Bloch). *Journal of Environmental Biology* 21: 121-123.

Scholander, P.F., L. van Dam and S.I. Scholander. 1955. Gas exchange in the roots of mangroves. *American Journal of Botany* 42: 92-98.

Scott, G.R., K.A. Sloman, C. Rouleau and C.M. Wood. 2003. Cadmium disrupts behavioural and physiological responses to alarm substance in juvenile rainbow trout (*Oncorhynchus mykiss*). *Journal of Experimental Biology* 206: 1779-1790.

Scott, G.R. and K.A. Sloman. 2004. The effects of environmental pollutants on complex fish behaviour: integrating behavioural and physiological indicators of toxicity. *Aquatic Toxicology* 68: 369-392.

Seymour, R.S., K. Christian, M.B. Bennet, J. Baldwin, R.M.G. Wells and R.V. Baudinette. 2004. Partitioning of respiration between the gills and air-breathing organ in response to aquatic hypoxia and exercise in the Pacific tarpon, *Megalops cyprinoides*. *Physiology and Biochemical Zoology* 77: 760-767.

Shelton, G., D.R. Jones and W.K. Milsom. 1986. Control of breathing in ectothermic vertebrates. In: *Handbook of Physiology*, A.P. Fishman, S.N. Cherniack, J.G. Widdicombe and S.R. Geiger (eds.), American Physiology Society, Madison,

Bethesda, *The Respiratory System,* Vol II, section 3, *Control of Breathing,* part 2, pp. 857-909.

Takita, T., A. Agusnimar and A.B. Ali. 1999. Distribution and habitat requirements of *Oxudercine gobies* (Gobiidae: Oxudercinae) along the Straits of Malacca. *Ichthyology Research* 46: 131-138.

Val, A.L. 1995. Oxygen transfer in fish: Morphological and molecular adjustments. *Brazilian Journal of Medical and Biological Research* 28: 1119-1127.

Val, A.L. 1996. Surviving low oxygen levels: Lessons from fishes of the Amazon. In: *Physiology and Biochemistry of the Fishes of the Amazon,* A.L. Val, V.M.F. Almeida-Val and D.J. Randall (eds.). Instituto Nacional de Pesquisas da Amazônia (INPA), Manaus, Brazil, pp. 59-73.

Val, A.L. 1997. Efeitos do petróleo sobre a respiração de peixes da Amazônia. In: *Indicadores Ambientais,* H.L. Martos and N.B. Maia (eds.). Pontifícia Universidade Católica. Sorocaba, São Paulo. pp. 109-119.

Val, A.L. 1999. Water-air-breathing transition in fishes of the Amazon. In: *Water/Air Transition in Biology,* A.K. Mittal, F.B. Eddy and J.S. Datta Munshi (eds.). Science Publishers, Inc., Enfield (NH), USA, pp. 145-161.

Val, A.L. 2000. Organic phosphate in the red blood cells of fish. *Comparative Biochemistry and Physiology* A125: 417-435.

Val, A.L. and V.M.F. Almeida-Val. 1995. *Fishes of the Amazon and their Environments. Physiological and Biochemical Features.* Springer-Verlag, Heidelberg.

Val, A.L. and V.M.F. Almeida-Val. 1999. Effects of crude oil on respiratory aspects of some fish species of the Amazon. In: *Biology of Tropical Fishes,* A.L. Val and V.M.F. Almeida-Val (eds.). Instituto Nacional de Pesquisas da Amazônia (INPA), Manaus, Brazil, pp. 277-291.

Val, A.L. and V.M.F. Almeida-Val. 2004. Crude oil, copper and fish of the Amazon. In: *Proceedings of the VI International Congress on the Biology of Fish. Behaviour, Physiology and Toxicology Interactions in Fish,* K. Sloman, C. Wood and D. MacKinlay (eds.). Manaus, Amazonas, pp. 1-6.

Val, A.L., E.G. Affonso, R.H.S. Souza, V.M.F. Almeida-Val and M.A.F. Moura. 1992. Inositol pentaphosphate in erythrocytes of an Amazonian fish, the pirarucu (*Arapaima gigas*). *Canadian Journal of Zoology* 70: 852-855.

Val, A.L., V.M.F. Almeida-Val and A.R. Chippari-Gomes. 2003. Hypoxia and petroleum: extreme challenges for fish of the Amazon. In: *Fish Physiology, Toxicology, and Water Quality. Proceedings of the Seventh International Symposium,* G. Rupp and M.D. White (eds.). Tallinn, Estonia, pp. 227-241.

Watters, J. V. and J.J. Cech Jr. 2003. Behavioral responses of mosshead and woolly sculpins to increasing environmental hypoxia. *Copeia* 2003: 397-401.

Yadu, Y. and M. Shedpure. 2002. Pinealectomy does not modulate the characteristics of 24-h variation in air-gulping activity of *Clarias batrachus. Biological Rhythm Research* 33: 141-150.

8

The Osmo-respiratory Compromise in Fish: The Effects of Physiological State and the Environment

Brian A. Sardella* and Colin J. Brauner

OVERVIEW AND BACKGROUND

The gills of fishes play a central role in gas exchange, ion regulation, acid-base balance and nitrogenous waste excretion. Consequently, there are trade-offs and compromises in gill design and function so that several processes can occur across a single structure to a satisfactory, and in some cases, optimal level. For example, conditions beneficial to gas exchange, such as large surface area, reduced diffusion distance and high blood and water flow rates, are the very characteristics that are detrimental to osmoregulation. This chapter will focus on the respiratory and osmoregulatory functions of the gill and the trade-offs that exist in gill

Authors' address: Department of Zoology, University of British Columbia 6270 University Boulevard, Vancouver BC V6T 1Z4, Canada.
**Corresponding author:* E-mail: sardella@zoology.ubc.ca

design— which has been defined as the osmo-respiratory compromise. We will start with a description of the basis for the osmo-respiratory compromise and how it has been quantified historically. This will be followed by a discussion of how short-term changes in the environment, such as salinity, temperature, and oxygen, can affect the osmo-respiratory compromise. Finally, we will conclude with a discussion of gross morphological changes in the gills associated with either development or long term exposure to different environments that reflect the nature and degree of the osmo-respiratory compromise.

(A) General Gill Design

The morphological design of the gill has been well reviewed previously (Laurent, 1984), and there is a clear regional separation of osmoregulatory and respiratory functions. Gas exchange takes place primarily across the lamellar epithelium, where oxygen and carbon dioxide move via simple diffusion across flattened squamosal cells called pavement cells (PVCs). While PVCs are found throughout the branchial epithelium, they are the dominant cell type on the surface of lamellae in the majority of fish species. In contrast, ion regulation occurs along the filamental epithelium, within specialized mitochondria-rich cells, or chloride cells (CCs). The filamental epithelium has a more mosaic design, containing CCs, mucous-secreting cells (goblet cells) and PVCs. Furthermore, it has been noted that CCs tend to be more localized to the afferent filamental epithelium, while the efferent filamental epithelium tends to be richer in mucous-cells. Chloride cells are functionally specialized according to the ionic concentration of the environment, undergoing morphological and physiological transformations during acclimation to different ionic compositions (i.e., freshwater or seawater).

(B) Gill Blood Flow

While the majority of attention has been paid to the cellular arrangements of the gill, the vascular anatomy and physiology of the gill was also reviewed by Olson (1991). The vasculature of the gill consists of two circuits, the arterioaterial and arteriovenous, or interlamellar, circulations. The arterio-arterial circulation is primarily involved in gas exchange, and is conducted via the following path; ventral aorta-afferent branchial artery-afferent filamental artery-afferent lamellar arteriole-lamellae-efferent lamellar arteriole-efferent filamental artery-efferent branchial artery-dorsal aorta. The thin-walled lamellae are held open by muscular pillar cells supporting columns of collagen that can be adjusted in length to control lamellar blood flow patterns (Olson, 1991).

In most fish, the gills are the only organ that receives the entire cardiac output, all of which must pass through the lamellae where it is potentially subjected to osmoregulatory perturbations associated with the environment. Some species such as eels possess a lamellar shunt, whereby the osmotic disturbances associated with lamellar perfusion can be eliminated under conditions of low oxygen demand (Olson, 1991).

Most fish are without lamellar shunts, thus the entire cardiac output potentially must pass through the lamellar circulation; however, alternative mechanisms exist that minimize lamellar flow when oxygen demand is low. Furthermore, the vasculature of the teleost gill is muscular and innervated with several nerve fiber types, so the possibility of redirecting blood flow within the gill via vasoconstriction/vasodilatation also exists. Recruitment of unused filaments and/or lamellae under conditions of oxygen demand has been observed (Randall and Stevens, 1967; Randall et al., 1967), so it can be speculated that down-regulation occurs when oxygen demand is low or when osmotic balance is threatened. Additionally, the blood flow pattern within the lamellae can be redirected to avoid areas of high diffusion lessen ion balance disturbances. The most medial sections of the lamellae are embedded within the filament and are not involved in gas exchange. Blood flow can be directed into these areas via pillar cells contraction/relaxation; essentially this mechanism functions as a shunt around the outer edges of the lamellae where diffusion forces are much greater (Olson, 1991).

THE COMPROMISE BETWEEN OSMOREGULATION AND RESPIRATION AT THE GILL

Oxygen and carbon dioxide are exchanged across the gills via simple diffusion, where exchange rate is a function of the partial pressure gradient for the respective gases, diffusional surface area, diffusion distance, and permeability of the barrier. Gas flux across the gills can be increased, under conditions such as exercise, by increasing water ventilation rate and the perfusion rate of blood through the gills. Increased perfusion leads to a recruitment of lamellae (Booth, 1978), which increases the functional surface area for diffusion and reduces the blood to water diffusion distance. In general, diffusion rate is maximal when the diffusion distance is low, the surface area and gradient are large, and the permeability is high. These characteristics are in direct contrast with those that minimize ion loss or gain across the gills. Net ionic movements are minimized by thick epithelial membranes and many large ion exchanging chloride cells. Alterations in both metabolic rate and salinity tolerance are potentially constrained due to compromises between maintaining sufficient gas transfer and defending osmotic balance.

The compromise between osmoregulation and respiration was first noted by Randall *et al.* (1972), although Nilsson (1986) first referred to it as the osmo-respiratory compromise. To test the hypothesis that enhanced gas transfer would increase ion diffusion, Randall *et al.* (1972) monitored the flux of radiosodium (Na^{22}) in fresh water (FW) rainbow trout (*Oncorhynchus mykiss*) that had been exercised, or injected with either noradrenalin or isoprenaline in order to elicit a catecholamine response. All three treatments increased sodium efflux, with the bout of exercise resulting in the greatest rate of loss. Catecholamine stimulation resulted in about 50% of the efflux observed during exercise. The hypothesis of Randall *et al.* (1972) was that the increase in sodium loss was catecholamine mediated, and resulted from alterations in blood flow patterns through the gills under increased oxygen demand. That work was followed up by Wood and Randall (1973), where multi-directional sodium fluxes were investigated during exercise and recovery in the same species. Activity was imposed upon the fish by chasing, and resulted in increased ventilation and perfusion of the gills. During activity, sodium efflux increased approximately 70%, resulting in a net sodium loss. Once again, changes in gill blood flow were thought to result in a recruitment of additional lamellae and a subsequent increase in diffusive ion loss to the freshwater environment. Results from the studies discussed above provided a framework within which the osmo-respiratory compromise could be further investigated.

QUANTIFICATION OF THE OSMO-RESPIRATORY COMPROMISE

(A) The Ion-Gas Ratio

One study in particular was instrumental in describing the relative changes in sodium efflux (J^{Na+}) and oxygen consumption (MO_2); the ion-gas ratio (IGR). Gonzalez and McDonald (1992) investigated the IGR of fresh water (FW)-acclimated rainbow trout under osmotic challenge, low ambient $[Ca^{2+}]$, and exercise. As predicted, increases in metabolic rate corresponded with increased ion efflux. The authors concluded that the increase in MO_2 was associated with an increase in gill functional surface area, a decrease in diffusion distance resulting from lamellar thinning, and an increased O_2 permeability mediated by catecholamines, all of which were expected to result in increased ion losses proportional to the increases in MO_2. Interestingly, J^{Na+} and MO_2 did not increase proportionally, and as a result there was an increase in the IGR anywhere from 1.6 to 10-fold. The disproportional increase in IGR was attributed to a distortion of paracellular tight junctions along the respiratory epithelium (Gonzalez and McDonald, 1992), which increases the rate of ion loss without affecting gas flux.

(B) Paracellular Tight Junctions

Several of the experimental treatments used by Gonzalez and McDonald (1992) resulted in a loss of tight junction integrity. Tight junctions provide the necessary adhesion between cells along an epithelium, turning a group of like cells into a membrane barrier. Because trans-cellular movement of solutes and water is nearly impossible without involving cellular processes, paracellular junctions largely determine the overall barrier capability of an epithelium. Tight junctions consist of proteinaceous strands, and the number of strands determine the depth and permeability characteristics of the junction. Typically, tight junctions found in the epithelia of FW-acclimated fishes are deep and contain multiple strands. Tight junctions within FW-acclimated epithelia are quite impermeable relative to their shallow, sea water (SW)-acclimated counterparts, where junctional leakiness plays an important role in Na^+ excretion. Negatively-charged amino acid residues in the areas where junctional strands overlap are often bound by Ca^{2+} ions, which in the Gonzalez and McDonald (1992) study, were most-likely removed due to exposure to a low $[Ca^{2+}]$ environment, making them more susceptible to sodium efflux. Additionally, junctional integrity was distorted by cell shrinkage resulting from exposure to hyperosmotic media, and by an increased hydrostatic pressure when perfusion rate was elevated. A combination of these factors led to the disproportional increase in ion efflux rate as metabolism increased. The results of Gonzalez and McDonald (1992) indicate that the capacity to maintain tight junction integrity has direct effects on the IGR in rainbow trout, and furthermore, may place a limit on metabolic scope.

(C) Active vs Non-active Fishes

If the increase in ion losses during exercise does limit the metabolic scope of a fish, it would be expected that the IGR is affected differently during exercise in active versus non-active species. This was investigated by Gonzalez and McDonald (1994) using nine species of FW fishes from a range of activity levels. In this study, the authors found that active fish had a much greater increase in MO_2 following exercise relative to those considered to be non-active. Interestingly, active fish were able to achieve a higher MO_2 relative to less active fish, while minimizing ion losses. Species with low routine MO_2 values lost more Na^+ per mole of O_2 absorbed, resulting in a negative correlation between IGR and MO_2; however, within a given species, increases in MO_2 over routine levels still resulted in increased ion loss. If the increase in IGR can be attributed to the distortion of tight junctions along the respiratory epithelium, then it can be hypothesized that more active fishes are able to somehow compensate for, or prevent, tight junction distortion to a greater degree than non-active fishes. Due to the drastically different mechanisms, this hypothesis is only supported for FW-acclimated fishes; however an investigation of tight junction structure and/or function between active and non-active fishes would prove very interesting. Gonzalez and McDonald (1994) further suggested that, regardless of the ability of active fishes to reduce tight junction distortion, ion losses are never fully alleviated, as ion exchanges are necessary for other functions such as acid-base balance. Clearly there are species specific differences that may act to reduce the magnitude of the osmo-respiratory compromise, which in this case likely reside in the characteristics of the gill epithelium. However, within a given species exposures to different environmental challenges over the short term are useful in illuminating the nature and degree of the osmo-respiratory compromise.

EFFECT OF SHORT TERM CHALLENGES ON THE OSMO-RESPIRATORY COMPROMISE

There are two major scenarios in which the osmo-respiratory compromise may interfere with osmotic homeostasis: (1) when oxygen demand is high, and gill perfusion must be increased to favor gas exchange at the expense of ion regulation; and (2) when the epithelium must be thickened to defend ion balance at the expense of gas exchange.

Stressors such as salinity, exercise, hypoxia, and temperature changes can all have a dramatic effect on the rate of oxygen consumption and ion balance. The short-term impacts of these different physiological states and environments are discussed below.

(A) Salinity

The metabolism of euryhaline species is often decreased when they are exposed to salinities above SW, even though the gradient and cost of osmoregulation should be greatly increased (Sardella et al., 2004a). The salinity at which metabolism becomes impaired most likely corresponds with the loss of ability to osmoregulate efficiently, as indicated by increased signs of osmoregulatory stress (Haney and Nordlie, 1997; Swanson, 1998; Sardella et al., 2004a). As a compensatory mechanism, a decrease in lamellar profusion to reduce the surface area for ion and water movement may partially alleviate osmoregulatory challenges, but also dramatically impairs gas exchange. Mozambique tilapia hybrids (Oreochromis mossambicus × O. urolepis hornorum) exposed to hypersaline conditions for two weeks exhibited up to a 38% decrease in resting MO_2 following 2 weeks of exposure to salinities ranging from 35 to 95 g/l (Sardella et al., 2004a) (Fig. 8.1). The reduced MO_2 correlated with increased plasma osmolality, $[Na^+]$, and $[Cl^-]$, and high levels of branchial Na^+, K^+-ATPase activity; there were few changes in these variables at salinities at or slightly above SW (35 g/l). A reduction in metabolic rate in response to elevated salinity has also been observed in the euryhaline milkfish (Chanos chanos; Swanson, 1998) where acclimation to 55 g/l salinity resulted in a 26% reduction in MO_2 relative to fish acclimated to 35 g/l (Fig. 8.1). Furthermore, as was observed with tilapia hybrids, the fall in MO_2 was correlated with a decline in osmoregulatory control, with a decrease in swimming activity also observed (Swanson, 1998). Swanson (1998) concluded that the reduction in swimming performance was a result of the need to defend osmotic balance, supporting the hypothesis of Gonzalez and McDonald (1992, 1994) that maintaining osmotic balance places limitations on metabolic scope. Swanson (1998) concluded that the high metabolism of fish at 35 g/l salinity was associated with higher spontaneous activity, while the depressed metabolism at 55 g/l represents the contrary.

The sheepshead minnow (Cyprinidon variegatus) is one of the most stress tolerant fishes, surviving exposures to extreme salinities and/or temperatures. Haney and Nordlie (1997) investigated the effects of

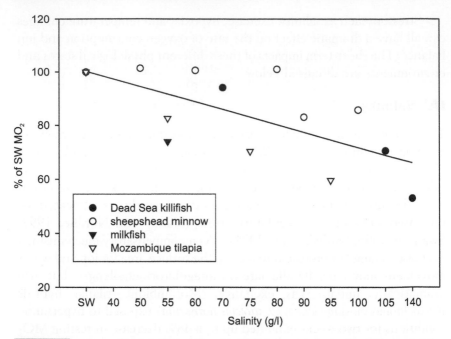

Fig. 8.1 Resting metabolic rates of four euryhaline fish species exposed to salinities greater than SW; values are expressed as a percentage of resting metabolic rate in SW.
• Dead Sea killifish (*Aphinius dispar*) from Plaut (2000); o sheepshead minnow (*Cyprinidon variegatus*) from Haney and Nordlie (1997); ▼ milkfish (*Chanos chanos*) from Swanson (1997); ▽ California Mozambique tilapia (*Oreochromis mossambicus* × *O. urolepis hornorum*). Adapted from Sardella *et al.* (2004a).

salinity on oxygen consumption and critical oxygen tension in this species. When exposed to 40 g/l salinity or greater, metabolic rate in this species fell steadily as salinity increases (Fig. 8.1). Additionally, Haney and Nordlie (1997) found that critical oxygen tension (P_{Crit}) increased with salinity. The (P_{Crit}) of blood represents the partial pressure of oxygen (PO_2) at which MO_2 begins to decrease as a function of PO_2. Whether the increase in P_{Crit} and reduction in MO_2 observed in these fish species exposed to hypersaline environments is a regulated process representing a compromise between gas exchange and osmoregulation, or whether it is a direct effect of elevated salinity is not known. In the case of the Mozambique tilapia, the reduction in metabolic rate at elevated salinities lasts for up to 2 weeks at 60 g/l salinity, returning to normal levels after 4 weeks exposure, possibly reflecting a strategy of initial resistance prior to acclimation (Sardella and Brauner, unpublished).

(B) Exercise

During exercise, increased O_2 and CO_2 flux across the gills is accomplished by an increase in ventilation volume, an increase in functional surface area (accomplished through lamellar recruitment), a reduction in diffusion distance, and increased perfusion of blood. While all these changes facilitate gas exchange across the gills, they increase the potential for osmoregulatory disturbances. Several studies have demonstrated osmoregulatory disturbances following exercise; salmonid species are often used as models in these studies as they are exceptional athletes and their anadromous natural history results in exposure to a large range of salinities during migration. Wood and Randall (1973) imposed one hour of exercise by chasing fish, while Gonzalez and McDonald (1992) exercised trout at 85% of critical swimming speed (U_{crit}) for 5-6 min. Both studies were conducted in FW and showed significant osmoregulatory disturbances (sodium losses and/or water gains). Wood and Randall (1973) observed a 70% increase in sodium efflux following exercise, while Gonzalez and McDonald (1992) observed a near four-fold increase. Postlethwaite and McDonald (1995) monitored changes in ion fluxes during a more sustainable exercise regime, using continuous aerobic exercise at 1.8 body lengths per second for up to 96 h. They found that sustained exercise resulted in decreased plasma $[Na^+]$ and $[Cl^-]$, due to effluxes of ions, influx of water, or both. Furthermore, they found that compensatory influxes of both ions increased as much as three-fold in rapid fashion (10-12 h following initiation of swimming). The authors concluded that the magnitude and time course of the compensatory influx indicates an increase in the number of transporters for both ions, likely activated from a pre-existing non-active state, as opposed to being newly synthesized. In this case, the osmotic imbalance was corrected for in the early stages of the exercise protocol. Thus, the rapid exercise-induced changes to satisfy metabolic demand at the gills initially affected ion homeostasis negatively; however, existing pathways could quickly be activated to compensated for this and bring the system back toward its pre-exercise state. Investigating the effects of exercise training, Gallaugher *et al.* (2001) swam both trained and untrained SW-acclimated Chinook salmon (*Oncorhynchus tshawytscha*) at U_{crit} and 80% U_{crit}, and measured the changes in plasma osmolality. They found that trained fish once again appeared to be able to achieve higher levels of MO_2, but had significantly lower plasma osmolality at both swimming speeds relative to their untrained counterparts. The beneficial effect of exercise training on osmoregulation in SW fish may have its basis at the level of the gut rather than the gill. In a SW-acclimated fish, the gut plays a large osmoregulatory role, as it actively absorbs ions from imbibed water, to drive water uptake. Trained fish show less of a reduction in gut blood flow during exercise than untrained fish, and thus trained fish may be more capable of water absorption across the gut to combat loss at the gill (Gallaugher et al., 2001). While the examples given represent the effects of exercise in salmonids, there has also been some limited work on other species. Farmer and Beamish (1969) swam Nile tilapia (Tilapia nilotica) in FW, 15 g/l, and 30 g/l. While there were no statistically significant changes in plasma osmolality, the authors did note a trend where fish exercised at these salinities had lower, equal, and higher osmolality, respectively, when compared with resting tilapia, suggesting that this effect of exercise on osmotic homeostasis is probably common among fishes.

(C) Temperature

Elevation in water temperature directly affects metabolism, where a Q_{10} of 2 is commonly assumed (Hochachka and Somero, 2002). The temperature-induced elevation in metabolism results in increased gill ventilation volume and gill blood flow (Barron et al., 1987) to enhance O_2 and CO_2 transport. As with exercise, this could have negative effects on osmoregulation depending upon the salinities in which fish reside. In general, there are few studies that have investigated the effect of temperature on osmoregulation in fish, especially at salinities greater than SW. In Mozambique tilapia hybrids directly transferred from SW to 43, 51, or 60 g/l salinity at 25 or 35°C there was no significant change in plasma osmolality following 24 h (Fig. 8.2), indicative of the incredible salinity tolerance of this species (Sardella et al., 2004b). However, in fish that were acclimated to SW at 25°C, and transferred to 43, 51, or 60 g/l salinity at 35°C, there was a dramatic increase in plasma osmolality after 24 h (Fig. 8.2). Thus, while the osmoregulatory mechanisms in 25°C acclimated fish appear to be sufficient to deal with large elevations in environmental salinity, when this is accompanied by an increase in temperature (and associated elevation in metabolic rate) osmoregulation is clearly impaired (Sardella et al., 2004b). The short duration (24 h) of exposure to elevated salinities and temperature in this study precludes acclimation, and helps to illustrate the negative interaction between gas exchange and ion

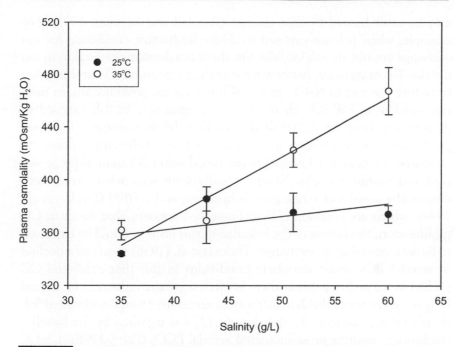

Fig. 8.2 The effect of direct salinity and temperature transfers on plasma osmolality (±SE) in California Mozambique tilapia (*Oreochromis mossambicus × O. urolepis hornorum*). Adapted from Sardella *et al.* (2004b).

regulation at the gills in these fish. Longer duration exposure and acclimation to combined salinity and temperature may alter this relationship; however, this is not known.

CHANGES IN GILL MORPHOLOGY THAT ALTER THE OSMO-RESPIRATORY COMPROMISE

Short-term changes in physiological state and the environment, as discussed above, are important in illuminating the nature of the compromise between gas and ion regulation at the gills. However, long-term changes in the environment and subsequent acclimation, as well as developmental changes, provide yet another view to help understand how the osmo-respiratory compromise operates.

(A) Reversible Changes in Gill Morphology

When fish are exposed to a new environmental condition, the level at which acclimation at the gill is observed is indicative of the process that may be most limited by that change (Randall and Brauner, 1991). For example, when fish are exposed to dilute freshwater, conditions for gas exchange are not altered initially, but there is a dramatic reduction in ion uptake. To compensate, rainbow trout undergo extensive CC proliferation to restore the rate of NaCl uptake. While CCs are generally absent from the lamellae in FW fish, their presence is extensive in fish exposed to dilute water. Greco et al. (1996) showed that diffusion distance increases predictably with an increase in CC fractional area. Following exposure to an ion-poor medium, a doubling of the blood-water diffusion distance was observed within 4 weeks. Similar results were seen when trout were chronically dosed with ovine growth hormone and cortisol (Bindon et al., 1994), which are common SW-acclimatory hormones that result in CC proliferation; thickening of the lamellae in this fashion would be expected to be detrimental to gas exchange. Thomas et al. (1988) observed a decline in arterial PO_2 under normoxic conditions in fish that exhibited CC proliferation, and in hypoxia there is often a greater reduction in arterial PO_2 at a given water PwO_2 in soft water- versus hard water-acclimated fish (Perry, 1998). Additionally, transfer of CO_2 was impaired by the lamellar thickening, resulting in an increased arterial PCO_2 (Perry, 1998). Loss of effective gas transfer associated with CC proliferation can be compensated for to some degree by increasing ventilation, and thus water flow over the gills; Greco et al. (1995) showed that the ventilation of rainbow trout gills increased following a thickening of the lamellar epithelium. Increasing gill ventilation may result in other problems, as ionic gradients can become further enhanced, and the muscular action of ventilation is energetically expensive. Greco et al. (1995) showed that ventilation reaches maximum levels when compensating for CC proliferation, even under conditions of normoxia; normally, during exposure to hypoxia, the resulting increase in ventilation leads to a respiratory alkalosis; however, this was not observed in trout experiencing a CC proliferation. If ventilation is maximal under normoxic conditions in fish experiencing CC proliferation, then any decrease in environmental oxygen, or increase in carbon dioxide could be detrimental. Thus, the increased CC proliferation to ensure adequate ion uptake in soft water comes at a cost to gas transfer, and the resultant changes during soft water exposure represent a compromise between these two, and likely other, processes.

The driving force for O_2 transport across the gills is accomplished by the ΔPO_2 between the water and the blood. Many hypoxia tolerant fishes, such as goldfish (*Carrassius auratus*) and crucian carp (*Carassius*

carassius), have very low whole blood P_{50} to safeguard O_2 uptake during exposure to low water O_2 tensions. This must be associated with adaptations at the tissue level to deal with the low P_{50}. In normoxia, however, the ΔPO_2 across the gills would be expected to be large relative to other fishes with higher whole blood P_{50}'s (assuming all else is equal), which would result in unnecessary costs associated with water and ion transfers. In both the crucian carp, and goldfish, it appears that in normoxia, functional gill surface area is minimized by the filling of interlamellar regions (Sollid *et al.*, 2003, 2005), resulting in a column like appearance of the filaments. Upon exposure to hypoxia, the interlamellar mass is reduced through apoptosis and cell cycle arrest, and the lamellae become exposed to the ambient environment. The final result is a 7.5 fold increase in respiratory surface area of the gills that correlates with a reduced critical oxygen tension from 1.0 to 0.5 mg O_2/l. The process appears to be temperature dependent, indicating that remodeling of the gills is likely driven by the animals' metabolic demand (Sollid *et al.*, 2005). The increase in gill surface area associated with exposure to hypoxia appears to come with an osmoregulatory cost in that fish with protruding lamellae have plasma [Cl⁻] that is significantly reduced relative to controls, and those with exposed lamellae have significantly elevated plasma [Cl⁻] following a 16 g/l salinity challenge (Sollid *et al.*, 2003). Both the crucian carp and goldfish are capable of producing ethanol during exposure to anoxia which may be a prerequisite of such large, and relatively slow, morphological changes associated with exposure to hypoxia.

In another species of carp, the scale-less carp (*Gymnocypris przewalskii*) from Lake Qinghai, China, similar effects of hypoxia on the gills are observed (Brauner and Matey, unpublished data) despite the absence of the ability to produce ethanol (Wang, unpublished data), but the magnitude of the change, and in particular the degree to which lamellae were embedded in normoxia, was greatly reduced relative to crucian carp. Again, protrusion of the lamellae associated with exposure to hypoxia resulted in a significant reduction in plasma Na^+ and Cl^- levels (Richards and Wang, unpublished data) indicating that the reduced functional surface area of the gills is beneficial in terms of osmoregulation. Remodeling of the gills during exposure to hypoxia is time consuming (up to 7 days for completion in the crucian carp at 8°C), and presumably costly. It is not clear why carp do not just reduce the functional surface area of the gills in normoxia by reducing the degree of lamellar perfusion as other fish do, but these gross morphological changes that are seen in the gills are convincing as to the existence and the magnitude of the osmo-respiratory compromise that must exist at the gills.

(B) Non-reversible Changes in Gill Morphology

In larval fish, such as the salmonids, the surface area of the body and yolk sac may be sufficient for gas exchange, however, with development, an increase in gill surface area is crucial for compensating for the reduction in body surface area to volume ratio (SVR). Interestingly, however, the gills develop long before larvae reach the size where gills are required for gas exchange (Rombough and Moroz, 1997), implying that gills are either developing in anticipation of the need for gas exchange, or that they are developing for another purpose. Rombough (1999) has proposed that initially gills develop for ion regulation and/or acid-base balance rather than gas exchange. This theory is based largely upon two observations: (1) that CCs appear on the gill filament of pre-hatch rainbow trout long before the lamellae develop, and (2) that by the time rainbow trout hatch, filamental CC density is similar to that measured in adult fish, despite the immature status of the lamellae. Thus, gill development may be most strongly influenced by the need for ion regulation and/or acid-base balance early on, and then with development, as total body SVR is reduced, additional constraints associated with gas exchange may result in changes in gill morphology that are superimposed upon the existing gills; in particular resulting in extensive lamellar development. Assuming gill development is not hard wired, exposure to hypoxia early in development would be expected to accelerate the onset of extensive lamellar development, as has been observed in arctic char (*Salvenius alpinus*; McDonald and McMahon, 1977). Thus, developing larvae have the ability to accelerate the onset of lamellar proliferation but "choose" not to, presumably due to the osmo-respiratory compromise.

The obligate air-breathing teleost, *Arapaima gigas*, represents yet another interesting model to investigate how selective pressures associated with osmoregulation and gas exchange influence gill morphology during development. Initially, A. *gigas* are obligate water-breathers up to about 8-9 days post-hatch (18 mm in length; see Graham, 1997), however, shortly thereafter they become air-breathers and drown without access to air within 10-20 minutes. The swimbladder is highly vascularized, and as air-breathers, responsible for about 80% of O_2 uptake at this time with the remaining 20% being taken up either across the skin or gills (Stevens and Holeton 1978; Brauner and Val, 1996). It is likely that the trajectory for gill development in A. *gigas* initially follows that described above for salmonids. That is, early on gills develop primarily for ion regulation and/or acid-base balance, and then as SVR decreases, the lamellae proliferate to satisfy gas exchange at the gills. With further development, however, the dependence upon the gills for O_2 uptake decreases and gill design will again be less influenced by conditions for O_2 transport. Consistent with the latter, large morphological changes are seen in the gills of A. *gigas*, from the time fish are about 10 g to 1 kg (Brauner *et al.*, 2004). At 10 g, A. *gigas* is an obligate air-breather, but slightly more dependent upon aquatic respiration than by the time they reach 1 kg. At 10 g, protruding lamellae from the filaments are readily visible (Fig. 8.3A) and are qualitatively similar to those of a closely related water-breathing fish. By the time fish reach 100 g (about 45 days later) the lamella become less visible, and by the time fish reach 1 kg (approximately 4-5 months later) the lamellae have completely disappeared and the interlamellar space becomes completely filled with cells, predominantly rich in mitochondria. Thus, as the constraints for O_2 uptake across the gills are reduced due to increased dependence upon aerial respiration, the gills become completely remodeled, appearing to be designed again for ion regulation and/or acid-base balance. Clearly the morphology of the gill changes throughout development and at any given stage of development morphology represents a compromise between gas exchange, ion regulation and likely other processes. Unlike the case in carp gills, these changes are not likely to be reversible, although experiments to validate this have not yet been conducted.

CONCLUSION

The gills are a multipurpose organ, and responsible for gas exchange and the majority of osmoregulation in most fish. The characteristics that maximize gas exchange, such as a large surface area, low diffusion thickness, and high water and blood flow through the gills, are the same characteristics that impair the ability to maintain plasma ion levels at consistent levels regardless of environmental salinity. Species-specific differences exist at the gills, such as modification of tight junctions, which reduce the extent to which an elevation in metabolic rate impairs osmoregulation. This may be especially important in more active species

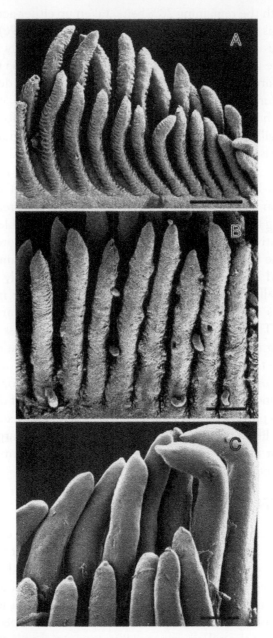

Fig. 8.3   Scanning electron micrographs (SEM) of the gills from three sizes of the obligate air-breather *Arapaima gigas*, showing the gradual loss of lamellae with age as fish become progressively more dependent upon air-breathing (**A**) 10 g, (**B**) 100g, and (**C**) 1 kg body mass. Scale bar = 500 μm. Adapted from Brauner *et al.* (2004).

that need a larger metabolic scope. Within a species, changes in environmental conditions, such as elevated temperature or salinity, and exercise, serve to unmask the nature and degree of the osmoregulatory compromise. In general, elevated metabolism associated with exercise and/or elevated temperature, result in a partial loss of osmoregulatory control. In other cases, exposure to conditions requiring an upregulation of osmoregulatory characteristics of the gills (e.g., exposure to dilute water) can result in morphological changes in the gills that impair gas exchange. Finally, there are several species of fish that exhibit large morphological changes in the gills as opposed to the mechanism of reducing gill surface area seen in several extreme euryhaline species. In some cases, morphological changes are reversible, such as in carp, where the transition from normoxia to hypoxia is associated with up to a seven fold increase in gill surface area to secure O_2 uptake at a cost to osmoregulation. In other species, such as *Arapaima gigas*, large reductions in total surface area of the gills are observed during development, as the fish become less dependent upon O_2 uptake across the gills due increased dependence upon aerial respiration. Under these conditions, the density of mitochondria rich cells increase dramatically, presumably indicating a transition in gill function.

In summary, there is a clear compromise between the osmoregulation and gas transfer in the gills of fishes. Like fishes in general, the mechanisms to minimize homeostatic disturbances in the face of this compromise are very diverse and depend on the natural history of the animal. In addition to the trade-offs between the two functions described throughout this chapter, compromises with other functions such as nitrogenous waste excretion and acid-base balance most likely exist as well, and would prove interesting for further study.

References

Barron, M.G., B.D. Tarr and W.L. Hayton. 1987. Temperature-dependence of cardiac output and regional blood flow in rainbow trout, *Salmo gairdneri* Richardson. *Journal of Fish Biology* 31: 735-744.

Bindon, S.D., K.M. Gilmour, J.C. Fenwick and S. F. Perry. 1994. The effects of branchial chloride cell proliferation on respiratory function in the rainbow trout (*Oncorhynchus mykiss*). *Journal of Experimental Biology* 197: 47-63.

Booth, J. H. 1978. The distribution of blood flow in the gills of fish: Application of a new technique to rainbow trout (*Salmo gairdneri*). *Journal of Experimental Biology* 73: 119-129.

Brauner, C.J. and A.L. Val. 1996. The interaction between O_2 and CO_2 exchange in the obligate air breather, *Arapaima gigas*, and the facultative air breather, *Lipossarcus pardalis*. In: *Physiology and Biochemistry of the Fishes of the Amazon*, A.L. Val, V.M.F. Almeida-Val and D.J. Randall (eds.). Instituto Nacional de Pesquisas da Amazônia (INPA), Manaus, Brazil, pp. 101-110.

Brauner, C.J., V. Matey, J.M. Wilson, N.J. Bernier and A.L. Val. 2004. Transition in organ function during the evolution of air-breathing; insights from *Arapaima gigas*, an obligate air-breathing teleost from the Amazon. *Journal of Experimental Biology* 207: 1433-1438.

Farmer, G.L. and F.W.H. Beamish. 1969. Oxygen consumption of *Tilapia nilotica* in relation to swimming speed and salinity. *Journal of the Fisheries Research Board of Canada* 26: 2807-2821.

Gallaugher, P.E., H. Thorarensen, A. Kiessling and A.P. Farrell. 2001. Effects of high intensity exercise training on cardiovascular function, oxygen uptake, internal oxygen transport, and osmotic balance in Chinook salmon (*Oncorhynchus tshawytscha*) during critical speed swimming. *Journal of Experimental Biology* 204: 2861-2872.

Gonzalez, R.J. and D.G. McDonald. 1992. The relationship between oxygen consumption and ion loss in a freshwater fish. *Journal of Experimental Biology* 163: 317-332.

Gonzalez, R. and D. Mcdonald. 1994. The relationship between oxygen uptake and ion loss in fish from divese habitats. *Journal of Experimental Biology* 190: 95-108.

Graham, J.B. 1997. *Air-Breathing Fishes: Evolution, Diversity and Adaptation*. Academic Press, San Diego.

Greco, A.M., K. Gilmour, J.C. Fenwick and S.F. Perry. 1995. The effects of soft-water acclimation on respiratory gas transfer in the rainbow trout (*Oncorhynchus mykiss*). *Journal of Experimental Biology* 198: 2557-2567.

Greco, A.M., J.C. Fenwick and S.F. Perry. 1996. The effects of soft water acclimation on gill structure in the rainbow trout (*Oncorhynchus mykiss*). *Cell and Tissue Research* 285: 75-82.

Haney, D. and F. Nordlie. 1997. Influence of environmental salinity on routine metabolic rate and critical tension of *Cyprinodon variegatus*. *Physiological Zoology* 70: 511-518.

Hochachka, P.W. and G.N. Somero. 2002. *Biochemical Adaptation: Mechanism and Process in Physiological Evolution*. Oxford University Press, New York.

Laurent, P. 1984. Gill internal morphology. In: *Fish Physiology*, W.S. Hoar and D.J. Randall (eds.). Academic Press, New York, Vol. 10A, pp. 73-183.

McDonald, D.G. and B.R. McMahon. 1977. Respiratory development in arctic char *Salvelinus alpinus* under conditions of normoxia and chronic hypoxia. *Canadian Journal of Zoology* 55: 1461-1467.

Nilsson, S. 1986. Control of gill blood flow. In: *Fish Physiology: Recent advances,* S. Nilsson and S. Holmgren (eds.). Croom Helm. London, pp. 87-101.

Olson, K.R. 1991. Vasculature of the fish gill: Anatomical correlates of physiological functions. *Journal of Electron Microscopy Technique* 19: 389-405.

Perry, S.F. 1998. Relationships between branchial chloride cells and gas transfer in freshwater fish. *Comparative Biochemistry and Physiology* A119: 9-16.

Postlethwaite, E.K. and D.G. McDonald. 1995. Mechanisms of Na^+ and Cl^- regulation in freshwater-adapted rainbow trout (*Oncorhynchus mykiss*) during exercise and stress. *Journal of Experimental Biology* 198: 295-304.

Randall, D. and C. Brauner. 1991. Effects of environmental factors on exercise in fish. *Journal of Experimental Biology* 160: 113-126.

Randall, D.J., G.F. Holeton and E.D. Stevens. 1967. The exchange of oxygen and carbon dioxide across the gills of rainbow trout. *Journal of Experimental Biology* 46: 339-348.

Randall, D.J. and E.D. Stevens. 1967. The role of adrenergic receptors in cardiovascular changes associated with exercise in salmon. *Comparative Biochemistry and Physiology* 21: 415-424.

Randall, D.J., D. Baumgarten and M. Malyusz. 1972. The relationship between gas and ion transfer across the gills of fishes. *Comparative Biochemistry and Physiology* A 41: 629-637.

Rombough, P.J. 1999. The gill of larvae. Is it primarily a respiratory or an ionoregulatory structure? *Journal of Fish Biology* 55: 186-204.

Rombough, P. and B. Moroz. 1997. The scaling and potential importance of cutaneous and branchial surfaces in respiratory gas exchange in larval and juvenile walleye. *Journal of Experimental Physiology* 200: 2459-2468.

Sardella, B., V. Matey, J. Cooper, R. Gonzalez and C.J. Brauner. 2004a. Physiological, biochemical, and morphological indicators of osmoregulatory stress in California Mozambique tilapia (*Oreochromis mossambicus* × *O. urolepis hornorum*) exposed to hypersaline water. *Journal of Experimental Biology* 207: 1399-1413.

Sardella, B.A., J. Cooper, R. Gonzalez and C.J. Brauner. 2004b. The effect of temperature on juvenile Mozambique tilapia hybrids (*Oreochromis mossambicus* × *O. urolepis hornorum*) exposed to full strength and hypersaline sea water. *Comparative Biochemistry and Physiology* A137: 621-629.

Sollid, J., P. De Angelis, K. Gundersen and G.E. Nilsson. 2003. Hypoxia induces adaptive and reversible gross morphological changes in crucian carp gills. *Journal of Experimental Biology* 206: 3667-3673.

Sollid, J., R.E. Weber and G.E. Nilsson. 2005. Temperature alters the respiratory surface area of crucian carp *Carassius carassius* and goldfish *Carassius auratus*. *Journal of Experimental Biology* 208: 1109-1116.

Stevens, E.D. and G.F. Holeton. 1978. The partitioning of oxygen uptake from air and from water by the large obligate air-breathing teleost pirarucu (*Arapaima gigas*). *Canadian Journal of Zoology* 56: 974-976.

Swanson, C. 1998. Interactive effects of salinity on metabolic rate, activity, growth osmoregulation in the euryhaline milkfish (*Chanos chanos*). *Journal of Experimental Biology* 201: 3355-3366.

Thomas, S., B. Fievet, G. Claireaux and R. Motais. 1988. Adaptive respiratory responses of trout to acute-hypoxia: effects of water ionic composition on blood acid-base status response and gill morphology. *Respiration Physiology* 74: 77-89.

Wood, C.M. and D.J. Randall. 1973. The influence of swimming activity on sodium balance in the rainbow trout (*Oncorhynchus mykiss*). *Journal of Comparative Physiology* 82: 207-233.

Dissolved Oxygen and Gill Morphometry

Marco Saroglia*, Genciana Terova and Mariangela Prati

INTRODUCTION

Fish gills are responsible for several physiological activities requiring anatomic and physiological compromises. Therefore, the different cell types forming gill epithelia, pavement, chloride and mucous cell distribution, plus morphology and morphometry, have been intensively investigated to understand the integration of several of their functions, such as gas exchange, ion and acid-base regulation, nitrogen excretion (Fernandes and Perna-Martins, 2001).

Randall *et al.* (1972) described the basic conflict between gas exchange and ion regulation in the gills of freshwater fishes. They indicated that a large permeable gill membrane is required for efficient gas transfer, but a small, impermeable epithelium is needed to minimize diffusive ion losses. They also demonstrated that rainbow trout accelerates Na^+ losses with an increase in oxygen consumption during exercise, which

Authors' address: Department of Biotechnology and Molecular Sciences, University of Insubria, via J.H. Dunant 3, 21100 Varese, Italy.

**Corresponding author:* E-mail: marco.saroglia@uninsubria.it

they attributed mainly to an increased functional surface area (FSA) of the gills during activity. As suggested by Nilsson (1986), the balance between 'need of oxygen' and 'need of osmotic regulation' was defined as *osmorespiratory compromise*.

Concerning the rate of gas transfer through fish gills, it is described by Fick's diffusion equation:

$$dV/dt = D \cdot S \cdot \Delta c/\Delta \chi$$

where the flow rate through the gill membrane (dV/dt) is directly proportional to the gas diffusion coefficient (D), to the total membrane surface (S), to the concentrations difference through the membrane (Δc) and inversely proportional to the membrane thickness ($\Delta \chi$).

When S is intended as the total respiratory surface area (RSA) and $\Delta \chi$ as the gas diffusion distance (GDD), the equation becomes:

$$dV/dt = D \cdot RSA \cdot \Delta c/GDD$$

The RSA and GDD values are not only a characteristic and vary with fish species, but they are also highly correlated with life-style and habitat of fish (Perry and McDonald, 1993). Also, GDD is reported to vary within the gill apparatus of a single animal (Hughes and Morgan, 1973) and with its onthogenetic development. Prasad (1986) has shown a positive correlation between body mass and GDD in *Esomus danricus*. Similarly, the surface area of the lamella and the number of lamellae vary within the same species, according to their size (Hughes, 1984b). Environmental factors such as water hardness can also affect GDD (Greco *et al.*, 1996). Water temperature and oxygen partial pressure (PO_2) are also reported to influence GDD (Randall and Daxboeck, 1984; Kisia and Hughes, 1992). In particular, GDD in gill of sea bass (*Dicentrarchus labrax*), has been found to reduce with an increase of water temperature, due to a natural reduction of PO_2 (Saroglia *et al.*, 2000), while RSA resulted inversely proportional to PO_2 (Saroglia *et al.*, 2002).

Moreover, changes in the anatomy of the gill respiratory surface area during body development appear to be related to routine metabolism in tilapia *Oreochromis niloticus* L. (Kisia and Hughes, 1993). Only a part of the total respiratory surface is perfused by blood during 'quiet' ventilation (Booth, 1978), compared with 'strong' ventilation (Nilsson, 1986), when the oxygen uptake rate is increased, and that also account for a modification of the FSA. Duthie and Hughes (1987), observed that in rainbow trout *Oncorhynchus mykiss* whose functional gill area had been reduced by cauterization, when forced to swim, showed a significant proportional reduction in maximum oxygen consumption. Oxygen consumption at rest and at subcritical swimming speed were not affected. The authors concluded that the total gill area is utilized for oxygen uptake only under conditions of maximum aerobic demand and a direct limit is imposed on oxygen uptake at the gills independent of environmental oxygen partial pressure (PO_2), when elevated above normoxia. Powell *et al.* (2000) have reported that Atlantic salmon (*Salmo salar*), previously infected with amoebic gill disease, developed physiological adaptations such as increases in gill perfusion or blood flow redistribution, when exposed to hypoxia. All these adaptation mechanisms in the gill permeability to gases may compensate for PO_2 fluctuations and for any impediment in gas exchange.

Several Reasons for Measuring *RSA* and *GDD*

Gills being not only responsible for gas exchange, several other physiological functions that are gill exchange related, may be expressed by comparing their rates to the *RSA*. In fact, gill being also the site of extrarenal ion excretion (Hartl *et al.*, 2000), their epithelia consist of several cell types. Pavement, chloride and mucous cell distribution and morphology have been intensively investigated to understand the integration of several of their functions, such as gas exchange, ion and acid-base regulation, and nitrogen excretion. These are surface area-dependent processes, and therefore accurate gill surface area estimates are essential for studying, on the top of the physiology of gas exchange, also ionic fluxes across the gills (Motais *et al.*, 1966; Hughes and Morgan, 1973). Such measurements are of particular significance in the physiology of estuarine fish that are frequently used as model species in studies of hydromineral regulation in fluctuating conditions (Potts and Eddy, 1973; Potts *et al.*, 1973; Hutchinson and Hawkins, 1990; Carrol *et al.*, 1995) and more generally, as sentinel organisms in water quality evaluations (Larson *et al.*, 1981).

In fact, measurements of the surface areas of the respiratory organs for different animals has become of increasing importance in relation to comparative quantitative studies of gaseous and ionic exchange (Hughes, 1984a). Gill area measurements have also been useful in relation to study of fish growth (Pauly, 1981) with potential applications in aquaculture. The studies have been of interest in relation to normal respiratory function but also with particular reference to the effects of change in environmental PO_2 level (Hughes *et al.*, 1978; Soivio and Tuurala, 1981; Saroglia *et al.*, 2000, 2002).

Other Factors Affecting Gills Morphometry

Gills morphometric studies are frequently reported in fish—together with morphological, histhological and electronic microscope observations—to monitor and evaluate quantitatively the effects due to hormones, water pollutants, water oxygenation regimes, pathological conditions and particular environmental occurrences. Beside the examples already reported and concerning, among others, the PO_2 regimes, some more cases will be offered below.

DNA containing the transgene salmon growth hormone attached either to appropriated protein promoters, has been used successfully to realize large increases in salmonid growth rate. In order to support their rapid growth, GH transgenic salmon consume food at a more rapid rate than control salmon. In addition, their oxygen uptake is about 60% more than that of controls during routine activity and during sustained swimming (Stevens *et al.*, 1998). So, in a study by Stevens and Sutterlin (1999), the hypothesis that some modifications to support the elevated oxygen uptake of the transgenic salmon might be reflected in aspects of their gill morphometry, particularly with respect to alterations that might increase respiratory exchange area, has been challenged.

Flounder (*Platichthys flesus*) exposed to TBTO (32 mg for 6 days) showed the localized character and three-dimensional extension of these lesions. The roughened surface of filaments and lamellae in the exposed animals show at the optic and scanning microscope, fusion of lamellae and respiratory epithelial proliferation (Grinwis *et al.*, 1998). The authors measured the functional surface area of gills and compared it with the controls, concluding for a markedly reduced functional surface area caused by the exposure to organo-tin compounds.

After exposing mosquitofish (*Gambusia holbrooki*) to mercury (300 nM treatment), Jagoe *et al.* (1996), reported that mean thickness of filaments epithelia tended to increase, becoming significantly greater than in control fish, even if exposure to lower concentrations (75 nM and 150 nM) did not generate significant differences from controls. On the same samples, there was no apparent trend in thickness of lamellae that did not differ significantly from the control values.

METHODS FOR GILL MORPHOMETRY

Basic methodologies to study gill morphometry were reported in some papers (Gray, 1954; Hughes, 1966, 1984b; Hughes and Perry, 1976). The same methods have been modified successively according with experience and with the improvement of available laboratory equipments.

On Hughes and Perry (1976), methods are described for the morphometric estimation of parameters of the gill system of trout which are relevant to is function in gas exchange. The methods have been used with 1 mm sections viewed under the light microscope. In particular the diffusion distance between water and blood were measured which, together with determination of gill area, provided figures for the morphometrically estimated diffusion capacity. The methods have been used to compare the diffusing capacity of gills from control fish and those treated in polluted waters. The concept of relative diffusing capacity (D_{rel}) was introduced, enabling comparisons to be made without the need to determine the absolute diffusing capacity. That is particularly convenient when gill surfaces from fish having had different environmental histories should be compared. Quantitative estimation of changes in relative volumes and surface areas of components of the secondary lamellae were determined by the authors, and employed to explain the possible anatomical causes of changes in D_{rel}, concluding that the same methods may be applied in the comparison of the gills of fish treated in different waters.

There are three essential measurements that have been reminded in Hughes (1984b), that are:

1. the total length of gill filaments (L, mm);
2. the number of lamellae (lamellae/mm) on both side of the filament (n); and
3. an estimate of the bilateral area (bl) of a lamellae which can be taken as representative of all lamellae of the particular gill system.

The product of the first two measurements gives the total number of lamellae. The total lamellar area is the product of all the measurements, i.e.:

$$\text{Total gill area} = L \cdot n \cdot bl$$

A projection microscope or *camera lucida* is necessary for the tracing of individual lamellae. The total filament length must be measured first of all, measuring each gill arch separately. Depending on the number of filaments on each arch, every 5th, 10th, or perhaps 20th filaments are measured and marked. In measuring filament length, carefulness must be paid as they may result bent after the fixation. The total length (L) is then obtained by extrapolating the addition of all the filaments lengths. Determination of the lamellae/mm (n) are best made by measuring the distance for ten lamellae (Fig. 9.1). The reciprocal of the average distance for each lamellae gives the frequency/mm ($1/d'$). Measurements are suggested on three parts (tip, middle and base) of the selected filament that is suggested to be from the 2nd arch. As Figure 9.1 refers to the lamellae of one side of the filament, a factor 2 must be introduced ($n = 2 \times 1/d'$). The product of n and total filaments length (L) gives the total number of lamellae (N).

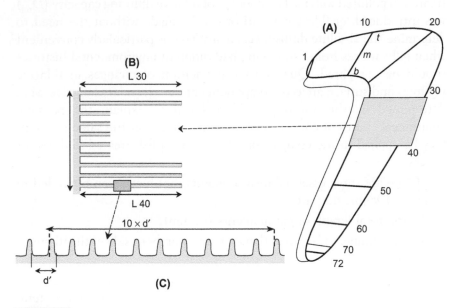

Fig. 9.1 **(A)** Diagram of a single hemibranch (72 filaments) of a teleost fish showing the position of filaments selected for measurement at regular intervals around the arch. Lenghts of the first and last filaments are also determined. The position of the lamellae selected for determination from the tip (t), middle (m) and base (b) of each selected filaments are indicated. **(B)** Diagram showing method of measurement of length of filaments 30 (l_{30}) and 40 (l_{40}) and the distance between filaments 30 and 40. **(C)** Diagram illustrating the method for measurement of secondary lamellar frequency ($1/d'$) by measuring the distance for ten secondary lamellae. Redrawn from G.M. Hughes, (1984b).

To obtain the average area of a lamella, some selection is requested. Hughes (1984b), together with other authors, suggests to choose only the lamellae of the second gill arch, selecting the filaments that were previously marked (5th, 10th, 20th). A lamella from the base, middle and tip of each selected filaments is taken. Measurement of its area may be done in different ways, as computerized surface analyzers may be utilized instead a projection microscope or a *camera lucida* as previous authors did.

Laser-scan microscopy has been adopted by Saroglia *et al.* (2002) on sea bass (*Dicentrarchus labrax*) exposed to different oxygen regimes, being hypoxia, normoxia and hyperoxia. While L, n and bl were measured, according to Hughes (1984b), utilizing light microscopy, bl was measured on laser-scan computerized microscopy. To determine L two successive steps were performed. Firstly, the total number of filaments of the superior hemibranch of all the four left gills was counted. Secondly, one filament in every 20 was sampled, and its length measured. The bilateral area bl of a representative lamella was calculated only in the second left gill arch and the value obtained was considered representative of the gill apparatus. Transverse sections of the sampled filament were calculated from the base, centre and apex. The transverse sections, showing two monolateral surfaces of two lamellae, were observed and the respiratory surfaces were measured using a computerized confocal laser-scan microscope (Zeiss LSM3) with image analyser software. Surface areas of 120 lamellae were measured from each of the fish sampled (Fig. 9.2A). By utilizing the same laser-scan microscope, Saroglia *et al.* (2000) measured gas diffusion distance. Twenty filaments of the first left gill of each fish were randomly sampled and isolated, cutting at their base. Then three lamellae, respectively from the base, center and apex of each filament, were measured. The thickness was measured at ten different points of each secondary lamella, not taking into consideration areas occupied by cell nucleus and overlapping cells. Thus, 600 measurements of thickness were performed altogether for each of the ten fish sampled under each experimental condition. Figure 9.2B shows the image of a secondary lamella obtained by the confocal laser-scan microscopy, in which the blood-to-water diffusion barrier is indicated between two crosses. Among the advantage of this method, related to other light microscopy or electron microscopy methods, resulted to be the high number of measures that can be done in a reasonable time span.

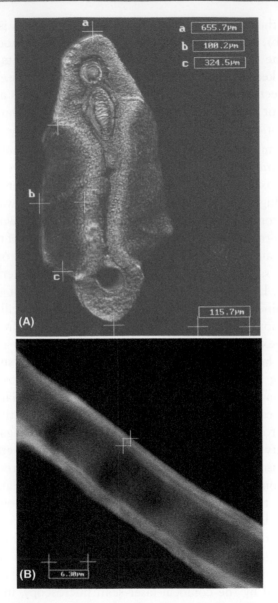

Fig. 9.2 (A) Transverse section of a sea filament, viewed by confocal laser-scan microscopy. The monolateral surfaces of two contiguous lamellae are shown. (B) Measure of gas diffusion distance in a fresh lamella. Adapted from M. Saroglia *et al.* (2000, 2002).

Hartl *et al.* (2000) performed the gills surface measurement on flounder (*Pleuronected flesus*) by applying a surface analyser software. All eight gill arches from each specimen were removed and preserved in cold

Bouin's fixative, following the protocol described by Hughes (1984b), who pointed to the fact that there is unavoidable shrinkage during fixation with Bouin's solution. In these studies, this shrinkage was quantified by extracting five individual filaments of different initial sizes that were placed in water on a cavity slide and filament length, interlamellar space and unilateral lamellar area were measured using the methodology described below. The water was replaced by Bouin's solution, and the slide covered and refrigerated. Further measurements were made after 24 and 72 h; 1, 2 and 6 weeks. All measurements presented have been corrected for shrinkage. The transfer of gill tissue from Bouin's to tap water showed no measurable distortion, compared with tissue transferred to an isotonic saline solution. The biometry of fresh and preserved material was determined by adapting the weighted method described by Hughes and Morgan (1973), making use of digital image analysis software (Sigma Scan Pro 4.0) so as to obtain the primary measurements under manual control. A single areal standard was required to calibrate the system, this was provided by the grid of a Neubauer haemocytometer; linear measurements were checked using a stage micrometer. Images were captured with an Olympus BH-2 light microscope linked by a Panasonic F10 CCD video camera to a desktop PC using Matrox Rainbow Runner (Fig. 9.3).

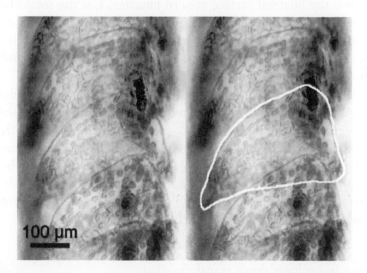

Fig. 9.3 Two micrographs of the same section of a filament (fish weight: 0·98 g). The unilateral lamellar area was measured on screen from grabbed digital images by hand tracing the shapes of intact lamellae (white line; right). The enclosed area was then integrated by a PC running "Sigma Scan Pro 4.0". Adapted from M.G. Hartl *et al.* (2000).

The length of each filament was measured from base to tip, taking the curvature of some filaments caused by fixation into account. The filaments of each hemibranch were grouped into several sections according to length. A medium-sized filament of each section was selected and every interlamellar space and the areas of one side of every lamella on both sides of individual filaments were measured, averaged and used to calculate the surface area of the particular group. This was achieved by separating the filament from the gill arch and placing it with one side facing upwards into a cavity slide filled completely with water and sealed with a cover slip. After measurements were made on the one side, the filament was turned over onto the other side.

Large filaments could be turned over with a dissecting needle but with very short filaments it was often enough to move the cover slip along the plane of the slide, the resulting movement of the water in the cavity then turned the filament onto the opposite side. Measurements had to be carried out on all eight gill arches because of the asymmetry of the upper and lower gill pouches. The total surface area (A) of a given hemibranch was determined by:

$$N = (L/d')2 \text{ and } A = 2bLn$$

where $2b$ is the bilateral surface area of lamellae, L is the total filament length, d' is the interlamellar spacing and n is the frequency of lamellae mm^{-1} filament. The surface area data of all the hemibranchs were summed to give the total gill surface area for each fish. The formulae of previous studies were applied to the present data set.

Gray (1954) averaged every 10th filament to calculate total filament length. Al-Kadhomiy (1985) calculated total filament length by dividing the filaments of a hemibranch into groups according to length and averaging the first and the last filament of each group. To demonstrate the effect of gill asymmetry on total gill area, the results from doubling the area of the upper gill pouch were compared with the area calculation derived from the sum of all eight gill arches. The gill area data (Y) was analysed in relation to body weight (W) with the allometric equation:

$$Y = aW^c$$

and using linear logarithmic transformation:

$$Y = \log a + c \log W$$

A number of authors adopted scan electron microscopy (SEM) to measure gills surface, with different levels of complexity. Fernandes and Perna-Martins (2001) measured lamellae surface in armored catfish, *Hypostomus plecostomus* from the Monjolinho Reservoir, São Carlos, SP, Brasil. The fish were anaesthetized with 0.01% benzocaine, killed and their gill arches excised and processed for light, scanning and transmission electron microscopy. For transmission electron microscopy (TEM), fixed pieces of individual filaments ($\sim$ 1 mm long) were post-fixed in 1% osmium tetroxide in 0.1 M phosphate buffer pH 7.3 at 4°C, dehydrated by a graded acetone series and embedded in Araldite 6005 (Ladd Research). Gill filaments from dorsal, middle and ventral portions of the each gill arch were cut off with a razor blade. Most of the arch tissue was removed but the anterior and posterior rows of filaments remained attached to the septum of the arch. Samples consisting of 1-5 gill filaments were fixed in buffered glutaraldehyde. Fixed tissue samples were dehydrated for SEM. Filament pairs were glued with silver paint onto the specimen stub, coated with gold in a vacuum sputter and examined under a DSM 940 ZEISS Scanning Microscope at 25 kV. Epithelial surfaces on the leading and trailing edges of the filaments near the base of the lamellae and from the lamella were randomly photographed with 3000-fold magnification (4 noncontiguous fields). The apical surface of individual chloride and mucous cells and their density on the filament and lamellar epithelia were determined by tracing cell perimeters using a morphometric software program (Sigma Scan, Jandel Scientific, Inc.). From these measurements, the mean chloride and mucous cell fractional area mm^{-1} epithelium were calculated. The same authors, measured the gas diffusion distance in armored catfish. Semi-thin sections were stained with toluidine blue and examined under an Olympus-Micronal photomicroscope. Ultra-thin sections were stained with uranyl acetate and lead citrate and examined with a JEOL 100 CX transmission electron microscope at 60 or 80 kV. The lamellae's water-blood barrier thickness (τ_h) were calculated from the harmonic mean intercept length (l_h) of random probing lines crossing the barrier, according to the equation $\tau_h = 2/3 \; l_h$. The l values were determined by superimposing a grid for layered structure on randomized electron micrographs (magnified $\times$ 3000) of lamellar cross sections.

Laurent and Hebibi (1989) studied the morphological parameters during adaptation of rainbow trout (*Oncorhynchus mykiss*) to different ionic environments. They found that the surface area of individual lamellae increased in trout acclimated to ion-poor water or seawater.

Conversely, the harmonic mean thickness of the lamellar epithelium decreased in seawater and to an ever greater extent in ion poor water.

In the study they also measured the surface covered by chloride cells. The authors utilized in their work, scanning electron microscopy (SEM), light microscopy (LM) and transmission electron microscopy (TEM). Dorsal, middle and ventral piece of arches I and II were cut from both sides of stunned trout. Each pieces contained a few pairs of filaments (Fig. 9.4a). Most of the arch tissue was removed with a razor blade, but the anterior and posterior rows of the filament still remained attached to septum of the arch. These pieces were immersed in the fixative, 5% glutaraldehyde buffered with 0.15 M sodium cocodylate, and subsequently rinsed in cocodylate buffer (0.15 M) (pH 7.4; osmotic pressure 292 mosmol) for 1 h at 4°C. Pieces of individual anterior and posterior filaments were removed at the place where filaments separate from the sectum (Fig. 9.4b).

These pieces (1 mm long), containing about 20 lamellae, were embedded separately in araldite in a way to allow the lamellae to be cross-sectioned (Fig. 9.3c). Thus each animal yielded 24 samples of filaments.

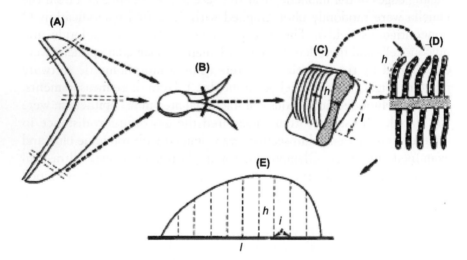

Fig. 9.4 Sampling and calculation of the lamellar surface area. (A) Gill arch; dorsal, middle and ventral region sampling. (B) Cross-sectioned piece of gill arch and region of filament from which piece are cut. (C) Filament sample bearing lamellae; *h*, height of lamella; *l*, length of lamella. (D) Cross section of the block C; the value of *h* are determined (see text). (E) Lamellae are reconstructed by placing the successive values of *h* along the length, *l*; *i*, interval between successive measured values of *h*. Adapted from P.L. Laurent and N. Hebibi (1989).

Ultra-thin sections were made in several regions of the lamella. They allowed subsequent measurement of the volume density and the harmonic mean thickness of the water-blood barrier.

Successive 1 μm thick section were photographed at × 1000 magnification, and the length of the lamellar profiles (h) was measured on a digitizer tablet, by taking into account the curvature, if any, of each lamella along the series of sections. The area of the lamella was computed according to the following algorithm:

$$S = i[h_0 + \tfrac{1}{2}(h_1 - h_0)] + [h_1 + \tfrac{1}{2}(h_2 - h_1)] + \cdots + [h_n + \tfrac{1}{2}(h_n - 1 - h_n)]$$

where i is the length of the interval between the successive profile, h, from h_0 to h_n, the length of n successive profiles. Thus, $n \cdot i$ is the length of the lamella.

The harmonic mean thickness of the lamella τ_h was calculated from the harmonic mean intercept length l_h of random probing lines traversing the barrier, according to the equation:

$$\tau_h = 2/3 \, l_h$$

Sollid *et al.* (2003) challenged crucian carp (*Carassius carassius*) with hypoxia and normoxia. The crucian carp, a North European freshwater fish often inhabits small ponds that—due to ice coverage—become hypoxic and finally anoxic for several months in winter. Its exceptional hypoxia and anoxia tolerance make it the sole fish species in this habitat. The resulting lamellae embedded under normoxia condition help the fish to face with osmoregulation, while when PO_2 strongly reduces, lamellae result protruded. The authors, among other tissutal and cellular analyses and observations, measured the area of the portion of the lamellae in contact with water, then the intralamellar cell mass (ILCM).

To do this, three measurements were taken on randomly selected gill filaments from four normoxic and four hypoxic fish (Fig. 9.5); lowercase letters denote measurements in normoxia, and uppercase letters denote measurements in hypoxia):

1. the basal length of the part of the lamellae in contact with water was measured using SEM (l and L, respectively);
2. the mean height of the lamellae in contact with water was measured using light microscopy on the sections obtained in the BrdU (5′-bromodeoxyuridine) and TUNEL experiments; the latter was done on cross-sections where one-third of the measurements were done

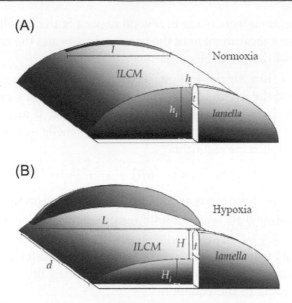

Fig. 9.5 Graphic illustration of the sites where lamellar diameter (*d*), height of protruding lamellae (*h* and *H*) and basal length of protruding lamellae (*l* and *L*) were measured in (**A**) normoxic and (**B**) hypoxic fish. Lowercase letters denote measurements in normoxia, and uppercase letters denote measurements in hypoxia. ILCM, interlamellar cell mass. Adapted from J. Sollid *et al.* (2003).

when the central venous sinus was visible and two-thirds were done when it was not; thus, one-third were from the central portion of the lamellae (h_c and H_c, respectively) and two-thirds were from the edges (h_e and H_e, respectively); and

3. the mean lamellae thickness (*t*) was also measured using light microscopy on BrdU and TUNEL sections.

The area of the normoxic lamella (*a*) was approximated by calculating the lamellar tip area, $a = pl$, where *l* is the protruding basal length and *p* is the ellipse perimeter formula divided by two:

$$p = \frac{2\pi\sqrt{\frac{1}{2}(r^2 + h^2)}}{2}$$

where $r = t/2$ and $h = (2h_e + h_c)/3$. The total lamellar area in hypoxia (A) was approximated by multiplying the mean protruding height minus the mean height of the lamella tip (obtained in the measurements of normoxic lamellae) by the mean basal length and finally adding the area of the lamellar tip:

$$A = pL + 2L(H - h)$$

where $H = (2H_e + H_c)/3$. The approximate volume (V) of the ILCM was obtained by measuring the distance between two adjacent lamellae (d) and multiplying by the measured height of the ILCM (h_i or H_i) and the lamellar basal length in hypoxia (L):

$$V = dh_iL$$

where $h_i = (2h_i, e + h_i, c)/3$. Note that most histological procedures are likely to distort the tissue, and comparisons between studies should be made with caution.

The exposure to hypoxia triggered a striking change in gill morphology (Fig. 9.6), where it is possible to see that the gills displayed protruding lamellae. The change was visible after 1 day of hypoxia and reached its greatest extent after 7 days. No further alteration was apparent after a further 7 days of hypoxia. The change was completely reversible, since, after an additional week in normoxia, the gills were indistinguishable from their initial state, showing no protruding lamellae. The changes occurred on all four gill arches and over the whole length of the gill filament.

Light microscopic examination of serially cross-sectioned gill arches showed that the lamellae were also present in normoxia but that the space between the lamellae was then completely filled by an interlamellar cell mass (ILCM). After 7 days in hypoxia, the mean area of the portion of the lamellae that was in contact with water increased by ~7.5-fold, from 1195 μm^2 per lamella in normoxia to 8898 μm^2 per lamella in hypoxia. This resulted from the reduction of the ILCM, which was reduced by ~52% from 2.01×10^5 μm^3 per interlamellar space in normoxia to 9.79 $\times 10^4$ μm^3 in hypoxia.

The reduction of the ILCM and the resultant exposure of a much larger lamellar area to water will greatly increase the part of oxygen transfer aided by convective water movements. However, some oxygen can be expected to diffuse through the ILCM to reach the blood inside the lamellae. It is not known how important this route is but it is likely to contribute to some of the oxygen uptake, especially in carp with embedded lamellae. The authors suggest that howewer, in those fish, it is possible that most of the O_2 is taken up by erythrocytes passing through the outer

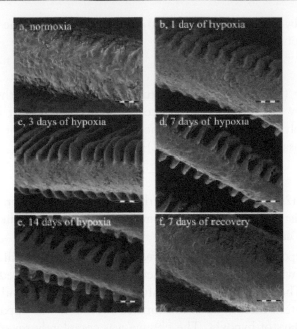

Fig. 9.6 Scanning electron micrographs from the 2nd gill arch of crucian carp kept in normoxic or hypoxic water. (a) In normoxia, the gill filaments have no protruding lamellae. (b) The morphology has already changed after 1 day of hypoxia exposure (0.75±0.15 mgO_2 l^{-1}). (c,d) The change progresses for up to 7 days in hypoxia, but (e) there were no further changes with subsequent exposure. (f) When the fish were moved to normoxic water, the morphological changes were reversed within 7 days.

Moreover, the authors suggest that the increased lamellar area must significantly increase the capacity for oxygen uptake, and it is also likely to increase water and ion fluxes between water and blood. Indeed, such effects were clearly indicated by the subsequent experiments described below. Scale bar, 50 μm. Adapted from J. Sollid *et al.* (2003).

marginal channel, which runs immediately inside the edge of the lamellae of fish gills. In the carp with embedded lamellae, about half of the lamellae had their outer edges not covered by ILCM cells.

Costa *et al.* (2001), considering that for fish gill morphometry no method exists that employs unbiased sampling and measuring techniques proposed an original method, which allows both the calculation of reference volume by the Cavalieri-method and stereological estimation of volume densities, surface densities and barrier thickness on the same sections. The gills of *Arapaima gigas* were fixed *in situ* in flowing glutaraldehyde (2.5% in phosphate buffer) and later removed to the same fixative. Trimmed arches from one side of the fish were cut into three pieces of approximately equal length, randomly sorted and orientated for embedding in glycol methacrylate. Three samples were sequentially embedded on top of one another in each of the four blocks. The length of the tissue perpendicular to the plane of sectioning was measured and the block was exhaustively sectioned. Ten 3-μm sections were adequate for Cavalieri volume estimation and for all subsequent stereological analysis. Results for one side of the fish were doubled to yield values for the whole fish. Systematic random sampling of the measurements of the water-blood barrier thickness in 0.5 μm sections were carried out in order to estimate the accuracy of the measurements previously performed in methacrylate embedded tissue.

The method has later been applied by Fernandes *et al.* (2004), to measure gill morphometry of the facultative air-breathing dipnoi fish, (*Lepidosiren paradoxa*). A perfusion method has been adopted to measure the gas diffusion capacity of gills, lungs and skin. On adult lungfish the heart was exposed and the fish was perfused through the ventral aorta with 0.1 M phosphate buffer (pH 7.4, 300 mOsM) containing Heparin. When the drained venous return was nearly colorless, the perfusion solution was changed to 2.5% glutaraldehyde in phosphate buffer as above (15-20 min). The animal was then decapitated and the entire gill apparatus was removed to the same fixative. The lungs were filled with fixative through the glottis and tied off, and the fish was left in fixative overnight at 4°C for complete fixation of lungs and skin. The right lung was used for light microscopy (LM) and the left was sampled for transmission electron microscopy (TEM). For skin analysis, the fixed fish was transected by 12 equidistant cuts, the location of the starting section being determined at random within the first increment. The sections were labeled and placed in glutaraldehyde storage solution, consisting of 0.5% glutaraldehyde in phosphate buffer. Stereological measurements were carried out according to standard techniques (Howard and Reed, 1998). The reference volume (V) of gills, lungs and skin were determined using the Cavalieri principle and the respiratory surface areas (S_R) were calculated using the vertical section method (Howard and Reed, 1998; Costa *et al.*, 2001). The barrier thickness (harmonic mean = τ_h) of gills and skin (water-blood distance = τ_h) were determined using LM and the diffusion barrier thickness of lungs (air-blood distance = τ_h) were determined using TEM. These determinations were employed in calculation of the anatomical diffusion factor (ADF = surface area/ harmonic mean barrier thickness) of each respiratory structure and the diffusing capacities for O_2 and CO_2, as the product of ADF and the appropriate Krogh's diffusion constant. Results are reported on Fig. 9.7 and Table 9.1.

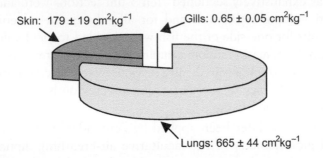

Skin: 179 ± 19 cm²kg⁻¹   Gills: 0.65 ± 0.05 cm²kg⁻¹

Lungs: 665 ± 44 cm²kg⁻¹

Fig. 9.7 The percentage of the total surface present in gills, skin and lungs, found in the air-breathing fish *L. paradoxa*. Adapted from M.N. Fernandes *et al.* (2004).

Table 9.1 Absolute surface area (S), harmonic average (τ_h) of the barrier diffusion distance between air-blood and or/water/blood, anatomic diffusion factor (ADF) and morphometric O_2 diffusion capacity of gills, skin and lungs of a 0.87 kg South American lungfish, *L. paradoxa*. Source: M.N. Fernandes *et al.* (2004).

Structure	S (cm²)	τ_h (μm)	ADF($cm^2\mu m^{-1}kg^{-1}$)	$cm^3min^{-1}mmHg^{-1}kg^{-1}$
Gills	0.6	110	$5.52 \cdot 10^{-3}$	$1.0 \cdot 10^{-6}$
Skin	404	138	3.46	$6.9 \cdot 10^{-4}$
Lungs	614	1.3	466.9	$1.1 \cdot 10^{-1}$

EFFECT OF DISSOLVED OXYGEN ON GILL MORPHOMETRY

Several methodologies have been reported in literature, to measure GDD as well as RSA in fish gills, and all of them enable to measure the effects of environmental factors, including PO_2. Nevertheless, advantages and disadvantages may characterize each of the different reported methodologies. The direct sample observation by laser-scan confocal microscopy enabled a rapid collection of data and easier sample preparation, in comparision to previous established methods (Hughes and Perry, 1976; Greco *et al.*, 1996) of lamellar surface measurement.

The reported data showed that water PO_2 affects RSA and GDD in all the species where either the effects of hypoxia or hyperoxia were studied. The morphometric analysis indicated that there were differences in the bilateral surface area of the lamella and filament length but not in lamella frequency (Saroglia *et al.*, 2002), although due to the sample size the means were not significantly different.

The RSA of the sea bass ranged from 212.90-351.99 $mm^2\ g^{-1}$ body mass, depending on PO_2. This is in the range of the majority of teleost species (100-400 $mm^2\ g^{-1}$ body mass) as reported by Perry and McDonald (1993). Applying a digital image analysis earlier described by McDonald *et al.* (1991), Hartl *et al.* (2000), analysed the gill surface area of European flounder *Pleuronected flesus* and concluded that data on the respiratory surface—as earlier reported in the literature—may be overestimated. Nevertheless, the methodology adopted here, an application of Hughes (1984b), would be less likely to have introduced errors because the plane of the lamellar surface not being perpendicular to the optical axis of the microscope. Some caution should be shown by all studies which depend on present-day computerized microscopic methods until the results are compared to traditional methods involving the dissection of individual lamella.

Also with the GDD, as measured in sea bass by Saroglia *et al.* (2000), clearly an adaptation occurs to PO_2 in water. In order to evaluate the masking effect of water temperature, the measures were carried out in early summer and in autumn, with fish bred under farm conditions, at different dissolved oxygen concentrations (DO): normoxia condition (80–100% of the saturation value) and "mild" hyperoxia condition (120–130% of the saturation value). There was a significant influence ($p <$ 0.001) of both dissolved oxygen concentration and environmental temperature on GDD. In summer it was 1.75 μm and 2.31 μm for fish reared under normoxia and hyperoxia, respectively, and in autumn 2.51 μm and 2.96 μm for fish reared under normoxia and hyperoxia, respectively. When DO was reduced at the higher temperatures, so did the GDD. Results lead to the conclusion that GDD increased with the increasing of DO, both due to reduced water temperature and to the mild oxygen hypersaturation following application of pure oxygen. Again, the advantage for fish may be found in the compromise between maximizing O_2 diffusion at the gills and ions/water intake/loss. However, when challenged with hypoxia, physiological mechanisms such as increases in gill perfusion or blood flow redistribution within the gill are likely to compensate for any impediment in gas exchange, so allowing the fish to compensate.

A strong compensation mechanism in crucian carp (*Carassius carassius*) has been reported by Sollid *et al.* (2003). They observed that crucian carp living in normoxic (aerated) water have gills that lack protruding lamellae, the primary site of O_2 uptake in fish. Such an unusual trait leads to a very small respiratory surface area. Histological examination showed that the lamellae of these fish were embedded in a cell mass (denoted embedded lamellae). When the fish were kept in hypoxic water, a large reduction in this cell mass occurred, making the lamellae protrude and increasing the respiratory surface area by ~7.5-fold. This morphological change was found to be reversible and was caused by increased apoptosis combined with reduced cell proliferation. Carp with protruding lamellae had a higher capacity for oxygen uptake at low oxygen levels rather than fish with embedded lamellae, but water and ion fluxes appeared to be increased, which indicates increased osmoregulatory costs. This is an important adaptive and reversible gross morphological change in the respiratory organ of an adult vertebrate in response to changes in the availability of oxygen.

After exposing Atlantic salmon *Salmo salar* with amoebic gill disease (AGD) to a graded hypoxia (135–40 mmHg water PO_2), Powell *et al.* (2000) analysed blood samples for respiratory gases and pH. Infected fish simulated a reduced gas exchange efficiency and there were no differences in the rate of oxygen uptake between infected and control fish. However, arterial PO_2, and pH were significantly lower in the infected fish, whereas PCO_2 was significantly higher in infected fish compared with controls prior to hypoxia and at 119 mmHg water PO_2. At 79.5 and 40 mmHg water PO_2 saturation, there were no significant differences in blood PO_2 or pH although blood PCO_2 was elevated in AGD affected fish at 50% hypoxia (79.5 mmHg water PO_2). The elevated levels of PCO_2 in fish affected by AGD, resulted in a persistent respiratory acidosis even during hypoxic challenge. These data suggest that even though the fish were severely affected by AGD, the presence of AGD while impairing gas transfer under normoxic conditions, did not contribute to respiratory failure during hypoxia. According to the authors, when challenged with hypoxia, physiological mechanisms such as increases in gill perfusion or blood flow redistribution within the gill are likely to compensate for any impediment in gas exchange, so allowing the fish to compensate.

In rainbow trout with a cautery-induced 30% reduction in functional gill area, Duthie and Hughes (1987) found a significant proportional reduction in maximum oxygen consumption (VO_{2max}), in comparison to controls, but oxygen consumption at rest and at subcritical swimming speed was not affected. This corroborates suggestion that the total gill area is utilized for maximum oxygen uptake only under condition of maximum aerobic demand. During swimming trials, hyperoxia ($PO_2 = 300$ mmHg) neither increased VO_{2max} in control fish or compensate for the reduced V_{O2max} apparent in fish with reduced gill area. They concluded that therefore a direct limitation of oxygen uptake at the gill is implied.

It can be assumed with Randall and Daxboeck (1984) that pillar cells are active in the changes in RSA, with a spatial modification. In fact, these authors have reported that the reduction of GDD is due to the contraction of pillar cells of lamellae. Hughes (1984a), who showed that the contraction of pillar cells is linked to the increase of ventral aorta blood pressure, came to the same conclusion. An increase in ventral aorta blood pressure giving rise to spatial modifications of pillar cells has been reported for hypoxia (Fritsche and Nilsson, 1993), exercise (Nilsson, 1986) and external hypercapnia (Perry *et al.*, 1999). It may be concluded that the direct effect of hypertension on the ventral aorta is a reduction of GDD together with an increase in the functional surface area (FSA) of the gills.

After studying the physiological adjustment to gas exchange in several fish species that differ in habitat and life-style, Gonzalez and McDonald (1994) concluded that the more active a species is, the more it relies on an increase in FSA, to meet the increased O_2 demand. Less active species that rarely exploit the upper limits during metabolism, rely more on haemodynamic adjustments, e.g., increasing blood flow through the gills and causing intralamellar blood pressure to rise.

It may be expected that sophisticated compensatory mechanisms have an energetic scope. The increase of FSA causes an increase in water loss and an ionic gain through the gills in fishes in a hypertonic environment (Evans, 1993). Fishes respond to this by increasing the water intake. This requires an increase in energy in order to excrete the excess ions through the gills.

Randall *et al.* (1972) described the basic conflict between gas exchange and ion regulation in the gills of freshwater fishes. They indicated that a large permeable gill membrane is required for efficient gas transfer, but a small, impermeable epithelium is needed to minimize the diffusive ion losses. They also demonstrated that rainbow trout accelerates

Na^+ losses with an increase in oxygen consumption during exercise, which they attributed mainly to an increased FSA of the gills during activity.

As suggested by Nilsson (1986), the balance between "need of oxygen" and "need of osmotic regulation" was defined as "osmorespiratory compromise". If this is the case, an adaptation of RSA and GDD may be expected, to compensate oxygen demand and PO_2 fluctuation in water. In fact, the positive correlation between environmental oxygen and GDD has been previously described (Saroglia et al., 2000). Sea bass reared under hyperoxia exibited significantly higher gas diffusion distance than fish reared under normoxia conditions.

References

Al-Kadhomiy, N.K. 1985. Gill development, growth and respiration of the flounder, Platichthys flesus (L.). Ph.D. thesis, University of Bristol, UK.

Booth, J.H. 1978. The distribution of blood flow in the gills of fish: Application of a new technique to rainbow trout (Salmo gairdneri). Journal of Experimental Biology 73: 119-129.

Carroll, S., N. Hazon and F.B. Eddy. 1995. Drinking rates and Na^+ effluxes in response to temperature-change in 2 species of marine flatfish—dab, Limanda limanda and plaice, Pleuronectes platessa. Journal of Comparative Physiology B164: 579–584.

Costa, O.F.T., S.F. Perry, A. Schmitz and M.N. Fernandes. 2001. Stereological analysis of fish gills: Method. Journal of Morphology 248: 219.

Duthie, G.G. and G.M. Hughes. 1987. The effect of reduced gill area and hyperoxia on oxygen consumption and swimming speed of rainbow trout. Journal of Experimental Biology 127: 349-354.

Evans, D.H. 1993. Osmotic and ionic regulation. In: The Physiology of Fishes, D.H. Evans (ed.). CRC Press, Boca Raton, pp. 315-341.

Fernandes, M.N. and S.A. Perna-Martins. 2001. Epithelial gill cells in the armored catfish (Hypostomus plecostomus) (Loricariidae). Revista Brasileira de Biologia 61: 69-78.

Fernandes, M.N., M.S.P.G. de Moraes, S. Höller, M.L. Glass and S.F. Perry. 2002. Partitioning of the morphometric diffusion capacity of respiratory organs in South American dipnoi, Lepidosirem paradoxa. In: The Tropical Fish: News and Reviews Symposium Proceedings, International Congress on the Biology of Fish, A. Val and D. Mackinlay (eds.). American Fisheries Society, Vancouver, pp. 105-113.

Fritsche, R. and S. Nilsson. 1993. Cardiovascular and ventilatory control during hypoxia. In: Fish Ecophysiology, J.C. Rankin and F.B. Jensen (eds.). Chapman and Hall, London, pp. 180-206.

Gonzalez, R.J. and D.G. McDonald. 1994. The relationship between oxygen uptake and ion loss in fish from diverse habitats. Journal of Experimental Biology 190: 95-108.

Gray, I.E. 1954. Comparative study of the gill area of marine fishes. Biological Bulletin of the Marine Biology Laboratory (Woods Hole) 107: 219–225.

Greco, A.M., J.C. Fenwick and S.F. Perry. 1996. The effects of soft-water acclimatation on gill structure in rainbow trout *Oncorhynchus mykiss*. *Cell and Tissue Research* 285: 75-82.

Grinwis, G.C.M., A. Boonstra, E.J. van den Brandhof, J.A.M.A. Dormans, M. Engelsma, R.V. Kuiper, H. van Loveren, P.W. Wester, M.A. Vaal, A.D. Vethaak and J.G. Vosa. 1998. Short-term toxicity of bis(tri-*n*-butyltin)oxide in flounder (*Platichthys flesus*): Pathology and Immune Function Aquatic Toxicology 42: 15–36.

Hartl, M.G.J., S. Hutchinson, L.E. Hawkins and M. Eledjam. 2000. The biometry of gills of O-group European flounder. *Journal of Fish Biology* 57: 1037-1046.

Howard, C.V. and M.G. Reed. 1998. *Unbiased Stereology. Three-dimensional Measurement in Microscopy*. Bios Scientific Publishers, Oxford.

Hughes, G.M. 1966. The dimensions of fish gills in relation to their function. *Journal of Experimental Biology* 45: 177–195.

Hughes, G.M. 1984a. General anatomy of the gills. In: *Fish Physiology*, W.S. Hoar and D.J. Randall (eds.). Academic Press, San Diego, Vol. 10 A, pp. 1-72.

Hughes, G.M. 1984b. Measurement of gill area in fishes: Practices and problems. *Journal of the Marine Biology Association of the United Kingdom* 64: 637-655.

Hughes, G.M. and M. Morgan. 1973. The structure of gills in relation to their respiratory function. *Biological Reviews* 48: 419-475.

Hughes, G.M. and S.F. Perry. 1976. Morphometric study of trout gills: A light-microscopic method suitable for the evaluation of pollutant action. *Journal of Experimental Biology* 64: 447-460.

Hughes, G.M., H. Tuurala and A. Soivo. 1978. Regional distribution of blood in the gills of rainbow trout in normoxia and hypoxia: a morphometric study with two fixatives. *Annales Zoologici Fennici* 15: 226-234.

Hutchinson, S. and L.E. Hawkins. 1990. The influence of salinity on water balance in O-group flounders, *Platichthys flesus*. *Journal of Fish Biology* 36: 751–764.

Kisia, S.M. and G.M. Hughes. 1992. Estimation of oxygen-diffusing capacity of different sizes of a tilapia, *Oreochromis niloticus*. *Journal of Zoology* (London) 227: 405-415.

Kisia, S.M. and G.M. Hughes. 1993. Routine oxygen consumption in different size of a tilapia, *Oreochromis niloticus* (Trewavas) using the closed chamber respiratory method. *Acta Biologica Hungarica* 44: 367-374.

Jagoe, C., A. Faivre and M.C. Newman. 1996. Morphological and morphometric changes in the gills of mosquitofish (*Gambusia holbrooki*) after exposure to mercury (II). *Aquatic Toxicology* 34: 163-183.

Larsson, A., B.-E. Bengsston and C. Haux. 1981. Disturbed ionic balance in flounder, *Platichthys flesus* L., exposed to sublethal levels of cadmium. *Aquatic Toxicology* 1: 19–35.

Laurent, P.L. and N. Hebibi. 1989. Gill morphometry and fish osmoregulation. *Canadian Journal of Zoology* 67: 3055–3063.

McDonald, D.G., J. Freda, V. Cavdek, R. Gonzalez and S.H. Zia. 1991. Interspecific differences in gill morphology of fresh-water fish in relation to tolerance of low-pH environments. *Physiological Zoology* 64: 124-144.

Moraes, M.F.P.G., S. Holler, O.T.F. Costa, M.L. Glass, M.N. Fernandes and S.F. Perry. 2005. Morphometric comparison of the respiratory organs of the South American lungfish, *Lepidosiren paradoxa* (Dipnoi). *Physiology and Biochemistry Zoology*. (In Press).

Motais, R., F.G. Romeu and J. Maetz. 1966. Exchange diffusion effect and euryhalinity in teleosts. *Journal of General Physiology* 50: 391-422.

Nilsson, S. 1986. Control of gill blood flow. In: *Fish Physiology: Recent Advances*, S. Nilsson and S. Holmgren (eds.). Croom Helm, London, pp. 86-101.

Perry, S.F. and G. McDonald. 1993. Gas Exchange. In: *The Physiology of Fishes.*, D.H. Evans (ed.). CRC Press, Boca Raton, pp. 251-278.

Perry, S.F., R. Fritsche, T.M. Hoagland, D.W. Duff and K.R. Olson. 1999. The control of blood pressure during external hypercapnia in the rainbow trout (*Oncorhynchus mykiss*). *Journal of Experimental Biology* 202: 2177-2190.

Potts, W.T.W. and F.B. Eddy. 1973. Gill potentials and sodium fluxes in the flounder *Platichthys flesus*. *Journal of Comparative Physiology* 87: 29-48.

Potts, W.T.W., C.R. Fletcher and B. Eddy. 1973. An analysis of the sodium and chloride fluxes in the flounder *Platichthys flesus*. *Journal of Comparative Physiology* 87: 21-23.

Powell, M.D., D. Fisk and B.F. Nowak. 2000. Effects of graded hypoxia on Atlantic salmon infected with amoebic disease. *Journal of Fish Biology* 57: 1047-1057.

Randall, D.J. and C. Daxboeck. 1984. Oxygen and carbon dioxide transfer across fish gills. In: *Fish Physiology*, W.S. Hoar and D.J. Randall (eds.). Academic Press, London, Vol. 10A, pp. 263-314.

Randall, D.J., D. Baumgarten and M. Malyusz. 1972. The relationship between gas and ion transfer across the gills of fishes. *Comparative Biochemistry and Physiology* A 41: 629-637.

Saroglia, M., S. Cecchini, G. Terova, A. Caputo and A. De Stradis. 2000. Influence of environmental temperature and water oxygen concentration on gas diffusion distance in Sea bass (*Dicentrarchus labrax* L.). *Fish Physiology and Biochemistry* 23: 55-58.

Saroglia, M., G. Terova, A. De Stradis and A. Caputo. 2002. Morphometric adaptations of sea bass gills to different dissolved oxygen partial pressure. *Journal of Fish Biology* 60: 1423-1430.

Soivo, A. and H. Tuurala. 1981. Structural and circulatory responses to hypoxia in the secondary lamellae of *Salmo gairdneri* at two temperatures. *Journal of Comparative Physiology* B145: 37-44.

Sollid, J., P. De Angelis, K. Gundersen and G.E. Nilsson. 2003. Hypoxia induces adaptive and reversible gross morphological changes in crucian carp gills. *Journal of Experimental Biology* 206: 3667-3673.

Stevens, D.E. and A. Sutterlin. 1999. Gill morphometry in growth hormone transgenic Atlantic salmon. *Environmental Biology of Fishes* 54: 405–411.

Stevens, E.D., A.M. Sutterlin and T. Cook. 1998. Respiratory metabolism and swimming performance in growth hormone transgenic Atlantic salmon. *Candian Journal of Fisheries and Aquatic Sciences* 55: 2028-2035.

Environmental Influences on the Respiratory Physiology and Gut Chemistry of a Facultatively Air-breathing, Tropical Herbivorous Fish *Hypostomus regani* (Ihering, 1905)

Jay A. Nelson[1],* Flavia Sant'Anna Rios[2], José Roberto Sanches[3], Marisa Narciso Fernandes[3] and Francisco Tadeu Rantin[3]

INTRODUCTION

Dissolved oxygen is an unpredictable resource for fishes in tropical freshwater ecosystems. Fish occupying these environments have evolved a suite of strategies to either avoid or exploit hypoxic waters (reviewed in

Authors' addresses: [1]Department of Biological Sciences, Towson University, Towson, Maryland 21252-0001, USA.
[2]Department of Cell Biology, Federal University of Paraná, Curitiba, PR, Brazil.
[3]Department of Physiological Sciences, Laboratory of Zoophysiology and Comparative Biochemistry, Federal University of São Carlos. Via Washington Luis, km 235, 13565-905–São Carlos, SP, Brazil.
Corresponding author: E-mail: jnelson@towson.edu

Val, 1996). Facultative air breathing is one hypoxia-avoidance strategy that has attracted substantial interest from both physiologists and ecologists. Physiologists are primarily interested in the mechanisms, metabolic cost and physiological consequences associated with switching between the respiratory modes of breathing water or air (Graham, 1997). Ecologists have primarily studied how interspecific interactions such as predation change when animals switch between these respiratory modes (e.g., Kramer et al., 1983). Despite this multidisciplinary interest, there is a lack of information on how environmental factors other than oxygen levels influence facultative air breathing.

Aquatic hypoxia can develop in tropical habitats due to various combinations of: (1) aquatic respiratory rates exceeding photosynthetic rates, (2) poor mixing at the aerial-aquatic interface, (3) poor light penetration due to shading or turbidity, and/or (4) isolation of water bodies during tropical dry seasons (Junk, 1984). The Neotropical catfish family Loricariidae is the most diverse siluriform family and occupies most hypoxia-prone habitats in the Neotropics. This family contains at least 108 genera, more than 692 described species (Isbrücker, 2002) and many more species awaiting description (Donald Stewart, SUNY-Syracuse USA personal communication). Although air-breathing is not synapomorphic in this family (Armbruster, 1998), most loricariids examined to date will facultatively breath air upon exposure to hypoxia using their gut as an air-breathing organ (ABO; Graham, 1997), or show morphological evidence of air-breathing capabilities (Armbruster, 1998). Cascudo (*Hypostomus regani*) are usually found on rocky substrates in well-oxygenated environments, but follow the predominant loricariid behavioral pattern and facultatively breath air when they are exposed to aquatic hypoxia (Mattias et al., 1998). While loricariids were among the first subjects of facultative air-breathing studies (Carter and Beadle, 1931) and studies of environmental influences on facultative air breathing in loricariids have continued sporadically (e.g., Graham and Baird, 1982; MacCormack et al., 2003), there is still limited knowledge about facultative air breathing in this speciose group and how the environment influences it.

Environmental temperature interfaces with air-breathing physiology of aquatic ectotherms because it influences the rates of metabolic processes as well as environmental oxygen availability and diffusivity. Environmental temperature has been studied with respect to how it influences facultative air breathing (e.g., Graham and Baird, 1982), but little attention has been given to the time course of temperature change. Animals may encounter waters of disparate temperature and dissolved oxygen content suddenly as they move between habitats or when flooding occurs. Water temperature and dissolved oxygen content also change more slowly with seasonal and climatic cycles. Thus, one aim of this study was to examine how temperature changes of different time course influence the physiological responses of cascudo to hypoxia.

The use of the gut as an ABO potentially compromises digestive function in herbivorous fishes. Many loricariids are herbivorous and presumably require an anaerobic gut to facilitate energy extraction from fermentative processes (Choat and Clements, 1998). Thus, the use of the gut as an ABO may oxygenate portions of the gut and diminish digestive performance. If this is the case, it may be manifest by a reluctance of fish with fibrous material in their guts to breath air under hypoxia. Conversely, the additional metabolic demand created by having food in the gut (specific dynamic action—SDA) could require a greater amount of facultative air breathing under hypoxic conditions. This would be manifest as increased surfacing behavior in well-fed fish under hypoxic conditions.

Finally, although hints are pervasive throughout the literature (e.g., Graham, 1997; MacCormack *et al.*, 2003), to our knowledge, the physiology corresponding to individual variance in air-breathing behavior at a given level of water oxygenation has not been analyzed. Thus, we will pay particular attention to whether variability in air-breathing behavior is related to individual variation in physiology under hypoxic water conditions.

MATERIALS AND METHODS

Experimental Animals

Adult specimens of *Hypostomus regani* (20.7 ± 2.0 SD cm total length; 202.2 ± 53.8 g for the dietary experiment; 243 ± 64.2 g SD for the temperature experiment) were collected in the Mogi Guaçu River Basin near Pirassununga, São Paulo State, Brazil by cast net. A group of eight animals of similar size from the same collection had 8.25 ± 2.32 SD growth rings on their sagittae otoliths (Nonogaki *et al.*, 2007). Since there is only a single rainy and dry season per year at this locale, each otolith ring is thought to be an annulus, suggesting that these animals are about 8 years old. In the laboratory, fish were maintained in 1000 L tanks supplied with dechlorinated, normoxic water (~140 mmHg) at 25 ± 1°C (acclimation temperature) for at least 3 weeks prior to experimentation.

Fish were fed *ad libitum* with commercial food pellets during acclimation. Food was withheld 2-3 days before an experimental trial in the temperature experiment. For the dietary manipulation experiment, animals were randomly assigned to either a fed or starved tank at least two weeks prior to experimentation. The "fed" tank contained natural algal growth that the animals foraged on and was supplemented every second day with summer squash (zucchini; *Cucurbita* sp.) and cucumbers (*Cucumis sativus*). The 'starved' tank was covered with black plastic to prevent algal growth and no additional food was provided.

Animal Preparation

Before surgery, fish were anaesthetized by immersion in a solution of benzocaine (0.1%) previously dissolved in 95% ethanol. Subsequently, animals were fixed to a surgical table with a continuous flow of a 0.08% benzocaine solution over the gills. Animals for the temperature experiment had electrocardiogram (ECG) electrodes, buccal and opercular cavity catheters inserted. Fish used in the diet experiment had ECG and a buccal cavity catheter installed. ECG electrodes were fashioned from 22 gauge syringe needles and inserted into the cleithra bones on either side of the heart. The electrodes were secured with atraumatic sutures on the ventral surface of the fish and again around the first dorsal fin spine. Sutures were further secured with acrylic cement. The buccal and opercular cavities were catheterized by inserting flared open-ended polyethylene tubing (PE90) across the tissue surrounding the cavity and gluing them into place (Hughes *et al.*, 1983). All wounds were dusted with antibiotic prior to placing the fish in the experimental chamber.

Following surgery, animals that weren't already ventilating were ventilated by hand until spontaneous breathing resumed and placed into either a flow-through respirometer (temperature experiment) or experimental chambers that allowed the animals to either breath water or air (diet experiment; Fig. 10.1). Animals were allowed a minimum of 16 h to recover from surgery. For the temperature experiment, oxygen consumption ($\dot{V}O_2$), respiratory frequency (f_R), heart rate (f_H) and the oxygen tension of inspired and expired water (P_IO_2, P_EO_2, respectively) were monitored during an experiment. For the diet experiment, f_R, f_H, air-breathing frequency (f_{AIR}) and air-breathing duration (t_{AIR}) were recorded. Water pressure was recorded with either a Telos® 4-327-I or a

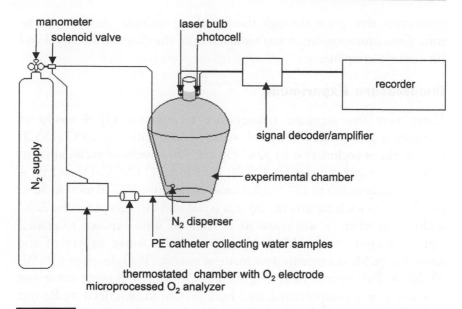

Fig. 10.1 Apparatus used to assess air breathing behavior during the diet experiment.

Narco® P-1000B pressure transducer. Electrical signals were appropriately amplified and either recorded on a Narco® Narcotrace 40 (Narco Bio-Systems, Houston, TX, USA) physiograph or digitized and captured with MacLab® hardware utilizing Chart™ data acquisition software running on a MacIntosh I-book computer.

Exposure to Hypoxia

Dietary experiment

The experimental design was 2×2 with two levels of water oxygen and either fed or starved fish as the experimental treatments and 25°C as the experimental temperature. Data were either recorded during 3 h of normoxia (air saturated) or for 3 h at a nominal P_wO_2 level of 20 mmHg, below the air-breathing threshold of 60 mmHg and P_cO_2 of 27 mmHg for this species at 25°C (Fernandes *et al.*, 1999). Hypoxia was induced by bubbling nitrogen into the water supplying the experimental chamber (Fig. 10.1). Fish in the hypoxia treatments were also recorded for 50 min under normoxia and during a 20 min reduction in P_wO_2 to the nominal level of 20 mmHg. At the end of the recording period, animals were anaesthetized until unresponsive (Stage III anesthesia) by introducing a solution of 1%

benzocaine retrograde through the breathing cannula. As soon as the animal was unresponsive, it was removed from the chamber for blood and gut content sampling.

Temperature Experiment

There were five separate temperature treatments: (1) 4 weeks of acclimation to 20°C (20°C), (2) 4 weeks of acclimation to 25°C (25°C), (3) 4 weeks of acclimation to 30°C (30°C), (4) 4 weeks of acclimation to 25°C with a rapid transfer ($\sim$ 1°C/10 min) to 20°C (20° C-T), and (5) 4 weeks of acclimation to 25°C with a rapid transfer ($\sim$ 1°C/10 min) to 30°C (30°C-T). For each treatment, (n) was equal to 8 fish except for the 20°C acclimation where n was equal to 7. Animals were exposed to graded aquatic hypoxia by bubbling N_2 gas into the water supply of the respirometer. Measurements were made at nominal P_wO_2 levels of 130, 90, 70, 50, 40, 30, 20 and 10 mmHg). The PO_2 of the inflowing water was monitored by a computerized feed back system as described by Rantin *et al.* (1998), and each nominal exposure level was maintained for at least 1 h.

Blood and Gut Chemistry

Fish from the diet experiment were sampled for blood and gut chemistry immediately following the 3 h hypoxic ($PwO_2 = 20$ mmHg) or a 3 h normoxic period. Animals in Stage III anesthesia were removed from the tank and rapidly killed by a blow to the head. The heart was exposed within seconds and blood withdrawn via cardiac puncture into "gas-tight" syringes. Immediately after the blood sample was taken, a small incision was made with artery scissors and a PE 190 cannula coated with mineral oil was advanced in one of three pre-determined gut regions in a randomly determined order as the gut contents were aspirated into a 'gas-tight' syringe. The three gut regions were: (1) foregut, immediately posterior to the stomach, (2) midgut, in the region of the ducting from an accessory liver lobe, and (3) hindgut, proceeding anterior from the gut's entrance to the cloaca. The PE 190 cannula approximated the diameter of the gut lumen such that surface tension and smooth muscle constriction formed a seal around the advancing cannula. Since one-half of the animals were starved, one-half were breathing air, and the gut length of this species is approximately twenty times the standard length of the animal (average 4 m for the animals in this study), there was some deviation from this prescribed protocol. If there was no fluid material or the material was too viscous, no sample could be taken. If the prescribed region was entirely filled with air, the sample was taken from the next adjacent region. As soon as enough material for a pH and PO_2 determination and a perchloric acid extract was removed, sampling moved to the next gut region.

Venous blood and gut content PO_2 were measured by injecting the fresh samples anaerobically into a thermostated cuvette housing an O_2 electrode (FAC 001—O_2, FAC—São Carlos, SP, Brazil), connected to an FAC-204A O_2 analyzer. The exponential decay of PO_2 was recorded on the PowerLab®/Chart™ data acquisition system. Extrapolation of the linear portion of the PO_2 decay curve back to the time of injection was used as the measure of the *in situ* PO_2. Venous blood and gut content pH were measured by injecting the sample anaerobically into a sealed cuvette containing the pH sensing portion of a Mettler Toledo # 405-M3-S7/60 pH electrode connected to a Quimis 400A pH meter (Quimis, Brazil). Care was taken not to use the portion of the sample at the air interface for either measurement.

A portion of the blood and gut samples were deproteinized by placing 0.1 ml of sample into 0.9 ml of 0.6N $HClO_4$ and subsequently neutralized with KOH and centrifuged. The extracted supernatant was divided into two fractions: for short chain fatty acid analysis (SCFA), deproteinized supernatants of blood and gut contents were diluted 1:1 and brought to pH 1.0 with HCl to keep all acids in the protonated form; the second fraction was frozen at neutral pH. Both supernatant fractions were immediately frozen at $-80°C$ for later analysis.

Hematocrit (Ht) was determined by centrifugation at $3000 \times g$ in microhematocrit capillary tubes. The red blood cell count (RBC#) was determined optically with a Neubauer chamber. The mean of seven replicate counts of 1 mm^2 was calculated. The hemoglobin concentration ([Hb]) was determined by a cyanomethemoglobin method that made the blood react with Drabkin's reagent and then the optical density was recorded at 540 nm. Mean cell volume (MCV), cell hemoglobin (MCH), and cell hemoglobin concentration (MCHC) were computed from the Ht, [Hb] and RBC# by standard methods.

In order to determine SCFA levels, acidified extracts of blood and gut contents were extracted with an equal volume of chromatography grade methanol and run on a Supelco SPB-1000 capillary column on a Hewlett-Packard 5973 helium flow gas chromatograph coupled to a Hewlett-

Packard 5973 mass spectrometer. Hewlett-Packard "Chemstation®" software (version B.02.00) controlled operations and coordination of the gas chromatograph with the mass spectrometer; peak detection and identification utilized Hewlett-Packard NIST98® spectral search software. Prepared standards of formic acid, acetic acid, propanoic acid, butyric acid, isobutyric acid, isovaleric acid, pentanoic acid, hexanoic acid and heptanoic acid were cleanly separated and identified by this system. The detection limit for non-polar compounds in aqueous media has been reported to be in the parts per billion range for this system (Li and Fingas, 2003).

Physiological Measurements

Oxygen uptake

Oxygen uptake ($_2$ - $mLO_2 \cdot kg^{-1} \cdot h^{-1}$) was measured by flow-through respirometry according to Rantin et al. (1992). The oxygen tension of incoming ($P_{IN}O_2$) and outgoing ($P_{OUT}O_2$) water was continuously monitored. This was accomplished by siphoning water samples via polyethylene tubing to O_2 electrodes (FAC 001—O_2, FAC—São Carlos, SP, Brazil) that were housed within temperature controlled cuvettes and connected to a FAC-204A O_2 analyzer. Oxygen uptake was calculated as:

$$\dot{V}O_2 = V_R \cdot \alpha \cdot (P_{IN}O_2 - P_{OUT}O_2) \cdot Wt^{-1},$$

where V_R represents the constant water flow rate through the respirometer ($L \cdot h^{-1}$) α denotes the solubility coefficient for O_2 in water ($mLO_2 \cdot L^{-1} \cdot mmHg^{-1}$) and Wt the body mass (kg). Flow rate was 300 $ml \cdot min^{-1}$ and measured manually. Critical oxygen tension was calculated by curve-fitting according to Yeager and Ultsch (1989).

Gill ventilation

Gill ventilation (- $mLH_2O \cdot kg^{-1} \cdot min^{-1}$) was measured and calculated according to Hughes et al. (1983). Permanently implanted PE catheters allowed continuous measurement of inspired (P_IO_2—buccal catheter) and expired (P_EO_2—opercular catheter) water O_2 tensions. Gill ventilation was calculated according to:

$$\dot{V}_G = V_R \cdot [(P_{IN}O_2 - P_{OUT}O_2)/(PIO_2 - PEO_2)] \cdot Wt^{-1}.$$

Respiratory frequency

Respiratory frequency (f_R—breaths $\cdot$ min^{-1}) was obtained from buccal pressure variations and calculated manually from physiograph (Narco, Narcotrace 40, Houston TX, USA) traces or automatically via the ratemeter function of the Chart™ software.

Ventilatory tidal volume

Tidal volume (V_T—mLH$_2$O$\cdot$kg$^{-1}\cdot$breath^{-1}) was calculated by dividing gill ventilation by the respiratory frequency ($\dot{V}_G \cdot f_R^{-1}$).

Oxygen extraction from the ventilatory current

The oxygen extraction by the gills from the ventilatory water current (EO_2 —%) was estimated according to the following equation (Dejours, 1981):

$$EO_2 \ (\%) = 100 \cdot (\ P_IO_2 - P_EO_2) \cdot P_IO_2^{-1}$$

Heart rate

Heart rate (f_H—beats min^{-1}) was obtained from ECG recordings and was either calculated manually from physiograph traces or automatically computed by the ratemeter function of the Chart™ software.

Air-breathing frequency and duration

Air-breathing frequency (f_{AIR}) and air breathing duration (t_{AIR}) were recorded automatically in the dietary experiment. A laser light was shone across the air portion of the air/water interface of the experimental tank (Fig. 10.1) activating photocells on the opposite side of the tank. Any surfacing by the fish (fish are benthic and only surface to breath air) was recorded as disruption of the signal from the photocell by the PowerLab®/ Chart™ data acquisition system (Fig. 10.1). At 1 h and 2 h of hypoxic exposure, the files were saved and recording re-initiated to void filling the computer's memory buffer. This process required 0.5–1 min, during which time any surfacing activity by the fish would not have been recorded. This may have resulted in slight underestimates of reported air-breathing frequencies.

Statistical Analysis

Significant variance among the various treatments was detected with repeated measures multiple analysis of variance (MANOVA). When significant variance was detected, differences between groups were analyzed with Scheffe's test. Relationships between variables were evaluated by least-squares linear regression and F-test. All statistical analyses were performed with Statistica® for microcomputers. ANOVA was used to determine the levels of significance of the physiological data, and the Tukey's test with 95% confidence limits was applied to compare the means whenever there was a significant difference (GraphPad InStat software, San Diego, CA).

RESULTS

Dietary Experiment

Blood and gut chemistry

Animals in the dietary experiment were statistically uniform for all measured variables except venous oxygen tension. Venous PO_2 was significantly lower in the animals exposed to three hours of hypoxia ($P<0.05$; Table 10.1). Venous blood pH was approximately 8 for all treatments and venous blood PO_2 was about 10 mmHg in hypoxic fish and 30 mmHg in normoxic animals (Table 10.1). All digestive tract pH values were near neutral and the gut was oxygenated under both air and water breathing conditions.

No SCFAs were detected in any blood or gut samples. The test had a conservative detection limit of 10 μM, so if free fatty acids were present in the blood or gut samples, they were present in nanomolar amounts, well below the levels found in fish suspected of having fermenting microorganisms present in their gut (Clements, 1997).

Erythrocyte data were pooled by environmental oxygen treatment (Table 10.2). This was because only a few normoxic animals were analyzed and there were no statistical differences between hypoxic animals based upon gut fullness. Hypoxic animals were characterized by having more,

Table 10.1 Animal sizes, blood and gut chemistry after either a three-hour normoxic or hypoxic recording period in fed and starved cascudo. Treatments significantly different from each other are distinguished by superscripted letters. Means ± 1 standard deviation are reported; 'nm' means "not measured".

		Treatment		
Parameter	Normoxia Fed	Normoxia Starved	Hypoxia Fed	Hypoxia Starved
n	5	4	6	6
Standard Length	19.0 ± 1.3	21.2 ± 1.0	21.6 ± 1.0	20.9 ± 3.0
Mass	166.6 ± 29.9	214.6 ± 32.2	217.1 ± 33.9	208.7 ± 86.1
Venous blood pH	8.032 ± 0.133	8.058 ± 0.336	8.134 ± 0.436	7.824 ± 0.631
Venous blood PO_2	24.8 ± 6.2[a]	29.1 ± 7.8[a]	12.5 ± 5.9[b]	8.61 ± 5.2[b]
Foregut pH	7.94 ± 0.20	7.48 ± 0.03	7.49 ± 0.17	7.53 ± 0.59
Foregut PO_2	18.2 ± 14.4	19.5 ± 8.0	33.3 ± 40.2	22.4 ± 19.5
mid/hindgut pH	8.00 ± 1.2	nm	7.30 ± 0.21	7.38 ± 0.37
mid/hindgut PO_2	35.3 ± 34.1	42.8 ± 24.7	57.8 ± 72.5	41.33

Table 10.2 Erythrocytic characteristics of cascudo after either a three-hour normoxic or hypoxic recording period. Treatments significantly different from each other are distinguished by superscripted letters. Means ± 1 standard deviation are reported.

	Treatment	
Parameter	Normoxia	Hypoxia
n	4	10
Hematocrit (%)	27.4 ± 5.0	27.1 ± 6.7
Hemoglobin (g/dL)	5.7 ± 1.8	6.7 ± 1.5
Erythrocyte count (#/mm^3 × 10^6)	0.89 ± 0.043	0.93 ± 0.21
Mean cell volume (μm^3)	308.6 ± 68.6	290.4 ± 25.0
Mean cell hemoglobin concentration (%)	19.8 ± 2.6[a]	24.8 ± 1.8[b]

smaller erythrocytes that contained proportionally more hemoglobin than normoxic animals; the only statistically distinct inter-group difference was the proportion of the cell taken up by hemoglobin ($P < 0.05$; Table 10.2).

Physiological Responses to Hypoxia

Dietary experiment (access to air)

Figure 10.2 demonstrates the heart rate response of fed and unfed cascudo to 3 hours of hypoxic exposure to a nominal P_wO_2 level of 20 mmHg or exposure to normoxic water. Hypoxia exposure immediately initiated an approximate 50% reduction in heart rate that was gradually ameliorated over time. Mean heart rate was significantly lower in the hypoxia-exposed animals throughout the exposure period (MANOVA P<0.001), although individual animals would briefly elevate their heart rate back to control levels or even higher when surfacing to breath air (see below). Fed animals tended to have a slightly higher heart rate in both treatments than unfed animals (significant interaction term MANOVA P < 0.05).

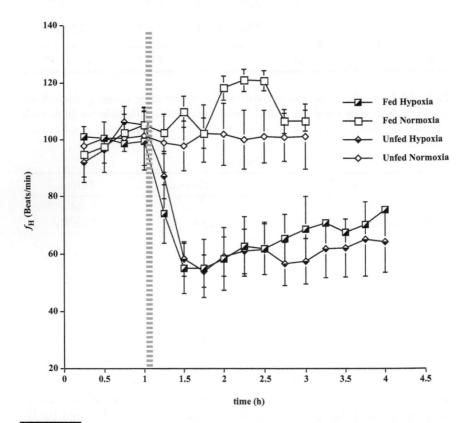

Fig. 10.2 Heart rate f_H in *Hypostomus regani* exposed to 20 mmHg PO$_2$ for three hours (half-closed symbols) or under normoxic conditions for three hours (open symbols). Each symbol represents the mean for that particular group ± 1 SE. Fed animals are represented by squares and unfed animals by diamonds. The striped bar designates hypoxia initiation.

Aquatic respiratory rate was quite variable among individuals (Fig. 10.3), but there was no discernible effect of hypoxia exposure on f_R, nor were there any apparent differences between the two dietary treatments (Fig. 10.3).

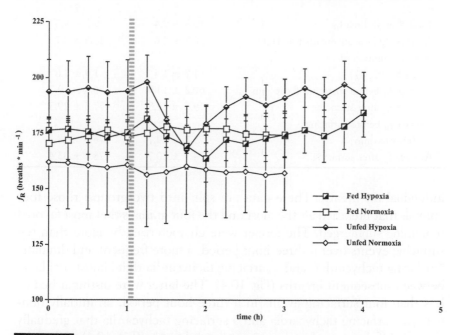

Fig. 10.3 Aquatic respiratory rate f_R in *Hypostomus regani* exposed to 20 mmHg PO_2 for three hours (half-closed symbols) or under normoxic conditions for three hours (open symbols). Each symbol represents the mean for that group ± 1 SE. Fed animals are represented by squares and unfed animals by diamonds. The striped bar designates hypoxia initiation.

Air-breathing behavior and physiology were remarkably consistent between groups of *Hypostomus regani* differing in their dietary status (Table 10.3); fed and unfed fish had virtually identical responses to hypoxia.

Individual analysis of air-breathing

Although dietary treatment did not affect mean air-breathing behavior and physiology, coincident monitoring of air-breathing activity and heart rate revealed substantial inter-individual variation in both behavior and physiology, and a connection between the two. Figure 10.4 plots the change in heart rate after an air-breathing episode as a function of the

Table 10.3 Air-breathing behavior and physiology of cascudo exposed to hypoxia (3 h at a nominal P_wO_2 of 20 mmHg); reported by dietary regime. Means ± 1 standard deviation are reported.

Parameter	Fed	Unfed
n	6	7
Total # of air breaths	10.8 ± 9.4	9.2 ± 4.5
Average P_wO_2 at air breath (mmHg)	20.5 ± 2.6	20.6 ± 3.7
Average duration of air breath (s)	1.7 ± 1.4	1.9 ± 1.8
Average f_H before surfacing (beat·min⁻¹)	60.2 ± 13.9	56.8 ± 19.7
Average f_H after surfacing	79.0 ± 22.8	77.5 ± 29.3
Average f_R before surfacing (breaths·min⁻¹)	175.7 ± 19.7	176.9 ± 28.1
Average f_R after surfacing	174.9 ± 20.7	178.1 ± 33.4

individual air breath. These data are split into two groups: those four animals that breathed air the most and the four animals that most favored remaining submerged. The former were characterized by more than ten surfacing events over a three-hour period, a more frequent and dramatic "surfacing tachycardia" and a surfacing tachycardia that almost oscillated between subsequent breaths (Fig. 10.4). The latter were distinguished by less than four surfacing events in a three-hour period, an initially non-existent surfacing tachycardia and a surfacing tachycardia that gradually increased if the animal took subsequent air breaths (Fig. 10.4). All of the 'frequent air-breathing' fish had an initial surfacing tachycardia of at least 20 bpm, whereas the infrequent air-breathing fish had virtually no surfacing tachycardia at their first air breath.

There was virtually no change in aquatic ventilatory rate (f_R) with air-breathing activity (Table 10.3). The mean change in f_R was $-0.36 ± 11.6$ SD breaths·min⁻¹ for 123 total air-breathing episodes. With a mean f_R around 176 breaths·min⁻¹, this is strong evidence that individual cascudo do not adjust aquatic ventilation when breathing air. No trend was observed for f_R to change as air breaths accumulated; nor was there a discernible difference in aquatic respiration between frequent air breathers and infrequent air breathers.

The periodicity of air breathing was highly dependent on the individual and no effects due to diet or the propensity for air breathing were found. One of the most regular air breathers took only 3 breaths at intervals of 63.6 ± 0.2 SD min (56.7 min would have been perfectly

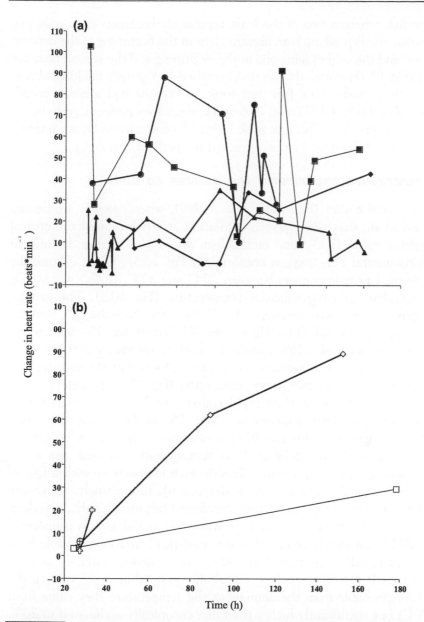

Fig. 10.4 Heart rate changes in individual *Hypostomus regani* exposed to 20 mmHg PO$_2$ for 3 h. Each symbol represents the time post-hypoxia initiation of an individual air-breathing episode for an individual plotted as the difference in heart rate immediately after the surfacing event minus the heart rate immediately before the surfacing event. (a) Upper panel (closed symbols) contains the four animals that most frequently surfaced (>10 times) (b) bottom panel (open symbols) those four animals that surfaced least (<4 times).

periodic), whereas two of the least regular air breathers took only two breaths: one fish taking both breaths right at the beginning of the hypoxic period and the other taking one at the beginning and the second near the end (Fig. 10.4b). Similarly, fish that breathed air regularly could be almost perfectly periodic (one fish that took 16 breaths had a mean breath interval of 9.6 ± 9.4 SD min; 10.6 min expected for perfectly periodic) or not at all periodic (a fish that took 10 breaths had a mean breath interval of 3.5 ± 1.7 SD min; 17 min expected for perfectly periodic).

Temperature experiment (no access to air)

As reported earlier (Fernandes *et al.*, 1999), when cascudo are denied access to air, they cease oxygen regulation at a point called the critical oxygen tension (P_cO_2) and metabolism drops with further declines in environmental PO_2 (oxygen conforming; Fig. 10.5). When chronically acclimated to temperatures between 20°C and 30°C, (P_cO_2) was largely independent of environmental temperature (Fig. 10.5). The critical oxygen tension was determined to be 27 mmHg for both the 25°C and the 30°C treatments and 33 mmHg for the 20°C treatment. However, when fish were acclimated to 25°C and then acutely transferred to 20°C or 30°C, the (P_cO_2) tended to increase irrespective of whether the transfer was made to colder or warmer temperature water (Fig. 10.5). This elevation of (P_cO_2) was 13 mmHg when rapid transfer to 30°C is compared with acclimation to that temperature, but 55 mmHg when the same comparison is made for the 20°C treatments. The metabolic rate of animals chronically acclimated to temperature followed the usual ectothermic pattern and varied directly with temperature with a Q_{10} of approximately 2.0 at PO_2s above the (P_cO_2). Interestingly, above the (P_cO_2), metabolic rates of acutely transferred fish were highly dependent upon which temperature the transfer was made to. Fish rapidly transferred to 30°C had a metabolic rate that was statistically indistinguishable from fish chronically acclimated to 30°C. In contrast, animals acutely transferred to 20°C had a metabolic rate that was statistically indistinguishable from the animals at the temperature they came from (25°C) but significantly higher than fish chronically acclimated to 20°C (Fig 10.5). The significant $PO_2 \times$ treatment interaction (MANOVA; $P<0.01$) shows that the change in metabolism with change in PO_2 was different from the corresponding chronic acclimation temperature for both 20°C and 30°C.

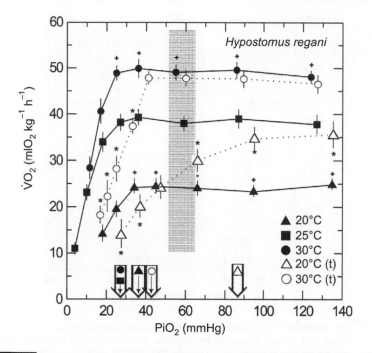

Fig. 10.5 Oxygen consumption of *Hypostomus regani* exposed to progressive hypoxia over a period of 9 hours and denied access to the surface. Closed symbols represent animals chronically acclimated to a temperature, whereas the opened symbols represent animals acutely transferred to the experimental temperature. Each symbol represents the mean for that specific group ± 1 SE. The stippled bar designates the range of PO_2s over which air breathing would normally commence. The arrows designate the calculated P_cO_2 for a treatment.

Analysis of gill ventilation in cascudo revealed a strong uncoupling effect due to rapid temperature change when exposed to hypoxia but denied access to surface (Fig. 10.6).

In animals chronically acclimated to 20°, 25° and 30°C, gill ventilation was temperature insensitive. Although metabolic rate was unaffected by rapid transfer to 30°C, fish rapidly transferred to 30°C had significantly elevated gill ventilation to achieve the same metabolic rate as animals that had acclimated to that temperature (Fig. 10.6). The significant elevation in metabolic rate in fish rapidly transferred to 20°C was also accompanied by an increase in gill ventilation. Greater gill ventilation in rapidly transferred fish was largely accomplished through increases in tidal volume V_T (Fig. 10.7b); Aquatic ventilatory rate (f_R) was relatively unaffected by whether transfer to temperature was acute or chronic at

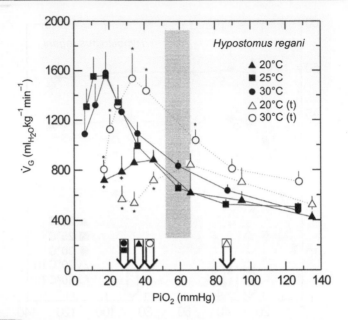

Fig. 10.6 Gill ventilation of *Hypostomus regani* exposed to progressive hypoxia over a period of 9 hours and denied access to the surface. Closed symbols represent animals chronically acclimated to a temperature, whereas the opened symbols represent animals acutely transferred to the experimental temperature. Each symbol represents the mean for that specific group ± 1 SE. The stippled bar designates the range of PO_2s over which air breathing would normally commence. The arrows designate the calculated P_cO_2 for a treatment.

normoxic PO_2s and was temperature-specific. However, upon exposure to hypoxia, both acute exposure groups saw an earlier decrease in both tidal volume and respiratory frequency compared to their chronically exposed counterparts as hypoxia progressed (Fig. 10.7). This is reflected in the higher P_cO_2 for these treatments (Fig. 10.5). Thus, acute change of temperature produced an earlier and more dramatic decline in gill ventilation during progressive hypoxia when compared to chronically acclimated fish (Fig. 10.6). Only two groups had tidal volumes (V_T) that differed significantly from the 25°C control temperature: (1) those acclimated to 30°C, and (2) those acutely transferred to 20°C.

Aquatic respiratory frequency (f_R) mirrored the oxygen consumption results and varied directly with temperature with a Q_{10} of 2.0 at PO_2s above the (P_cO_2). As in the dietary experiment, hypoxia had little effect upon f_R, particularly if the animals were chronically acclimated to temperature. Tidal volumes were substantially increased in both acutely

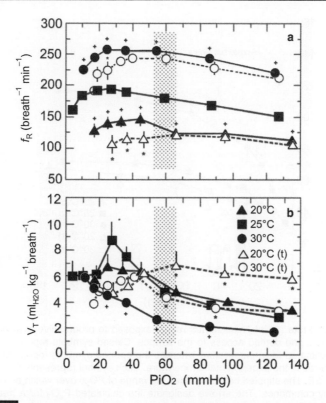

Fig. 10.7 Respiratory frequency (f_R; a) and tidal volume (V_T; b) of *Hypostomus regani* exposed to progressive hypoxia over a period of 9 hours and denied access to the surface. Closed symbols represent animals chronically acclimated to a temperature whereas the opened symbols represent animals acutely transferred to the experimental temperature. Each symbol represents the mean for that group ± 1 SE. The stippled bar designates the range of PO_2 over which air breathing would normally commence.

transferred groups of fish when compared with chronically acclimated animals at the same temperature ($P<0.01$; Fig. 10.7b).

Heart rate in chronically acclimated animals generally increased with temperature at PO_2s above the (P_cO_2), but with a substantially larger Q_{10} (2.6) between 25°C and 30°C than between 20°C and 25°C ($Q_{10}=1.2$; Fig. 10.8). Heart rate of acutely transferred fish was highly dependent upon which temperature the transfer was made to, above the P_cO_2. Acute transfer to 30°C caused a significant elevation of heart rate compared to animals acclimated to that temperature, whereas transfer to 20°C caused a significant depression of f_H when compared to acclimated animals (Fig. 10.8). There was also a significant interaction between temperature and

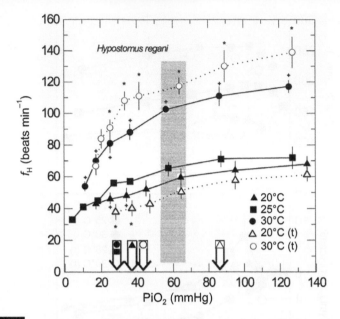

Fig. 10.8 Heart rate (f_H) of *Hypostomus regani* exposed to progressive hypoxia over a period of 9 hours and denied access to the surface. Closed symbols represent animals chronically acclimated to a temperature whereas the opened symbols represent animals acutely transferred to the experimental temperature. Each symbol represents the mean for that group ± 1 SE. The stippled bar designates the range of PO_2s over which air breathing would normally commence. The arrows designate the calculated P_cO_2 for a treatment.

f_H with respect to the bradycardia that developed with progressive hypoxia. Animals acutely transferred to 30°C developed a more substantial bradycardia than chronically acclimated animals at that temperature so that animals from the two treatments at 30°C had identical heart rates at 20 mmHg PO_2. In contrast, animals acutely transferred to 20°C developed a similar bradycardia to animals chronically acclimated to that temperature (Fig. 10.8). The hypoxic bradycardia was significantly more established at 30°C than at either 20°C or 25°C.

DISCUSSION

Individual Analysis

The results from the dietary experiment show that air-breathing behavior and its physiological support varied substantially between individuals in the laboratory. The results from the temperature experiment indicated that cascudo are similar to other facultative air breathers capable of oxygen conforming when faced with hypoxic conditions and only breathing water (Graham, 1997; Fig. 10.5). When fish are given access to air, as in the dietary experiment, the "choice" to keep breathing water and oxygen conform or to breath air and oxygen regulate varies considerably among individuals. Air-breathing frequency among hypoxia-exposed individuals varied from only 1 breath to at least 27 breaths over the same 3 h hypoxic period. These variable breath numbers corresponded to total surface times of less than a second in the least frequently surfacing fish to almost a minute in the fish most likely to surface. Some of this large variance in surfacing behavior may be due to a differential response to the cannulae or the experimental chamber. Graham (1997) discussed the likelihood that air-breathing behavior changed due to cannulation in the obligate air-breathing species *Electrophorus* and *Arapaima*. A second possible explanation for these results is that the frequently surfacing animals had explored the tank during the acclimation period (16 h) and were aware of the air access, whereas the infrequently surfacing animals may have been first discovering the air access during the hypoxic exposure. Several other authors have commented on the large individual variability in air-breathing frequency among conspecific facultative air-breathers exposed to hypoxia (e.g., Gee, 1976; Graham and Baird, 1982; Graham, 1997; MacCormack *et al.*, 2003) suggesting that this phenomenon may be widespread. Considering the additional exposure to predation surfacing entails (Kramer *et al.*, 1983), one can easily envision opposing selection pressures creating the diversity of air-breathing behavior found in cascudo.

The behavioral difference in surfacing behavior between individual *Hypostomus regani* was somewhat connected to the pattern of surfacing tachycardia (Fig. 10.4). Animals that surfaced frequently had a large initial surfacing tachycardia that appeared to oscillate over subsequent surfacing episodes (Fig. 10.4a). Conversely, infrequently surfacing animals had no initial surfacing tachycardia, but then exhibited an increasing surfacing tachycardia with subsequent breaths, if taken (Fig. 10.4b). Surfacing tachycardia, although not universal, has been observed frequently in air-breathing fishes (Table 6.5 in Graham, 1997). In this study, an average 34% increase in heart rate was observed with surfacing in cascudo (Table 10.3); this result is similar to the 36% surfacing tachycardia reported by Graham (1983) for the facultatively air-breathing cofamiliar *Ancistrus*. To our knowledge, ours is the first study to show such large individual variance in surfacing tachycardia. Individual tachycardia varied from 4.4 ± 8.8 SD beats·min^{-1} (post-air-breath cardiac rate–pre-air-breath cardiac rate) from 8 breaths for the animal that exhibited the least tachycardia to three animals that had a mean surfacing tachycardia of around 50 beats·min^{-1} for as many as 16 breaths. Surfacing tachycardia was not dependent upon the total number of air breaths taken, the dietary treatment or f_H immediately preceding the air breath. Thus the oscillatory nature of the tachycardia exhibited by frequently air-breathing fish (Fig. 10.4a) was not merely a result of cardiac rate being relatively high from a previous air breath. This large variation in a physiological measurement, purportedly under just autonomic control (Taylor, 1992), in animals exposed to identical environmental conditions, clearly merits further study. The parsimonious explanation for these results is that a differential stress response to the experimental conditions (cannulation, human presence, foreign tank upon arousal from anaesthesia, etc.) modulated the physiological response to surfacing in *Hypostomus regani*. However, from these data, we cannot exclude the possibility of higher brain center coordination between cardiac regulation and air-breathing behavior in this species.

Our conclusion that air breathing during hypoxia is an individualized event in cascudo was also supported by the periodicity of air breathing. Some animals exhibited highly periodic air breathing whereas other animals breathed air in a manner that could only be described as chaotic (Fig. 10.4). Air breathing in fish has been described by most investigators as arrhythmic, so it was not surprising to find that for these 13 fish, air breathing was not significantly periodic. Interestingly, four of the thirteen animals exhibited very periodic breathing for substantial portions of the three-hour exposure period. Periodicity of air breathing has been demonstrated in *Amia*, but only after long-term recordings and sophisticated statistics were employed by Hedrick *et al.* (1994). We conjecture from the behavior of these four animals that longer-term studies under more natural conditions may have evoked rhythmic air-breathing in cascudo. MacCormack *et al.* (2003) studied air-breathing behavior of *Glyptoperichthyes gibbceps*, another air-breathing loricariid, under simulated natural conditions. They found this species more likely to surface at night-time. Although the chamber in the present experiment was darkened, it is possible that less variant surfacing behavior might have been observed had the experiments been run at night.

Chemistry of Hypoxia Exposure

Despite dramatic differences in air-breathing behavior, blood and gut oxygen and pH levels were maintained at relatively constant levels across treatments in the dietary experiment. The only significant difference between the hypoxic and normoxic treatments was the lower venous oxygen tension in the hypoxic group. Whether this was due to greater oxygen extraction or diminished oxygen uptake by these animals is unknown because arterial blood samples were not taken. The hypoxic group generally had higher gut oxygen tensions, but not significantly so. Since these animals had each taken at least one air breath, finding the gut to be well oxygenated was not surprising. Indeed, fish that breathed air frequently during the hypoxic period had air bubbles throughout their four-meter digestive tracts. Finding well-oxygenated gut contents in the control fish, coupled with the lack of volatile short chain fatty acids (SCFAs), if confirmed, is suggestive of novel digestive mechanisms in loricariids. The gut and blood samples were stored at –80°C for 1 year in sealed containers as acidified perchloric acid extracts, so decomposition should have been minimal. However, the possibility exists that some SCFAs were originally present but were later lost in handling. Oxygenated guts at neutral pH levels, although consistent with the flora culturable from loricariid guts (Nelson *et al.*, 1999), are not suitable for any currently known mechanism of recalcitrant carbon bond breakage in fishes (Choat and Clements, 1998); future studies of digestive mechanisms in loricariids and interactions with air-breathing could prove very interesting.

Hypoxic cascudo were characterized by having more, smaller erythrocytes that contained statistically more hemoglobin per erythrocyte than normoxic animals (Table 10.2). This result is most likely due to the hypoxic animals releasing immature erythrocytes from the spleen to enhance oxygen transport. These results are very similar to those reported by Fernandes *et al.* (1999) for cascudo exposed to graded hypoxia at 25°C for a longer period and Val *et al.* (1990) also reported a higher cell hemoglobin concentration in a loricariid exposed to hypoxia. However, Weber *et al.* (1979) reported cell swelling and decreased cell hemoglobin concentrations in loricariids exposed to hypoxia for 4-7 days, suggesting that the cascudo response is not a generalized loricariid one.

Physiology of Hypoxia Exposure

Metabolic and ventilatory responses of *Hypostomus regani* to hypoxia and temperature have been described previously by Fernandes *et al.* (1999)

and Mattias *et al.* (1998) and will not be repeated here. The focus of this study has been to differentiate between rapid transfer to temperature and acclimation to that temperature in determining the response to hypoxia.

Rapid transfer to a new temperature dramatically altered an animal's metabolic response to progressive hypoxia when compared to animals chronically held at a temperature (Fig. 10.5). Fish rapidly transferred to a 5°C cooler temperature, despite having a respiratory rate entirely appropriate for the temperature they were at (Fig. 10.7a), had a significant elevation of metabolic rate compared with conspecifics acclimated to 20°C. This higher oxygen consumption was accomplished through a significant elevation of aquatic tidal volume (Fig. 10.7b). Fish rapidly transferred to a 5°C warmer temperature also had a significant elevation of gill ventilation when compared to chronically acclimated animals (Fig. 10.6), that was also accomplished entirely through changes in V_T (Fig. 10.7b). As these changes in ventilatory parameters did not translate into elevated metabolic rates at 30°C, this suggests that either O_2 uptake or utilization is compromised by rapid transfer to a 5°C warmer temperature. Graham and Baird (1982) studied the air-breathing response to hypoxia in another facultatively air-breathing loricariid (*Ancistrus chagresi*) and found that at acclimation times intermediate to the chronic and acute times used here, air-breathing frequency changed little between 20°C and 25°C but dramatically increased in animals at 30°C. Graham and Baird (1982) also found that P_cO_2 changed little between 20°C and 30°C when animals are acclimated, supporting our findings for *Hypostomus*. The mechanism behind the elevation of P_cO_2 in both groups of rapidly transferred animals is unknown, but we speculate that a rapid change of temperature is stressful to cascudo, more so when the change is towards colder temperatures. Furthermore, changes in ventilation immediately adjacent to air breaths, common in other air breathers (Graham, 1997) were not apparent in cascudo (Table 10.3). An interesting future experiment would be to determine if the rapid transfer to a different temperature further altered the already variable air-breathing behavior.

Bradycardia is a universal response to hypoxia in *Hypostomus regani* regardless of temperature or dietary treatment (Figs. 10.2 and 10.8). Comparison of fish from the dietary experiment (Fig. 10.2) with the 25°C treatment of the temperature experiment (Fig. 10.8) suggests that the sudden onset of hypoxia elicits a more severe bradycardia than does progressive hypoxia. The development of a significant bradycardia during progressive hypoxia, when metabolic rate remained unchanged, suggests that the cardiac response of cascudo to hypoxia may be somewhat similar to that for carp (*Cyprinus carpio*) at 15°C (Stecyk and Farrell, 2002), where falling f_H was somewhat compensated for by increases in stroke volume. The hypoxic bradycardia observed in cascudo is presumably the same generalized, although not universal, vagally mediated reflex bradycardia response to hypoxia (Taylor, 1992). Interestingly, some fish that do not exhibit hypoxic bradycardia are also loricariids (MacCormack *et al.*, 2002). Although f_H generally tracked metabolic rate in animals chronically acclimated to temperature (Figs. 10.5 and 10.8), a rapid transfer of 5°C had an interesting effect upon f_H. Rapid transfer to 30°C caused animals to have a higher f_H than conspecifics chronically acclimated to that temperature, whereas rapid transfer to 20°C produced a lower cardiac rate than in the comparison group. Rapid transfer did not alter the development of hypoxic bradycardia from that observed in acclimated fish at the same temperature. Bradycardia development was, however, temperature dependent and mirrored metabolic demand. Bradycardia was greatest in fish at 30°C followed by fish at 25°C and was least in animals at 20°C. Presumably the significantly higher cardiac rate in food supplemented fish when compared with starved fish (Fig. 10.2) was due to a higher metabolic rate from specific dynamic action, although SDA is difficult to detect in herbivorous catfishes (Nelson, 2002) and without actual $\dot{V}O_2$ measurements, this is speculative.

In conclusion, dietary treatment had little influence on the air-breathing behavior or physiology of cascudo except for a modest tachycardia in fed animals under both normoxia and hypoxia when compared with starved conspecifics. In contrast, rapid transfer of animals to a 5°C different temperature dramatically altered their response to progressive hypoxia, suggesting that rapid exposure to hypoxic water of different temperature may be the most significant environmental physiological challenge faced by tropical loricariids. The partitioning of respiratory demand between air and water in animals challenged with hypoxia in the laboratory appears to be a highly individualized phenomenon, and may be connected to individual variance in physiology. Presumably the increased susceptibility to avian predation experienced by surfacing loricariids has produced this large variance in air-breathing behavior, but that hypothesis needs to be tested in the field.

Acknowledgements

We wish to thank Nelson A.S. Matos for his excellent technical assistance. Dr Joy E.M. Watts provided helpful comments on the manuscript. Funded by NSF INT-0086474 to JAN and from FAPESP Proc. 96/1691-0 and CNPq Proc. 522904-96-2 to MNF.

References

Armbruster, J.A. 1998. Modifications of the digestive tract for holding air in Loricariid and Scoloplacid catfishes. *Copeia* 1998: 663-675.

Carter, G.S. and L.C. Beadle. 1931. The fauna of the swamps of the Paraguayan Chaco in relation to its environment II. Respiratory adaptations in the fishes. *Journal of the Linnean Society (Zoology)* 37: 327-368.

Choat, J.H. and K.D. Clements. 1998. Vertebrate herbivores in marine and terrestrial environments: A nutritional ecology perspective. *Annual Review of Ecology and Systematics* 29: 375-403.

Clements, K.D. 1997. Fermentation and gastrointestinal microorganisms in fishes. In: *Gastrointestinal microbiology. Gastrointestinal Ecosystems and Fermentations*, R.I. Mackie and B.A. White (eds.). Chapman and Hall, New York, Vol. 1, pp. 156-198.

Dejours, P. 1981. *Principles of Comparative Respiratory Physiology*. 2nd edition. Elsevier/ North-Holland, Amsterdam.

Fernandes, M.N., J.R. Sanches, M. Matsuzaki, L. Panepucci and F.T. Rantin. 1999. Aquatic respiration in facultative air-breathing fish: Effects of temperature and hypoxia. In: *Biology of Tropical Fishes*, A.L. Val and V.M.F. Almeida-Val (eds.). Instituto Nacional de Pesquisas da Amazônia (INPA), Manaus, pp. 341-352.

Gee, J.G. 1976. Buoyancy and aerial respiration: factors influencing the evolution of reduced swim-bladder volume of some Central American catfishes (Trichomycteridae, Callichthyidae, Loricariidae, Astroblepidae). *Canadian Journal of Zoology* 54: 1030-1037.

Graham, J.B. 1983. The transition to air breathing in fishes II. Effects of hypoxia acclimation on the bimodal gas exchange of *Ancistrus chagresi* (Loricariidae). *Journal of Experimental Biology* 102: 157-173.

Graham, J.B. 1997. *Air-breathing Fishes: Evolution, Diversity and Adaptation*. Academic Press, San Diego.

Graham, J.B. and T.A. Baird. 1982. The transition to air breathing in fishes. *Journal of Experimental Biology* 96: 53-67.

Hedrick, M.S., S.L. Katz and D.R. Jones. 1994. Periodic air breathing behaviour in a primitive fish revealed by spectral analysis. *Journal of Experimental Biology* 197: 429-436.

Hughes, G.M., C. Albers, D. Muster and K.H. Götz. 1983. Respiration of carp, *Cyprinus carpio* L., at 10 and 20°C and the effect of hypoxia. *Journal of Fish Biology* 22: 613-628.

Isbrücker, I.J.H. 2002. Nomenclature of the 108 genera with 692 species of the mailed catfishes, family Loricariidae Rafinesque, 1815 (Teleostei, Ostariophysi). *Cat chat* 3: 11-30.

Junk, W.J. 1984. Ecology of the varzea, floodplain of Amazonian whitewater rivers. In: *The Amazon. Limnology and Landscape Ecology of a Mighty Tropical River and its Basin*, H. Sioli (ed.).W. Junk, Dordrecht, pp. 215-244.

Kramer, D.L., D. Manley and R. Bourgeois. 1983. The effect of respiratory mode and oxygen concentration on the risk of aerial predation in fishes. *Canadian Journal of Zoology* 61: 653-665.

Li, K. and M. Fingas. 2003. Evaluation of the HP 6890/5973 bench-top gas chromatograph/mass selective detector for use in mobile laboratories. *Journal of Hazardous Materials* 102: 81-91.

MacCormack, T.J., R.S. McKinley, R. Roubach, V.M.F. Almeida-Val, A.L. Val and W.R. Driedzic. 2003. Changes in ventilation, metabolism, and behaviour, but not bradycardia, contribute to hypoxia survival in two species of Amazonian armoured catfish. *Canadian Journal of Zoology* 81: 272-280.

Mattias, A.T., F.T. Rantin and M.N. Fernandes. 1998. Gill respiratory parameters during progressive hypoxia in the facultative air-breathing fish, *Hypostomus regani* (Loricariidae). *Comparative Biochemistry and Physiology* A120: 311-315.

Nelson, J.A., D.J. Stewart, M.E. Whitmer, E.A. Johnson and D. Wubah. 1999. Wood-eating catfishes and their aerobic, cellulolytic gut symbionts: ecological and evolutionary implications. *Journal of Fish Biology* 54: 1069-1082.

Nelson, J. A. 2002. Metabolism of three species of herbivorous loricariid catfishes: influence of size and diet. *Journal of Fish Biology* 61: 1586-1599.

Nonogaki, H., J.A. Nelson and W.P. Patterson. 2007. Dietary histories of herbivorous loricariid catfishes: Evidence from ^{13}C values of otoliths. *Environmental Biology of Fishes* 78: 13-21.

Rantin, F.T., A.L. Kalinin, M.L. Glass and M.N. Fernandes. 1992. Respiratory responses to hypoxia in relation to mode of life of two erythrinid species (*Hoplias malabaricus* and *Hoplias lacerdae*). *Journal of Fish Biology* 41: 805-812.

Rantin, F.T., C.D.R. Guerra, A.L. Kalinin and M.L. Glass. 1998. The influence of aquatic surface respirtion (ASR) on cardio-respiratory function of the serrasalmid fish *Piaractus mesopotamicus*. *Comparative Biochemistry and Physiology* A119: 991-997.

Stecyk, J.A.W. and A.P. Farrell. 2002. Cardiorespiratory responses of the common carp (*Cyprinus carpio*) to severe hypoxia at three acclimation temperatures. *Journal of Experimental Biology* 205: 759-768.

Taylor, E.W. 1992. Nervous control of the heart and cardiorespiratory interactions. In: *Fish Physiology*, W.S. Hoar, D.J. Randall and A.P. Farrell (eds.). Academic Press, San Diego, Vol. 12B, pp. 343-389.

Val, A.L. 1996. Surviving low oxygen levels: Lessons from fishes of the Amazon. In: *Physiology and Biochemistry of Fishes of the Amazon*, A.L. Val, V.M.F. Almeida-Val and D.J. Randall (eds.). Instituto Nacional de Pasquisas da Amazônia (INPA). Manaus, pp. 59-73.

Val, A.L., V.M.F. de Almeida-Val and E.G. Affonso. 1990. Adaptive features of Amazon fishes: Hemoglobins, hematology, intraerythrocytic phosphates and whole blood Bohr effect of *Pterygoplichthys multiradiatus* (Siluriformes). *Comparative Biochemistry and Physiology* B97: 435-440.

Weber, R.E., S.C. Wood and B.J. Davis. 1979. Acclimation to hypoxic water in facultative air-breathing fish: blood oxygen affinity and allosteric effectors. *Comparative Biochemistry and Physiology* A62: 125-129.

Yeager, D.P. and G.R. Ultsch. 1989. Physiological regulation and conformation: A BASIC program for the determination of critical points. *Physiological Zoology* 62: 888-907.

11

Osmoregulatory and Respiratory Adaptations of Lake Magadi Fish (*Alcolapia grahami*)

Daniel W. Onyango* and Seth M. Kisia

INTRODUCTION

Alcolapia grahami inhabits an extremely stressful aquatic environment of Lake Magadi, Kenya. The lake water temperature is often above 40°C, sometimes as high as 46°C, osmolality close to 600 mOsm/L, alkalinity above 350 mmol/L, pH 10 and very low carbon dioxide content, reaching a high of 200 mmol/L (Johansen *et al.*, 1975, Wood *et al.*, 1989; Johnston *et al.*, 1994; Maina *et al.*, 1996; Narahara *et al.*, 1996) (see Table 11.1).

The lake milieu is derived from discharges of hot alkaline springs containing very high levels of HCO_3^- and CO_3^{2-} which themselves are products of intense volcanic activity characteristic of this region. This fish is the only known vertebrate inhabiting this harsh environment. Perhaps due to these environmental pressures, they are generally small in size

Authors' address: Department of Veterinary Anatomy, University of Nairobi, P.O. Box 30197, Nairobi, Kenya.
*Corresponding author: E-mail: dwo@uonbi.ac.ke

Table 11.1 Summary of the aquatic conditions in which *A. grahami* subsists.

Temperature (°C)	Osmolality (mOsm/L)	pH	CO_2	Author(s)
30-36.5	525	10	180	Wood *et al.* (1989)
43	600	9.6-10.5	200	Johansen *et al.* (1975)
42.8-44.8	–	10.35	–	Johnston *et al.* (1994)
32-40*	525-600**	–	–	*Maina, 1996; **Narahara *et al.* (1996)

(Fig. 11.1), weighing up to 54 g (Maina *et al.*, 1996). *A. grahami* feeds on algae at the bottom of the lake and *Cynobacteria* found in abundance in the lake water. The fish can gradually adapt to diluted water environments but sudden transfers lead to significant mortalities (Wood *et al.*, 2002).

Fig. 11.1 An adult lake Magadi tilapia (*A. grahami*). Mature *A. grahami* tilapia are small in size often weighing up to 54 g.

OSMOREGULATION

Significant salt gain and water losses in fish living in a hyperosmotic environment occur through the gill surface and gastro-intestinal tract since the skin and scales offer relatively impermeable surfaces. Fish living in marine environments, due to the high concentration of salts in the surrounding water, tend to lose water and gain salts while those in freshwater environments lose salts and gain water. Successful survival of fish in, either marine or freshwater environments require strategic structural as well as ecophysiological adaptations.

Since *A. grahami* lives in a hyperosmotic environment and drinks highly alkaline water at a relatively high rate, the design of the esophagus, stomach and intestines ensures bypassing of an empty stomach by the alkaline water (Bergman *et al.*, 2003). Such a design ensures that this tilapia is able to maintain a low stomach pH for digestion as well as processing of alkaline water imbibed for osmoregulation. Some of the osmoregulatory and respiratory adaptive changes in *A. grahami* have been discussed below.

Urea Excretion

Due to the hostile environment in which *A. grahami* lives it has developed a remarkable capacity to circumvent these adverse environmental conditions. The excretion of nitrogenous waste is fundamentally different from that of freshwater fish. Whereas in the latter the predominant by-product is ammonia; in the former, urea is the by-product, a feature that makes them entirely ureotelic (Randall *et al.*, 1989; Wood *et al.*, 1989, 1994). Ureotelism is a phenomenon in which an organism excretes its nitrogenous waste in the form of urea. It commonly occurs in terrestrial vertebrates, elasmobranchs and the coelacanth but this is one of the very rare instances of complete ureotelism in an entirely aquatic teleost fish (Wood *et al.*, 1989). Ureotelism also occurs in other fishes but only during special conditions; for example, it occurs in South American and African lungfishes during estivation and also, in embryos and larvae of the African cat-fish (*Clarias gariepinus*) and gulf toadfish. In terrestrial vertebrates, the liver is the central organ where, through ornithine-urea cycle, urea is produced. Indeed the first enzyme in the urea cycle, carbamoyl-phosphate synthetase I (CPSase), is predominantly found in the liver of these vertebrates (Mommsen and Walsh, 1989). Conversely, in *A. grahami* a similar enzyme, CPSase III believed to be the evolutionary precursor of CPSase I (Mommsen and Walsh, 1989), has been shown to occur abundantly in muscles (Lindley *et al.*, 1999) suggesting that the muscle plays a significant role in urea excretion. The removal of nitrogenous waste in this manner enables this fish to survive the alkaline conditions of Lake Magadi (Randall *et al.*, 1989) but, apparently, does not seem to play a role in the regulation of acid-base balance (Wood *et al.*, 1994). Ureagenesis is controlled by the quantity of ammonium (NH_4^+) ions to be removed (Halperin *et al.*, 1986).

Urea excretion in *A. grahami* occurs largely across the gills although small quantities may also be excreted through the bile and urine (Wood *et al.*, 1994). The permeability of *A. grahami* gills to urea is many times higher than that of other urea permeable teleosts making it the most important organ in urea excretion. For example, urea excretion in *A. grahami* gills is 5 times higher than that of gulf toadfish (Walsh *et al.*, 2001). Across the gills, the excretion is continuous and its transport bi-directional, occurring mainly through pavement cells by a special urea transporter, mtUT (Walsh *et al.*, 2001; Wilkie, 2002). Urea trafficking in these cells, perhaps, occurs in the form of dense-cored vesicles found, as in toadfishes which also inhabit highly alkaline environments (Laurent *et al.*, 2001), between the well developed Golgi cisternae and apical membranes of epithelial cells surrounding gill filaments and lamellae.

Ionic Regulation

As a result of the extremes of conditions under which *A. grahami* survives, this fish has adopted a delicate and dextrous means of balancing the ionic concentrations in the body. In seawater adapted fish, ionic exchange generally occurs across the gills (Girard and Payan, 1980; McGeer and Eddy, 1998) mainly as a function of a special type of cells, the chloride cells. Like marine fish, *A. grahami* also possess these cells where they act as important channels for excretion of excess monovalent ions as the fish consumes a lot of alkaline water in order to be in osmotic balance with the environment. The cells are mitochondrion-rich with extensive tubular reticulum and sometimes possess apical pores; features which not only reflect their capacity to act as ion pumps but also help them intercommunicate within the surrounding water (Sardet *et al.*, 1979; Maina, 1990, 1991). The ion pump possibly operates as a branchial Na^+ plus HCO_3^-/CO_3^{2-} active excretion scheme. Excess alkaline water consumed by this fish is handled by the unique arrangement of the oesophagus, intestines, stomach and pyloric sphincter which affords a bypass of the empty stomach by highly alkaline water into the intestine raising its pH level. The gut wall quickly absorbs Na^+, titratable base and Cl^- using, to a large extent, Na^+ and base co-transport system (Bergman *et al.*, 2003). Water absorption in the gut follows that of Na^+. In this way, this unique teleost is able to survive the extremes of osmolality and alkalinity prevalent in its natural habitat.

RESPIRATION

Gaseous Exchange

The aquatic environment of Lake Magadi provides unique respiratory challenges to the inhabitants. Firstly, the oxygen content of the water is highly variable, ranging from hyperoxia during the day, due to the photosynthetic activity of the algae, to virtual anoxia during the night as a result of the activities of *Cynobacteria* (Narahara *et al.*, 1996; Maina, 2000). Secondly, the free carbon dioxide content of the surrounding medium is extremely low, resulting into similarly low carbon dioxide levels in the fish blood, hence, very high pH values unparalleled in other vertebrate species (Johansen *et al.*, 1975). In order to survive in this extreme situation, this fish has evolved gills with a very large surface area (seen as long and numerous gill filaments, Fig. 11.2) and a thin water-blood barrier compared to other fresh-water tilapiine species (Maina *et al.*, 1990, 1996). This is ostensibly to facilitate fast and efficient gaseous exchange in an environment that has, sometimes, very extreme levels of these gases. In addition, the swimbladder in this species acts as an accessory respiratory organ and controls the buoyancy of the fish (Maina, 1996; Maina *et al.*, 2000). Thus, the columnar epithelial cells lining the lumen of the bladder frequently show disruptions of the cell membranes over their nuclei associated with discharge of gas-bubbles.

Fig. 11.2 *A. grahami* gills. Notice the presence of numerous gill filaments providing a large surface area for gaseous exchange.

In hot springs where temperatures are 40°C or more, these fish frequently swim to the surface to "gulp air" (Franklin *et al.*, 1995; Maina *et al.*, 1996). The fish resort to aerial respiration when there is high oxygen demand occasioned either by the increased level of activity (exercise) or decreased partial pressures of dissolved oxygen in the water due to increase in temperature. The overall objective of this aerial respiration is to pass air through the buccal cavity into the swimbladder which acts as an air-breathing accessory organ (Maina *et al.*, 1996).

Tissue Respiration

It is uncertain as to how A. *grahami,* inhabiting this environment with extremes of temperature, still manages to retain a reasonable level of activity. Perhaps, this thermal tolerance has been achieved through acclimatization and acclimation although evidence concerning tissue respiration in this species is scanty. Red muscle mitochondria—for which information is available—only exhibit modest evolutionary adjustments in the maximal rate of mitochondrial respiration (Johnston *et al.*, 1994). The mitochondria seem to utilize pyruvate as a respiratory substrate more than glutamate, an observation that may reflect the unusual pattern of nitrogen metabolism in this fish. Obviously, further investigations on tissue respiration in this species are desirable. This will, most likely, bring forth the tissue peculiarities in this fish with regard to thermal tolerance.

References

Bergman, A.N., P. Laurent, G. Otiang'a-Owiti, H.L. Bergman, P.J. Walsh, P. Wilson and C.M. Wood. 2003. Physiological adaptations of the gut in the Lake Magadi tilapia, *Alcolapia grahami*, an alkaline- and saline-adapted teleost fish. *Comparative Biochemistry and Physiolology* A136: 701-715.

Franklin, C.E., I.A. Johnston, T. Crockford and C. Kamunde. 1995. Scaling of oxygen consumption of Lake Magadi tilapia, living at 37°C. *Journal of Fish Biology* 46: 829-834.

Girard, J.P. and P. Payan. 1980. Ion exchange through respiratory and chloride cells in freshwater- and seawater-adapted teleosteans. *American Journal of Physiology* 238: R260-R268.

Halperin, M.L., C.B. Chen, S. Cheema-Dhadli, M.L. West and R.L. Jungas. 1986. Is urea formation regulated primarily by acid-base balance *in vivo*? *American Journal of Physiology* 250: F605-F612.

Johansen, K., G.M. Maloiy and G. Lykkeboe. 1975. A fish in extreme alkalinity. *Respiration Physiology* 24: 159-162.

Johnston, I.A., H. Guderley, C. Franklin, T. Crockford and C. Kamunde. 1994. Are mitochondria subject to evolutionary temperature adaptation? *Journal of Experimental Biology* 195: 293-306.

Laurent, P., C.M. Wood, Y. Wang, S.F. Perry, K.M. Gilmour, P. Part, C. Chevalier, M. West and P.J. Walsh. 2001. Intracellular vesicular trafficking in the gill of urea-excreting fish. *Cell and Tissue Research* 304: 197-210.

Lindley, T.E., C.L. Schreiderer, P.J. Walsh, C.M. Wood, H.L. Bergman, A.N. Bergman, P. Laurent, P. Wilson and P.M. Anderson. 1999. Muscles as the primary site of urea cycle enzyme activity in an alkaline lake-adapted tilapia, *Oreochromis alcalicus grahami. Journal of Biological Chemistry* 274: 29858-29861.

Maina, J.N. 1990. A study of the morphology of the gills of an extreme alkalinity and hyperosmotic adapted teleost *Oreochromis alcalicus grahami* (Boulenger) with particular emphasis on the ultrastructure of the chloride cells and their modification with water dilution: A SEM and TEM study. *Anatomy and Embryology* 181: 83-98.

Maina, J.N. 1991. A morphometric analysis of chloride cells in the gills of the teleosts *Oreochromis alcalicus* and *Oreochromis niloticus* and a description of presumptive urea excreting cells in *O. alcalicus. Journal of Anatomy* 175: 131-145.

Maina, J.N. 2000. Functional morphology of the gas-gland cells of the air-bladder of *Oreochromis alcalcus grahami* (Teleostei: Cichlidae): An ultrastructural study on a fish adapted to a severe, highly alkaline environment. *Tissue and Cell* 32: 117-132.

Maina, J.N., S.M. Kisia, C.M. Wood, A.B. Narahara, H.L. Bergman, P. Laurent and P.J. Walsh. 1996. A comparative allometric study of the morphometry of the gills of an alkalinity adapted cichlid fish, *Oreochromis alcalicus grahami*, of Lake Magadi, Kenya. *International Journal of Salt Lake Research* 5: 131-156.

McGeer, J.C. and F.B. Eddy. 1998. Ionic regulation and nitrogenous excretion in rainbow trout exposed to buffered and unbuffered freshwater of pH 10.5. *Physiological Zoology* 71: 179-190.

Mommsen, T.P. and P.J. Walsh. 1989. Evolution of urea synthesis in vertebrates: the piscine connection. *Science* 243: 72-75.

Narahara, A.B., H.L. Bergman, P. Laurent, J.N. Maina, P.J. Walsh and C.M. Wood. 1996. Respiratory physiology of the lake Magadi tilapia (*Oreochromis alcalicus grahami*), a fish adapted to a hot, alkaline, and frequently hypoxic environment. *Physiological Zoology* 69: 1114-1136.

Randall, D.J., C.M. Wood, S.F. Perry, H. Bergman, G.M. Maloiy, T.P. Mommsen and P.A. Wright. 1989. Urea excretion as a strategy for survival in a fish living in a very alkaline environment. *Nature* (London) 337: 165-166.

Sardet, C., M. Pisam and J. Maetz. 1979. The surface epithelium of teleostean fish gills. Cellular and junctional adaptations of the chloride cell in relation to salt adaptation. *Journal of Cell Biology* 80: 96-117.

Walsh, P.J., M. Grosell, G.G. Goss, H.L. Bergman, A.N. Bergman, P. Wilson, P. Laurent, S.L. Alper, C.P. Smith, C. Kamunde and C.M. Wood. 2001. Physiological and molecular characterization of urea transport by the gills of the Lake Magadi tilapia (*Alcolapia grahami*). *Journal of Experimental Biology* 204: 509-520.

Wilkie, M.P. 2002. Ammonia excretion and urea handling by fish gills: present understanding and future research challenges. *Journal of Experimental Zoology* 293: 284-301.

Wood, C.M., S.F. Perry, P.A. Wright, H.L. Bergman and D.J. Randall. 1989. Ammonia and urea dynamics in the Lake Magadi tilapia, a ureotelic teleost fish adapted to an extremely alkaline environment. *Respiration Physiology* 77: 1-20.

Wood, C.M., H.L. Bergman, P. Laurent, J.N. Maina, A. Narahara and P. Walsh. 1994. Urea production, acid-base regulation and their interactions in the Lake Magadi tilapia, a unique teleost adapted to a highly alkaline environment. *Journal of Experimental Biology* 189: 13-36.

Wood, C.M., P. Wilson, H.L. Bergman, A.N. Bergman, P. Laurent, G. Otiang'a-Owiti and P.J. Walsh. 2002. Obligatory urea production and the cost of living in the Magadi tilapia revealed by acclimation to reduced salinity and alkalinity. *Physiology Biochemistry and Zoology* 75: 111-122.

Wright, P.A. and M.D. Land. 1998. Urea production and transport in teleost fishes. *Comparative Biochemistry and Physiology* A119: 47-54.

Respiratory Function of the Carp, *Cyprinus carpio* (L.): Portrait of a Hypoxia-tolerant Species

Mogens L. Glass[1] and Roseli Soncini[2]

INTRODUCTION

The carp, *Cyprinus carpio* (L.), belongs to the family Cyprinidae, order Cypriniformes. A population of wild carp in the Donau River basin possibly gave rise to the form used in aquaculture. The species has a distribution from China across India and Siberia to Western Europe and was also introduced to the Americas. The unusually wide distribution of this species invites a discussion of its respiratory function. Aside from trout, the carp might be the best known teleost fish from the point of view of respiratory control and gas transport. Carp is a benthic herbivore of slowly moving or stagnant waters. In this chapter we discuss some specific respiratory characteristics of carp that correlate with this type of habitat. In particular, we have studied specimens from Brazilian fish cultures with

Authors' addresses: [1]Department of Physiology, Faculty of Medicine of Ribeirão Preto, University of São Paulo, Avenida Bandeirantes 3900, 14049-900 Ribeirão Preto, SP, Brazil. E-mail: mlglass@rfi.fmrp.usp.br
[2]Department of Biological Science, Federal University of Alfenas, Rua Gabriel Monteiro da Silva 714, 37130-000 Alfenas, MG, Brazil. E-mail: soncinir@yahoo.com.br

elevated water temperatures. To highlight these adaptations of gas transport, we compare carp to dourado (*Salminus maxillosus*) and traira (*Hoplias malabaricus*).

Temperature and O_2 levels of the aquatic environment change on a daily and annual basis. This influences gill ventilation, circulatory function, O_2 transport and blood acid-base status. In this context, carp has acute and efficient ventilatory responses to ambient hypoxia (Glass *et al.*, 1990). Most importantly, carp possess high affinity hemoglobin, which permits O_2 transport at very low ambient O_2 levels (Albers *et al.*, 1983; Soncini and Glass, 2000). Above all, carp is a very hypoxia-tolerant species. In regard to temperature and hypoxia, it becomes interesting to consider the broad distribution of the species. A question is whether or not the basic characteristics of carp are the same within the wide range.

GILL VENTILATION AND ITS CONTROL

Ventilatory Responses to Hypoxia and Hypoxemia

Carp ventilates the gills in a periodic manner, when in normoxic conditions. Bursts of gill movements alternate with short periods of apnea. Gill ventilation becomes continuous during exposure to hypoxia and/or high temperature (Lomholt and Johansen, 1979; Glass *et al.*, 1990). Largely, the studies agree as to the magnitude of ventilation and arterial blood gases in relation to temperature, hypercarbia, hypoxia and hyperoxia. See Table 12.1.

At 20°C and under normoxic conditions, the gill ventilation was 195 $mL \cdot kg^{-1} \cdot min^{-1}$ in the study by Lomholt and Johansen (1979) and 241-253 $mL \cdot kg^{-1} \cdot min^{-1}$ (Glass *et al.*, 1990). In both reports, rather small decreases of PO_2 provoked substantial increases of gill ventilation. For example, a small reduction of water PO_2 from 150 mmHg to 110 mmHg was accompanied by significant increases of gill ventilation. The responses were significant at 10 and 20°C (Glass *et al.*, 1990; Soncini and Glass, 2000). The values for increases of ventilation were also divided by the normoxic resting gill ventilation:

$$(V_{Gt} \text{ hypoxia}/V_{Go} \text{ normoxia}) \cdot 100\%$$

Expressed is this way, the relative increases were practically the same at 10, 20 and 25°C. In other words, increased temperature equally scaled up the control value and the ventilation in hypoxia by the same factor

Table 12.1 Gill ventilation and arterial PO_2 and acid-base status in *Cyprinus carpio*.

Mass (kg)	Temperature (°C)	Pw_{O_2} (mmHg/kPa)	$\dot{V}_G$ (mL kg^{-1} min^{-1})	PaO_2 (mmHg/kPa)	$PaCO_2$ (mmHg/kPa)	pHa	$[HCO_3^-]pl$ (mmol L^{-1})	Method for measurement of ventilation	Reference
0.17	20	High	327	...	...	...	...	Estimation by Fick principle	Saunders (1962)
0.17	20	Low	3065	...	...	...	...		
3.2	10	~ 130/17.3	...	8.0/1.07	3.5/4.7	7.91	12.7		Garey (1967)
0.56	25	141/18.8	244	24.8/3.3	3.9/0.52	7.89	...	Estimation by Fick principle	Itazawa and Takeda (1978)
0.56	25	52/6.9	917	18.7/2.5	1.7/0.23	...	...		
0.56	25	25/3.3	2908	5.4/0.72	<1.0/0.13	...	...		
1	20	>100/13.3	195*	...	...	...	...	Direct, by electromagnetic probes	Lomholt and Johnsen (1979)
1	20	<40/5.3	1122	...	...	...	...		
1.5-3.1	15	135/18	...	73.2/9.76**	3.47/0.46	7.86	9.43	...	Ultsch et al. (1981)
1.78	15	~ 150/20	...	44.4/58.92	4.8/0.64	7.87	13.8	...	Clabiborne and Heisler (1987)
1.73	15	~ 150/20	...	24.0/3.2	3.0/0.4	8.01	12.2	...	Jensen et al. (1988)

(Table 12.1 Contd.)

(Table 12.1 Contd.)

...	20	~150/20	...	43.1/5.74	5.0/0.67	7.78	...	...	Fuchs and Albers (1988)
1.1	15	130/17.3	...	26.3/3.58	3.2/0.42	8.02	...	...	Jensen (1990)
2.34	10	~150/20	50-57	15.3/2.04	3.3/0.44	8.07	14.2	Direct, by electromagnetic probes	Glass et al. (1990)
2.34	10	758/10	132[+]	8.9/1.18	1.7/0.23	8.27	12.3		
2.34	20	~50/20	241-253	14.4/1.92	4.0/0.53	8.01	13.4		
2.34	20	75/10	659[+]	11.4/1.52	1.9/0.25	8.19	10.5		
0.55-0.77	25	130/17.3	258	23.2/3.09	3.9/0.52	7.95	...	Estimation by Fick principle	Takeda (1990)
0.55-0.77	25	313/41.7	104.8	31/4.1	9.4/1.2	7.73	...		
0.55-0.77	25	459/61.2	90.6	38.3/5.13	11.3/1.5	7.64	...		
108	15	~150/20	...	23.3/3.1	3.61/0.48	7.90	...	...	Williams, Glass and Heisler (1992)
1.89	15	75/10	...	13.6/26	2.41/0.32	8.03	...		
0.53-0.61	25	141/18.8	211	25.9/3.45	4.04/0.53	7.94	...	Estimation by Fick principle	Takeda (1993)
0.53-0.61	25	128***/17	1113	43.9/5.85	4.16/0.55	7.93	...		
0.66	25	127/16.9	296	31/4.1	3.03/0.4	7.81	...	Direct method by van Soncini and Dam (1938)	Glass (2000)
0.66	25	97/12.9	296	31/4.1	3.03/0.4	7.81	...		
0.66	25	89/11.8	853	25/3.33	5.39/0.72	7.85	...	Modified by Burggren and Cameron (1980)	
0.66	25	245/32.6	175	39/5.2	8.77/1.17	7.59	...		

*Carp pre-acclimated to normoxia, [+]4h-values.
**The authors explained that these high PO_2 values reflected disturbances. The PO_2 of undisturbed specimens were close to 25 mm Hg (t = 15°C).
***Exercise stress.

(Fig. 12.1). This makes sense, since O_2 uptake increases with temperature, while gill ventilation scales up demands for oxygen.

This situation contrasts with the responses of endothermic tetrapods, which lack fixed adjustments of the gain relative to normoxic control, when at low temperature. The secondary importance of O_2 regulation is evident from the depression or disappearance of hypoxia-related responses at low temperature (Kruhøffer et al., 1987). Fish are exposed to ever changing levels of O_2 and temperature, and the regulatory strategy of carp and other teleosts makes sense. The ventilatory responses to hypercarbia are weak or absent in teleosts (Dejours, 1981), while O_2-related drives continue at low temperature.

Moffitt and Crawshaw (1983) studied acute temperature changes in carp that were acclimated to 19°C. Their study in New York State gave a heart rate was 18 beats·per min^{-1} with $f_R = 21.8$ min^{-1}. It is striking that carp from various parts of the world may have some very similar characteristics: Carp at 20°C had a respiratory frequency of 16.1 min^{-1} in Göttingen, Germany (Glass et al., 1990) and 18.7 min^{-1} in Japan (Takeda, 1993).

Oxygen-sensing receptors are located in the gill arches of teleost fish. Part of these receptors screens the water, while some monitor O_2 of the blood (Milsom and Brill, 1986; Burleson and Milsom, 1990; 1995a,b). This provides the fish with two defense lines when confronted with hypoxia. In carp, gill ventilation increased by 58% as soon as the external O_2 receptors were exposed to light hypoxia. In this situation, there was no significant change in PaO_2 and acid-base status (Soncini and Glass, 2000). By contrast, land vertebrates possess only a single line of defence. Peripheral chemoreceptors screen PaO_2, but a reduction of this pressure is needed to elicit ventilatory responses.

Smith and Jones (1982) subjected rainbow trout (*Oncorhynchus mykiss*) to various combinations of normoxia, hypoxia, hyperoxia, and hypercarbia, which all changed O_2 content of the blood ($[O_2]a$), and they concluded that gill ventilation increased systematically with reductions of $[O_2]a$. Based on this, Randall (1982) suggested that $[O_2]a$ is the specific O_2-stimulus in teleost fish. By contrast, it was not possible to obtain such a straight-forward relationship in carp, which indicates that more components are involved. An example is provided by the external gill arch receptors that screen the water and not $[O_2]a$.

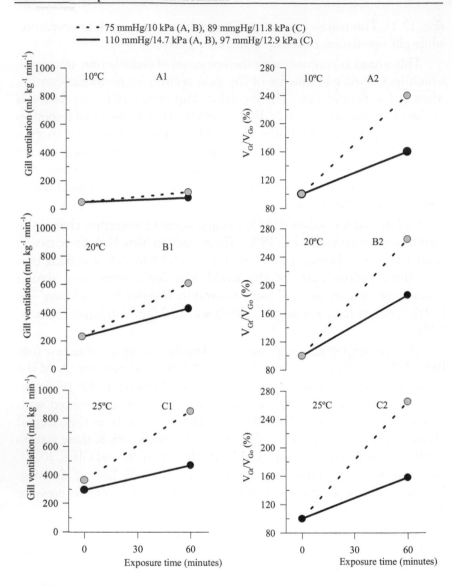

Fig. 12.1   Acute exposure of carp at 10, 20 and 25°C. Gill ventilation was directly measured using mask techniques based on flow probes (Glass *et al.*, 1990) or a modified van Dam technique (van Dam, 1938; Burggren and Cameron, 1980; Soncini and Glass, 2000). The left side depicts the responses to various levels of hypoxia and at the three temperatures. Continuous lines represent light hypoxia, while broken lines represent an increased but moderate level of hypoxia. The right side shows the same responses transformed to relative values (see text).

There is, however, evidence that reduction of $[O_2]$a is one of various stimuli to gill ventilation in carp. To detect possible O_2-content receptors in carp, we reduced $[O_2]$a and $Hb-O_2$ saturation by application of CO. With this procedure the $[O_2]$a fell to 60% of the previous value. In turn, this evoked a significant increase of gill ventilation (Soncini and Glass, 2000). In this context, it should be pointed out that the stimulus modality is O_2 partial pressure in amphibians and lungfish (Wang et al., 1994; Branco and Glass 1995; Sanchez et al., 2001).

Nitrite in the water reduces $Hb-O_2$ content, due to methaemoglobin formation. Jensen et al. (1987) pointed out that adult carp quickly accumulated nitrite in the blood and, moreover, pointed out that blood values exceeded the ambient concentration. Thus, NO_2^- is taken up against a gradient, suggesting an active transport by competition with Cl^- uptake (Jensen, 1990).

Williams et al. (1992) exposed carp to hypoxia (75 mmHg) combined with application of nitrite, which caused formation of methaemoglobin and reduced arterial $Hb-O_2$ saturation from 83.2% (control) to 47.5% (24 h of exposure). This provoked hyperventilation, accompanied by decreases of $PaCO_2$ and elevated pHa. Meanwhile, PaO_2 that was constant throughout the exposure. This raised questions concerning the regulated variable and feedback mechanisms. Reduced $[O_2]$a could be a factor and external O_2 receptors might have responded to aquatic hypoxia. More information is, however, needed to sort out the rather complex questions concerning the stimuli to ventilation. As pointed out above, reduction of $Hb-O_2$ saturation by application of CO provoked hyperventilation. Likewise, methaemoglobin formation and reduced $Hb-O_2$ content might have stimulated gill ventilation.

Ventilatory Responses to pH and CO₂

Central pHa/CO_2-receptors are located in the brainstem of tetrapods (amphibians, reptiles, birds and mammals) (Milsom, 2002) and are also present in lungfish (Sanchez et al., 2001). Evidence of such receptors is arguable in relation to teleost fish (Hedrick et al., 1991). Recently, Wilson et al. (2000) found that the isolated brain stem of the long-nosed gar pike (*Lepisosteus osseus*) increased the frequency of respiratory motor output from the in vitro brainstem, equilibrated to low pH solutions. This certainly represents a major progress, but the questions remain concerning differences between in vitro and in vivo conditions after isolation.

A classic study on the tench (Hughes and Shelton, 1962) supported the existence of non-branchial receptors since the IV and X cranial nerves were cut to avoid receptor signals from the gill. In spite of this, the respiratory pattern continued at a lower frequency combined with larger volume for each gill movement. Even then, ventilation could be stimulated by hypercarbia. As pointed out by Milsom (2002), the post-denervation responses to hypercarbia are very species-specific and a general statement is not possible at this point.

When blood CO_2 levels increase, the Hb-O_2 dissociation curve becomes right-shifted, and this is known as the classical Bohr effect. Less known is the Root effect, which characterizes teleost fish. The Root effect causes a right-shift but, in addition, the O_2 capacity becomes reduced, which stimulates content-sensitive O_2 receptors, screening the blood. In this situation, the hypercarbia would stimulate ventilation even if CO_2/pH receptors were absent. This presents a difficulty for interpretation of the stimuli, underlying the responses. Thus, significant ventilatory responses to hypercarbia were obtained in carp, but only when $[O_2]$a became significantly reduced. Therefore, it was not possible to identify the specific stimulus (O_2 or pH/CO_2). Certainly, O_2-oriented respiratory drives have priority, while acid-base regulation is achieved largely by modulations of plasma bicarbonate levels (Heisler, 1984; Claiborne and Heisler, 1986).

GILL VENTILATION AND CARDIAC FUNCTION

Normoxic carp ventilates the gills in an intermittent manner (Hughes et al., 1983), which is synchronized with increased cardiac frequency (f_H) during ventilation (Peyroud and Serfaty, 1964). Consistently, Glass et al. (1991) measured a f_H of 49 beats $\cdot$ min^{-1} during ventilation, compared to 26 beats $\cdot$ min^{-1} during apnea. This cardiorespiratory synchrony persisted during hypoxia-induced ventilatory and cardiovascular responses. Reduction of PwO_2 to 5 mmHg increased the respiratory frequency to as much as 105 breaths $\cdot$ min^{-1} (7-fold the normoxic value). Meanwhile, the cardiac frequency increased slightly down to a PwO_2 of 40 mmHg, while even lower PwO_2 induced a severe bradycardia, accompanied by altered ventricular depolarization (Glass et al., 1991).

The ventricular muscle of teleosts, including the carp, consists of an inner spongy layer and an outer compact layer. Force of contraction is primarily developed by the latter (Temma et al., 1987; Chugun et al., 1999). In carp, the weight of the compact layer of the ventricular muscle accounts for 37% of total weight (Bass et al., 1973; Chugun et al., 1999), which correlates with its rather sedentary behavior. The development of the compact layer relates to the swimming activity and mode of life. Highly active teleosts have large compact layers as exemplified by the Black skipjack tuna (51% of the total) and Bigeye tuna (74% of the total).

Itazawa and Takeda (1978) reported ventilatory responses in hypoxia-exposed carp at 24.5°C. Their extensive series of variables permitted to apply conductance equations for respiratory variables. The equation for gill ventilation is:

$$VO_2 = Vg \cdot \alpha O_2 \cdot (P_I O_2 - P_E O_2),$$

in which the conductance is $Gv = Vg \cdot \alpha O_2$, where Vg is gill ventilation and αO_2 is the solubility of O_2 in water. A decrease of inspired PO_2 from 141 to 25 mmHg provoked an 11-fold increase of this conductance, whereas the conductance for perfusion increased 8-fold. Concomitantly, PaO_2 fell from 25 to 5.4 mmHg, while the Hb-O_2 saturation decreased from 73 to 53%. This suggests a P_{50} of about 5 mmHg, which is consistent with the literature (cf. Albers et al., 1983). Hypoxia-exposed teleost fish release catecholamines, that activate a Na^+/H^+ exchanger and, in turn, lead to an increased pHa and reduced NTP. Both effects increase Hb-O_2 affinity, which permits O_2 transport to take place at extremely lower PO_2 (Holk and Lykkeboe, 1995). Moreover, the activation of the Na^+/H^+ exchanger increases intracellular $[Na^+]$, which reduces $[ATP]$ via effects on oxidative phosphorylation (Ferguson and Boutilier, 1989; Nikinmaa, 1990). In this context, Jensen et al. (1990) reported a linear decrease of NTP/Hb with increases of pH. In particular, the ratio NTP/Hb decreased steeply with reduction of PO_2 from 20 to 5 mHg.

Taken together, the modulations of ventilatory and cardiovascular responses confirm the carp is a very efficient oxyregulator, when confronted with moderate or severe hypoxia. Hughes et al. (1983) emphasized that carp can endure a wide scope of environmental conditions, ranging from near-freezing water and severe hypoxia to during the winter to high temperatures in the summer and O_2-supersaturation of the water (Garey and Rahn, 1970). Thus, Takeda (1993) reported that increases of $P_I O_2$ from 130 to 313 mmHg and, finally, to 459 mmHg reduced f_R from 22.4 to 9.7 min^{-1} (extreme hyperoxia), which reduced gill ventilation from 258 to 96 $mL \cdot kg^{-1} \cdot min^{-1}$. Concomitantly, PaO_2 increased from 23 to 38 mmHg, while pHa fell by about 0.3 units. The limited increase of PaO_2 relates to a highly reduced ventilation and, possibly, ventilation-perfusion mismatching. These and other respiratory adjustments apparently protected from excessive levels of hyperoxia.

Recently, the cardiorespiratory responses of carp were studied under deep hypoxia at three temperatures (Stecyk and Farrell, 2002). The authors found a profound down regulation of cardiac function and gill ventilation. The time tolerated under these strenuous conditions was inversely related to temperature. There was a certain tendency to increase recover cardiovascular function during the exposure, which, once again, stresses high tolerance to hypoxic stress.

BLOOD GAS TRANSPORT AND ACID-BASE STATUS

Figure 12.2 compares the oxygen dissociation curves (ODC) of three teleost fish: the carp, the dourado (*Salminus maxillosus*) and traira (*Hoplias malabaricus*). We constructed $Hb\text{-}O_2$ dissociation curves based on the mixing method developed by Haab *et al.* (1960). In this procedure, blood withdrawn from catheters was equilibrated to identical PCO_2, using two tonometers. One received 30% (fully oxygenated blood), while the other was equilibrated with N_2 (anoxic blood). The desired saturation is then achieved by quantitative mixing of the sample, after which the PO_2 was measured by means of an O_2 electrode. The curve is then derived from the PO_2 values at various saturations.

The ODC of carp has frequently been reported (Lykkeboe and Weber, 1978; Albers *et al.*, 1983; Soncini and Glass, 2000). Consistently, the curve is hyperbolic rather than sigmoid. As an example, Albers *et al.* (1983) reported temperature effects on the ODC of the temperature acclimated carp with the following characteristics: (a) $10°C$: $P_{50} = 7.6$ mmHg, $n_{Hill} = 1.14$ and Bohr factor $- 0.97$. (b) $20°C$: $P_{50} = 9.4$ mmHg, $n_{Hill} = 1.32$ and Bohr factor $- 0.98$. At $25°C$, Soncini and Glass (2000) reported $n_{Hill} = 1.1$ and a Bohr factor of $- 0.81$. These values are representative, but an even lower P_{50} value (4.8 mmHg at pH $= 7.8$ and $15°C$) was reported by Ultsch *et al.* (1981).

In addition, carp has a root effect that reduces the O_2 content $[O_2]a$ of the sample (Ultsch *et al.*, 1981; Soncini and Glass, 2000). Thus, at constant temperature ($25°C$), a decrease of pH from 7.92 to 7.09 reduced $[O_2]a$ from 7.9 to 5.4 vol.%.

These characteristics of the ODC imply that carps can survive exposures to extremely low oxygen levels. On the other hand, the O_2

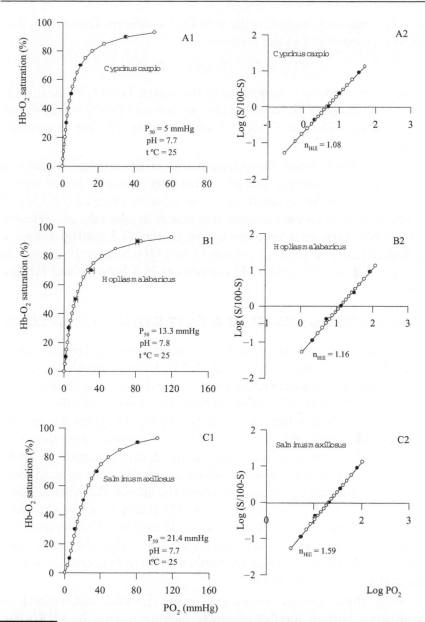

Fig. 12.2 Hemoglobin-oxygen binding curves for selected teleost fish. The composite figure is based on unpublished data for *Hoplias malabaricus*, while the curves for *Cyprinus* and *Salminus* have been published individually in a different form (Soncini and Glass, 2000; Salvo Souza *et al.*, 2001). The left side compares the ODCs. Notice the change of scale of the X-axis. The high-affinity curve of carp contrasts with the low-affinity position of the ODCs of more active species. The right side represents the corresponding Hill plots. Low n_{Hill} values characterize most teleosts, but the value for the active *Salminus* is rather high.

delivery to tissues is limited by the high O_2-Hb affinity. Thus the O_2 flux (VO_2) is proportional to the gradient:

$$VO_2 = Dt \cdot (PcapO_2 - PicO_2)$$

where Dt = diffusive conductance of the tissue; $PcapO_2$ = mean PO_2 during the passage of blood within the capillary and $PicO_2$ = PO_2 within the cell tissue. The low $PcapO_2$ clearly limits the O_2 flux to systemic tissues.

As explained above, the principal role of the respiratory drive is regulation of O_2 uptake, and gill ventilation primarily linked to O_2 transport. Teleost fish compensate hypercarbia by elevated [HCO_3^-] pl levels that depend on ion exchange that mainly involves the gills (Heisler 1984, 1986). Long-term exposure to hypercarbia (7.5 mmHg) caused a gradual increases of [HCO_3^-] pl from 13 to 25.9 mM after 19 days. This corresponded to 80% plasma pH compensation (Claiborne and Heisler, 1986).

COMPARISON OF CARP TO A FAST SWIMMING SPECIES

The dourado, *Salminus maxillosus*, is a fast swimming species of lotic waters of Southern Brazil (Salvo-Souza *et al.*, 2001). Unlike carp, this species is little resistant to ambient hypoxia. At $25°$ C, the Hb-O_2 dissociation curve of carp has a P_{50} (pH =7.7) of about 5 mmHg or less, while the P_{50} (pH = 7.7) of dourado is as high as 21 mmHg (See Fig. 12.2). For this reason, dourado fails to maintain adequate O_2 transport below a critical threshold PO_2 tension of 40 mmHg. Below this threshold, carp continues to obtain adequate supplies of O_2 down to about 20 mmHg (Ott *et al.*, 1980). Importantly, the curve positions determine the arterial PO_2-values. Thus, the *in vivo* PaO_2 of carp is about 15 mmHg (Glass *et al.*, 1990), while dourado maintains a PaO_2 of no less than 92 mmHg, which represents a six-fold difference (Salvo-Souza *et al.*, 2001). The PaO_2 of dourado is close but slightly lower than values for rainbow trout (*Oncorhynchus mykiss*) (Randall and Cameron, 1973).

As a third species, we include traira (*Hoplias malabaricus*), a carnivorous bottom dweller of slowly moving waters. Its ODCs are characterized by an intermediate position relative to values for carp and dourado. The Hill plots are linear for the three species, which indicates the presence of only one type of hemoglobin. The Hill plot involves the relationship: $n_{Hill} = \Delta log_{10} [SO_2 /(100\% - SO_2)]/\Delta log_{10} PO_2$. The higher the n_{Hill} value, the more S-shaped becomes the ODC. The n-values of the Hill plots are low in carp (Tan *et al.*, 1972) and in traira, which indicates a weak interaction between oxygen sites in the protein molecule. By contrast, a more sigmoid curve form in dourado indicates stronger interaction between the sites.

Acknowledgements

This research was supported by FAPESP (Fundação de Amparo à Pesquisa do Estado de São Paulo); Proc 98/06731-5, CNPq (Conselho Nacional de Desenvolvimento Científico o Tecnológico); Proc. 520769/93-7, FAEPA (Fundação de Apoio ao Ensino, Pesquisa e Assistência do Hospital das Clínicas da FMRP-USP).

References

Albers, C., R. Manz, D. Muster and G.M. Hughes. 1983. Effects of acclimation temperature on oxygen transport in the blood of the carp, *Cyprinus carpio*. *Respiration Physiology* 52: 165-179.

Bass, A., B. Ostadal, V. Pelouch and V. Vitek. 1973. Differences in weight parameters, myosin-ATPase activity and enzyme pattern of energy supplying metabolism between the compact and spongious cardiac musculature of carp (*Cyprinus carpio*) and turtle (*Testudo horsfieldi*). *Pflügers Archives* 343: 65-77.

Branco, L.G.S. and M.L. Glass. 1995. Ventilatory responses to carboxyhaemoglobinaemia and hypoxic hypoxia in *Bufo paracnemis*. *Journal of Experimental Biology* 198: 1417-1421.

Burggren, W.W. and D.J. Cameron. 1980. Anaerobic metabolism, gas exchange, and acid-base balance during hypoxia exposure in the channel catfish, *Ictalurus punctatus*. *Journal of Experimental Zoology* 213: 405-416.

Burleson, M.L. and W.K. Milsom. 1990. Propanol inhibits O_2-sensitive chemoreceptor activity in trout gills. *American Journal Physiology* 27: R1089-R1091.

Burleson, M.L. and W. K. Milsom. 1995a. Cardio-ventilatory control in rainbow trout: I. Pharmacology of branchial oxygen-sensitive chemoreceptors. *Respiration Physiology* 100: 231-238.

Burleson, M.L. and W. K. Milsom. 1995b. Cardio-ventilatory control in rainbow trout: II. Reflex effects of exogenous neurochemicals. *Respiration Physiology* 101: 289-299.

Chugun, A., T. Oyamada, K. Temma, Y. Hara and H. Kondo. 1999. Intracellular Ca^{2+} storage sites in the carp heart: comparison with the rat heart. *Comparative Biochemistry Physiology* A 123: 61–67.

Claiborne, J.B. and N. Heisler. 1984. Acid-base regulation and ion transfers in the carp (*Cyprinus carpio*): during and after exposure to environmental hypercapnia. *Journal of Experimental Biology* 108: 25-43.

Claiborne, J.B. and N. Heisler. 1986. Acid-base regulation and ion transfers in the carp (*Cyprinus carpio*): pH compensation during graded long and short-term environmental hypercapnia, and the effect of bicarbonate infusion. *Journal of Experimental Biology* 126: 41-61.

Dejours, P. 1981. *Principles of Comparative Respiratory Physiology.* 2nd edition. Elsevier/ North Holland/ Amsterdam.

Ferguson, R.A. and R.G. Boutilier. 1989. Metabolic membrane coupling in red blood cells of trout: The effects of anoxia and adrenergic stimulation. *Journal of Experimental Biology* 143: 149-164.

Fuchs, D.A. and C. Albers. 1988. Effects of adrenaline and blood gas condictions on red cell volume and intraerythrocytic electrolytes in the carp, *Cyprinus carpio. Journal of Experimental Biology* 137: 457-477.

Garey, W.F. 1967. Gas exchange, cardiac output, and blood pressure in free swimming carp (*Cyprinus carpio*). Ph.D. thesis, University of New York at Buffalo.

Garey, W.F. and H. Rahn. 1970. Gas tensions in tissues of trout and carp exposed to diurnal changes and oxygen tension of the water. *Journal of Experimental Biology* 52: 575-582.

Glass, M.L., N.A. Andersen, M. Kruhøffer, E.M. Williams and N. Heisler. 1990. Combined effects of environmental PO_2 and temperature on ventilation and blood gases in the carp *Cyprinus carpio* L. *Journal of Experimental Biology* 148: 1-17.

Glass, M.L., F.T. Rantin, R.M.M. Verzola, M.N. Fernandes and A.L. Kalinin. 1991. Cardio-respiratory synchronization and myocardial function in hypoxic carp. *Journal of Fish Biology* 39: 143-149.

Haab, P.E., J. Piiiper and H. Rahn. 1960. Simple method for rapid determination of an O_2 dissociation curve of the blood. *Journal of Applied Physiology* 15: 1148-1149.

Hedrick, M.S., D.R. Burleson, D.R. Jones and W.K. Milsom. 1991. An examination of central chemosensitivity in an air-breathing fish (*Amia calva*). *Journal of Experimental Biology* 155: 165-174.

Heisler, N. 1984. Acid-base regulation in fishes. In: *Fish Physiology*, W.S. Hoar and D.J. Randall (eds.). Academic Press, Orlando, Vol. 10A, pp. 315-401.

Heisler, N. 1986. Acid-base regulation in fishes. In: *Acid-base Regulation in Animals*, N. Heisler (ed.), Elsevier, Amsterdam, pp. 309-356.

Holk, K. and G. Lykkeboe. 1995. Catecholamine-induced changes in oxygen affinity of carp and trout blood. *Respiration Physiology* 100: 55-62.

Hughes, G.M. and G. Shelton. 1962. Respiratory mechanisms and their nervous control in fish. *Advances in Comparative Physiology and Biochemistry* 1: 275-364.

Hughes, G.M., C. Albers, D. Muster and K.H. Götz. 1983. Respiration of the carp, *Cyprinus carpio* L., at 10 and 20°C and the effects of hypoxia. *Journal of Fish Biology* 22: 613-628.

Itazawa, Y. and T. Takeda. 1978. Gas exchange in the carp gills in normoxic and hypoxic conditions. *Respiration Physiology* 35: 263-269.

Jensen, F.B. 1990. Nitrite red cell function in carp: control factors for nitrite entry, membrane potassium ion permeation, oxygen affinity and methaemoglobin formation. *Journal of Experimental Biology* 152: 149-166.

Jensen, F.B., N.A. Andersen and N. Heisler. 1987. Effects of nitrite exposure on blood respiratory properties, acid-base and electrolyte regulation in the carp (*Cyprinus carpio*). *Journal of Comparative Physiology* B157: 533-541.

Jensen, F.B., N.A. Andersen and N. Heisler. 1990. Interrelationships between red cell nucleoside triphosphate content, snad blood pH, O_2-tension and hemoglobin concentration in carp, Cyprinus carpio. Fish Physiology and Biochemistry 8: 459-464.

Kruhøffer, M., M.L. Glass, A.S. Abe and K. Johansen. 1987. Control of breathing in an amphibian Bufo paracnemis: Effects of temperature and hypoxia. Respiration Physiology 69: 267-275.

Lomholt, J.P. and K. Johansen. 1979. Hypoxia in carp — How it affects O_2 uptake, ventilation extraction from water. Physiological Zoology 52: 38-49.

Lykkeboe, G. and R.E. Weber. 1978. Changes in the respiratory properties of the blood in the carp, Cyprinus carpio, induced by diurnal variation in ambient oxygen tension. Journal of Comparative Physiology 128: 117-125.

Milsom, W.K. 2002. Phylogeny of CO_2/H^+ chemoreception in vertebrates. Respiration Physiology and Neurobiology 131: 29-41.

Milsom, W.K. and R.W. Brill. 1986. Oxygen-sensitive afferent information arising from the first gill arch of the yellowfin tuna. Respiration Physiology 66: 193-203.

Moffitt, B.P. and L.I. Crawshaw. 1983. Effects of acute temperature changes on metabolism, heart rate, and ventilation frequency in the carp Cyprinus carpio L. Physiological Zoology 56: 397-403.

Nikinmaa, M. 1990. Vertebrate Red Blood Cells. Adaptations of Function to Respiratory Requirements. Springer-Verlag, Berlin.

Ott, M.E., N. Heisler and G.E. Ultsch. 1980. Re-evaluation of the relationship between temperature and critical oxygen tension in fresh water fishes. Comparative Biochemistry and Physiology A67: 337-340.

Peyroud, C. and A. Serfaty. 1964. Le rythme respiratoire de la Carpe (Cyprinus carpio L.). Hydrobiologia 23: 165-178.

Rahn, H. 1966. Aquatic gas exchange: theory. Respiration Physiology 1: 1-12.

Randall, D.J. 1982. The control of respiration and circulation in fish during exercise and hypoxia. Journal of Experimental Biology 100: 275-288.

Randall, D.J. and J.N. Cameron. 1973. Respiratory control of arterial pH as temperature changes in rainbow trout Salmo gairdneri. American Journal of Physiology 225: 997-1002.

Salvo-Souza, R.H., R. Soncini, M.L. Glass, J.R. Sanches and F.T. Rantin. 2001. Ventilation, gill perfusion and blood gases in dourado, Salminus maxillosus Valenciennes (Teleostei, Characidae), exposed to graded hypoxia. Journal of Comparative Physiology B171: 483-489.

Sanchez, A., R. Soncini, T. Wang, P. Koldkjaer, E.W. Taylor and M.L. Glass 2001. The differential cardio-respiratory responses to ambient hypoxia and systemic hypoxaemia in the South American lungfish, Lepidosiren paradoxa. Comparative Biochemistry and Physiology A130: 677-687.

Saunders, R.L. 1962. The irrigation of gills in fishes. II. Efficiency of oxygen uptake in relation to respiratory flow, activity and concentrations of oxygen and carbon dioxide. Canadian Journal of Zoology 40: 817-862.

Smith, F.M. and D.R. Jones. 1982. The effect of changes in blood oxygen-carrying capacity on ventilation volume in the rainbow trout (Salmo gairdneri). Journal of Experimental Biology 97: 325-334.

Soncini, R. and M.L. Glass. 2000. Oxygen and acid-base related drives to gill ventilation in carp. *Journal of Fish Biology* 56: 528-541.

Stecyk, J.A.W. and A.P. Farrell. 2002. Cardiorespiratory responses of the common carp (*Cyprinus carpio*). *Journal of Experimental Biology* 205: 759-768.

Takeda, T. 1990. Ventilation, cardiac output and blood respiratory parameters in the carp, *Cyprinus carpio*. *Respiration Physiology* 81: 227-239.

Takeda, T. 1993. Effects of exercise-stress on ventilation, cardiac output and blood respiratory parameters in the carp, *Cyprinus carbio*. *Comparative Biochemistry Physiology* A106: 277-283.

Tan, A.L., A. Young and R.W. Noble. 1972. The pH dependence of the affinity, kinetics, and co-operativity of ligand binding to carp hemoglobin, *Cyprinus carpio*. *Journal of Biological Chemistry* 247: 2493- 2498.

Temma, K., H. Nagatomi, H. Hirano, T. Kitazawa and H. Kondo. 1987. Carp (*Cyprinus carpio*) heart has a high sensitivity to the positive ionotropic effect of strophanthidin despite negative force–frequency relationships. *General Pharmacology* 18: 617–622.

Ultsch, G.R., M.E. Ott and N. Heisler. 1981. Acid-base and electrolyte status in carp (*Cyprinus carpio*) exposed to low environmental pH. *Journal of Experimental Biology* 93: 65-80.

Wang, T., L.G.S. Branco and M.L. Glass. 1994. Ventilatory responses to hypoxia in the toad (*Bufo paracnemis* Lutz) before and after reduction of HbO_2 concentration. *Journal of Experimental Biology* 186: 1-8.

Williams, E.M., M.L. Glass and N. Heisler. 1992. Blood oxygen tension and content in carp, *Cyprinus carpio* L., during and methaemoglobinaemia. *Aquaculture and Fisheries Management* 23: 679-690.

Wilson, R.J., M.B. Harris, J.E. Remmers and S.F. Perry. 2000. Evolution of air-breathing and central CO_2/H^+ respiratory chemosensitivity: new insights from an old fish? *Journal of Experimental Biology* 203: 3505-3512.

Blood Gases of the South American Lungfish, *Lepidosiren paradoxa*: A Comparison to Other Air-breathing Fish and to Amphibians

Jalile Amin-Naves[1], A.P. Sanchez[1], M. Bassi[1], H. Giusti[1], F.T. Rantin[2] and M.L. Glass[1,*]

INTRODUCTION

Equipped with real lungs and reduced gills, the lungfish (Dipnoi) are a probable sister group relative to the land vertebrates (Tetrapoda) (Meyer and Dolven, 1992; Yokobori *et al.*, 1994; Tohyama *et al.*, 2000). The Australian lungfish, *Neoceratodus*, ventilates the lung at a very low frequency when in well-aerated water (Grigg, 1965; Johansen *et al.*, 1967; Fritsche *et al.*, 1993). In contrast, the lepidosirenid lungfish (*Lepidosiren*

Authors' addresses: [1]Department of Physiology, Faculty of Medicine of Ribeirão Preto, University of São Paulo, Avenida Bandeirantes 3900, 14049-900, Ribeirão Preto, SP, Brazil. [2]Laboratory of Zoophysiology and Cowparotive Biochemistry, Department of Physiological Sciences, Federal University of São Carlos, via Washington Luis, Km 235, 13565-905–São Carlos, SP, Brazil.
Corresponding author: E-mail: mlglass@rfi.fmrp.usp.br

parodoxa in South America and *Protopterus* sp. in Africa) are obligatory air-breathers (Johansen and Lenfant, 1967) with well-developed lungs and rudimentary gills. In *Lepidosiren*, the lung accounts for no less than 99% of the total morphological diffusing capacity, while the capacity of the gills is negligible (0.0013% of total), and the rest is contributed by the skin (de Moraes *et al.*, 2005). In spite of this distribution, the lung eliminates about 50% of the total CO_2 produced at 25°C, while the other half is eliminated to the water (Amin-Naves *et al.*, 2004). This reflects that aquatic gas exchange is important during prolonged dives.

Earlier studies on *Protopterus aethiopicus* established that this species has a low pHa, combined with high values for $PaCO_2$ and $[HCO_3^-]pl$ (Lenfant and Johansen 1968; DeLaney *et al.*, 1977). In the Australian lungfish *Neoceratodus forsteri* has $PaCO_2$ values that are only slightly higher than those for water breathing teleost fish (Lenfant *et al.*, 1966/1967). We hypothesized that *Lepidosiren* would resemble *Protopterus* as to set-points for acid-base status and PaO_2, since both genera are highly dependent on lung ventilation (Sawaya, 1946). According to Dejours (1981): "the more an animal depends on air breathing, the higher its PCO_2". We discuss this statement in relation to blood gas data for lungfish and other relevant groups of ectothermic vertebrates.

MATERIALS AND METHODS

Animals

Samples were obtained from 54 specimens of *L. paradoxa* (Fitzinger) (Weight: 300 to 800 g). The animals were collected close to the city of Cuiabá, Mato Grosso State, and transported to the Faculty of Medicine, University of São Paulo, Ribeirão Preto, and São Paulo State. They were kept in 1000-L tanks, containing dechlorinated water at 25°C. Animals were fed weekly on chicken liver, but food was withheld 48 h before experimentation. All blood gas samples were obtained during the non-aestivating period of the animals (October to the end of April). The animals represent 3 periods of collection and subsequent measurements: Group A (n=14): October 15, 2000 to April 01, 2001; Group B (n=15): February 01, 2002 to May 01, 2002 (n=15); Group C: October 01 to December 01, 2003 and March 01, 2004 to May 01, 2004 (n=24). The measurements comprise all our data, including pilot experiments.

Surgical procedures

Fish were immersed into a benzocaine solution (1 $g \cdot L^{-1}$) and became anesthetized in less than 10 min. During surgery, fish were maintained anesthetized with a weaker benzocaine solution (0.25 $g \cdot L^{-1}$). An incision was made to access the caudal part of the dorsal aorta, located about 1 cm behind the lung. Then, the vessel was separated from surrounding tissue and a PE50 catheter was inserted into the lumen of the vessel. Subsequently, the catheter was fixed to the vessel and adjacent tissue, exteriorized and secured to the sutures, which closed the incision. The animals recovered quickly, when placed into benzocaine free aerated water. Catheters were checked and flushed with heparinized Ringer solution 1 h before measurements (100 $IU \cdot ml^{-1}$).

Blood gas analysis

Blood samples (1 ml) were withdrawn without any visual or acoustic disturbance to the animal and always analyzed immediately. The electrodes were kept at the temperature of the animal (25°C), and PO_2 was measured using a FAC 001 O_2 electrode coupled to a 204 O_2 analyzer (FAC Instr., São Carlos, SP, Brazil), while PCO_2 and pH were measured using a Cameron Instr. CO_2 electrode and a Mettler Toledo (Mettler, Switzerland) pH microelectrode. The O_2 electrode was calibrated with N_2 for zero and with air-equilibrated water (PO_2 = 145 mmHg – elevation 600 m; P_B ~715 mmHg). The blood-gas sensitivity ratio was taken into account as described by Siggaard-Andersen (1974). Moreover, the reliability of the readings was checked by equilibration of blood samples to air (Dual equilibrater, MEQ1, Cameron Instr.). The red cells of *Lepidosiren* are large and nucleated, which leads to a slow linear decline due to O_2 consumption of the cells. This O_2 consumption was corrected for by linear regression to zero time.

The CO_2 electrodes was calibrated using a GF3/MP Gas Flow Meter (Cameron Instr., Port Arkansas, Texas, USA), providing 1% or, alternatively, 6% CO_2 at a flow rate of 100 $ml \cdot min^{-1}$, while the pH meter was calibrated using high precision buffer solutions (Queel, São Paulo, SP, Brazil). The CO_2 and the pH-electrodes were coupled to a BGM 200 Cameron Analyzer. Total plasma CO_2 was obtained by means of a Capni-Con 5 Analyzer, calibrated with standard bicarbonate solutions, and $[HCO_3^-]pl$ was obtained as: $[HCO_3^-]pl = [CO_2]tot - \alpha CO_2 \cdot PaCO_2$, where the value for αCO_2 was derived from Reeves (1976). Hematocrit was measured by sampling into microtubes that were centrifuged using a HERMLE Z.200 M/H.

Statistics

Data are presented as mean values ± SEM. One way ANOVA was applied to detect differences between the three groups. This was followed by Bartlett's normality test that was followed by Bonferroni's test in case of normal distribution.

Results and Discussion

Calculated for all animals (n = 54), the acid-base status of *Lepidosiren* at 25° C was characterized by a low arterial pH (7.51), a high $PaCO_2$ (21.6 mmHg), and a high $[HCO_3^-]$pl level (23.6 mM), and a high PaO_2 (80.9 mmHg). See Table 13.1. The resulting pK_1' value was 6.079 (αCO_2 derived from Reeves, 1972). As stated above, data on blood gases were obtained during the non-aestivating period of the year (October to May), and the blood gases were consistent with one exception. The animals of group A had a significantly lower $PaCO_2$ than in groups B and C (P< 0.05). Group A was studied around the turn of the year, which corresponds to the hot and rainy season, during which rivers expand. Apart from this, the blood gases were not significantly different for the groups. This is not trivial, since some teleost fish may exhibit large variations. As an example, PaO_2 and pHa in carp, *Cyprinus carpio*, may range from 44.4 mmHg with pHa = 7.87, depending on the specific group studied (Claiborne and Heisler, 1986) to PaO_2 = 14.4 mmHg and pHa = 8.01) (Glass *et al.*, 1990).

Comparison to other studies on Lepidosiren

Johansen and Lenfant (1967) measured blood gases in small specimens of *Lepidosiren* (104 to 212 g) with traces of external gills. They obtained a mean PaO_2 value of 31.5 mmHg which, according to their data, would saturate only 80% of the hemoglobin (Johansen and Lenfant, 1967). Our specimens (500-700 g) had a PaO_2 of 81 mmHg, which would saturate as much as 90-95% (Bassi *et al.*, in press). These differences as to PaO_2 may reflect a size dependence of the gas exchange surfaces (Johansen *et al.*, 1974). As a less attractive alternative, the small animals might have been inadequate for implantation of catheters and subsequent blood sampling.

Table 13.1 The results for three separate groups and for all the animals of this study (n = 54). Below can be seen the values from other blood gas studies on lungfish. Mean values ± SEM.

Animal	Body weight (kg)	Temp. (°C)	Vessel	n-value	$PaCO_2$ (mmHg)	pHa	$[HCO_3^-]pl$ (mM)	PaO_2 (mmHg)	Hct (%)
Lepidosiren (1a)	0.4-0.6	25	Dorsal aorta	14	16.9 ± 1.5	7.46 ± 0.05	24.2 ± 1.5	72.5 ± 12.6	22.7 ± 2.3
(1b)	0.4-0.8	25	-----	15	22.2 ± 2.3	7.53 ± 0.05	21.5 ± 3.1	80.5 ± 5.6	24.2 ± 4.2
(1c)	0.3-0.8	25	-----	25	24.3 ± 4.4	7.54 ± 0.08	26.1 ± 0.8	82.1 ± 8.8	19.8 ± 2.9
Lepidosiren (2)	0.8-2.0	25	Bulbus (via the fifth gill arch)	9	------	7.38 ± 0.04	------	62	36
Lepidosiren (3)	0.10-0.21	23	Dorsal aorta	8	------	------	------	31.6 ± 2.8	-----
Protopterus aaethiopicus (4)	2-12	25	Dorsal aorta	10	26.4 ± 0.4	7.60 ± 0.01	31.4 ± 0.4	------	-----
Protopterus aaethiopicus (4)	Not stated	20	Dorsal aorta	4	25.7 ± 1.6	------	------	27.0 ± 3.4	25
Protopterus dolloi (5)	0.174 ± 0.16	25	Dorsal aorta	16	34.8 ± 3.2	7.48 ± 0.04	------	49.2 ± 4.6	15 ± 2.4
Neoceratodus forteri (6)	4.2-8.8	18	Dorsal aorta	3 to 13	3.6 ± 0.1 n = 9	7.64 ± .02 n = 3	------	38.9 ± 1.2 n = 13	31 ± 1.2

(1) The present study, including three separate groups. The overall mean is presented below. (2) Sanchez *et al.* (2001), (3) Johansen and Lenfant (1967), (4) Delaney *et al.* (1977), (5) Lenfant and Johansen (1968) and (6) Lenfant *et al.* (1966/1967).

Large specimens (0.8 to 2.0 kg) of *Lepidosiren* were studied by Sanchez *et al.* (2001), while blood was taken from a catheter inserted into the bulbus. As expected, the values for pH and PO_2 were lower than for the arterial blood, which was consistent with higher PO_2-values in larger animals (Table 13.1).

Comparison with other lungfish

The blood gases of *Lepidosiren* were close to those of *Protopterus aethiopicus* (DeLaney *et al.*, 1977), although pHa and {HCO_3^-]pl are somewhat higher in the latter (7.602 and 31.4 mM). Our values for *Lepidosiren* are in better agreement with data *P. dolloi* (pHa = 7.48; $PaCO_2$ = 34.8; PaO_2 = 49 mmHg). Table 13.1. Moreover, the pK'_1 in *Lepidosiren* was close to the values for non-aestivating *Protopterus aethiopicus* (6.142 at 25°C and pH = 7.6). For *Protopterus*, Lenfant and Johansen (1968) reported a $PaCO_2$ of 27 mmHg, which is very reasonable. This can not be said about the corresponding PaO_2 of 25.7 mmHg (Table 13.1). With this PO_2, Hb-O_2 saturation would be as low as 65%, which seems unlikely.

Neoceratodus has a strikingly low $PaCO_2$, which beyond any doubt reflects its crucial dependence on aquatic respiration (Lenfant *et al.*, 1966/67). In this particular case, the low dependence on aerial respiration reduces $PaCO_2$ (cf. Dejours, 1981).

Comparison to other vertebrates

Table 13.2 compares the ratio (pulmonary CO_2-elimination/total CO_2 elimination · 100%) to arterial blood gases and pHa of selected teleosts, amphibians and lungfish at 25°C (for further values, see Ultsch 1996; Ultsch and Jackson, 1996). In this context, it becomes clear that the air-breathing bowfin (*Amia calva*) and the snakehead fish (*Channa argus*) have high $PaCO_2$-values relative to the exclusively water breathing carp. In this case, the presence of aerial respiration certainly increased $PaCO_2$. The data for *Cryptobranchus* are also consistent with the predictions stated by Dejours (1981) (See Introduction). The pulmonary CO_2-output accounts for a tiny fraction of total CO_2-elimination, this fraction is sufficient to reach a $PaCO_2$ of 5.7 mmHg, a value that falls between falls between those for carp and *Channa*.

There are some complications for the use of the principle: (1) The higher the activity level and/ temperature of the animal, the larger the role

Table 13.2 Data for selected species, including blood gases and the ratio: (CO_2 eliminated by the lung/total CO_2 elimination) · 100%.

Species	$V_LCO_2/V_{tot}CO_2$ (%)	$PaCO_2$ (mmHg)	pHa	$[HCO_3^-]pl$ (mM)	PaO_2 (mmHg)	Reference
Teleost: Carp (Cyprinus carpio)	0	1.7	8.08	6.5	31	1
Urodele: Hellbender (Cryptobranchus alleganiensis)	2	5.7	7.78	9.3	27.4	2
Teleost: Snakehead fish Channa argus	13	15.1	7.52	15.7	26	3
Anuran amphibian: (Xenophus laevis)	14	17.5	7.73	—	—	4
Anuran amphibian: (Bufo paracnemis)	35	7.7	7.75	13.7	61	5
Air-breathing teleost: Bowfin (Amia calva)	39	5.5	7.3	8.4	—	6
Lungfish: Pirambóia (Lepidosiren paradoxa)	50*	21.0 ± 1.3	7.51± 0.06	23.6 ± 2.1	80.9 ± 8.3	7
Anuran amphibian: Bullfrog (Rana catesbeiana)	59	9.6	7.96	28.2	—	8

1. Soncini and Glass (2000), 2. Boutilier and Toews (1981), 3. Ishimatsu and Itazawa (1983), 4. Boutilier (1984), Emilio and Shelton (1980), 5. Wang et al. (1998), 6. Randall et al. (1981), and 7. This study, including 54 animals – mean values ± SEM. *Based on Amin-Naves et al. (2004) and 8. Gottlieb and Jackson (1976).

of the lung for CO_2 elimination (2) A certain amount of CO_2 can as large gas volume of gas (low CO_2 concentration) or as a small gas volume (high CO_2 concentration). This will be explained in relation to blood gases and lungfish.

Activity and temperature

Cutaneous gas exchange is mainly diffusion dependent and K_{CO_2} (the Krogh diffusion constant for CO_2) increases by only 10%/10°C (Jackson, 1978). Due to this limitation, a larger fraction of total CO_2 output must be eliminated by the lung, whenever metabolism increases. Such increases can occur due to higher activity level (Jackson, 1978; Wang et al., 1998).

This also applies to Lepidosiren, since its gills are highly reduced (de Moraes et al., 2005). At 15°C, the lung of Lepidosiren eliminated as little as 14% of the total CO_2 produced. An increase to 35°C completely changed this picture, since as much as 73% was eliminated via the lung (Amin-Naves et al., 2004). Likewise, elevated metabolic demands increase the role of the lung for O_2 uptake and CO_2-elimination, which again reflects the limited capacity of the skin (Gottlieb and Jackson, 1976). Therefore, both activity level and temperature must be taken into consideration

High or low ventilation level in relation to CO_2 output

Comparing Rana catesbeiana, to Lepidosiren it turns out that $PaCO_2$ of Lepidosiren is about twice the value for Rana catesbeiana. This result was not expected from the principle of Dejours (1981), in particular because R. catesbeiana eliminates 59% of total CO_2 output by the lung, while the lungfish eliminates 50% (Table 13.2).

Values for effective ventilation of the lung are available for the toad B. schneideri (Wang et al., 1998) and for Lepidosiren. The effective lung ventilation of Bufo paracnemis was as high as 22.6 ml BTPS·kg^{-1}·min^{-1} (25-27°C), while the value for Lepidosiren was as low as 5.2 ml BTPS·kg^{-1}·min^{-1} (Bassi et al., 2005). In part, the high $PaCO_2$ levels in Lepidosiren reflect that this animal dives for prolonged periods, which leads to a high O_2 extraction from the lung and to high blood PCO_2-values (Bassi et al., 2005). In contrast, Bufo has a much higher effective ventilation, which lowers $PaCO_2$ and O_2-extraction from the lung (Wang et al., 1998). Obviously, if the same amount of CO_2 is distributed within a large volume the PCO_2 will be low as seen in Rana and Bufo. If the same amount of CO_2 is distributed within a small volume, then PCO_2 will be high, which is the case for *Lepidosiren*.

Clearly, this reflects mode of life and it is not to be excluded that distinct strategies and evolutionary trends underlie characteristics such as to pHa/PCO_2-values and bicarbonate levels (Bassi *et al.*, 2005). In spite of the limitations discussed above, the statement by Dejours (1981) is highly valuable as an overall rule for respiratory control in vertebrates. Fish inhabit an ever-changing O_2 availability and their respiratory control is O_2-oriented, whereas land vertebrates developed an acid-base oriented respiratory control.

Acknowledgements

This research was supported by FAPESP (Fundação de Ampora á Pesquisa do Estado de São Paulo); Proc 98/06731-5, CNPq (Conselho Nacional de Desenvolvimento Científico o Tecnológico); Proc. 520769/93-7, FAEPA (Fundação de Apoio ao Ensino, Pesquisa e Assistência do Hospital das Clínicas da FMRP-USP).

References

Amin-Naves, J., H. Giusti and M.L. Glass. 2004. Effects of acute temperature changes on aerial and aquatic gas exchange, pulmonary ventilation and blood gas status in the South American lungfish, *Lepidosiren paradoxa* (Fitzinger). *Comparative Biochemistry and Physiology* A 138: 133-139.

Bassi, M., W. Klein, M.N. Fernandes, S.F. Perry and M.L. Glass. 2005. Pulmonary diffusing capacity of the South American lungfish *Lepidosiren paradoxa*: Physiological values by the Bohr method. *Physiological and Biochemical Zoology* 78: 560-569.

Boutilier, R.G. and D.P. Toews. 1981. Respiratory, circulatory and acid-base adjustments to hypercapnia in a strictly aquatic and predominantly skin-breathing urodele, *Cryptobranchus alleganiensis*. *Respiration Physiology* 46: 177-192.

Boutilier, R.G. 1984. Characterization of the intermittent breathing pattern in *Xenophus laevis*. *Journal of Experimental Biology* 110: 291-309.

Claiborne, J.B. and N. Heisler. 1986. Acid-base regulation and ion transfers in the carp (*Cyprinus carpio*): pH compensation during graded long- and short-term environmental hipercarbia, and the effect of bicarbonate infusion. *Journal of Experimental Biology* 126: 1-17.

Dejours, P. 1981. *Principles of Comparative Respiratory Physiology*. Elsevier/North Holland/ Amsterdam. 2nd edition.

DeLaney, R.G., S. Lahiri, R. Hamilton and A.P. Fishman. 1977. Acid-base balance and plasma composition in the aestivating lungfish (*Protopterus*). *American Journal of Physiology* 232: R10-R17.

de Moraes, M.F.P.G., S. Höller, O.T.E. da Costa, M.L. Glass, M.N. Fernandes and S.F. Perry. 2005. Morphological diffusing capacity of the South American lungfish, *Lepidosiren paradoxa*. *Physiological and Biochemical Zoology* 78: 560-569.

Emilio, M.G. and G. Shelton. 1980. Carbon dioxide exchange and its effects on pH and the bicarbonate equilibria in the blood of an amphibian, *Xenophus laevis*. *Journal of Experimental Biology* 85: 253-262.

Fritsche, R., M. Axelsson, C.E. Franklin, G.C. Grigg, S. Holmgren and S. Nilsson. 1993. Respiratory and cardiovascular responses to hypoxia in the Australian lungfish. *Respiration Physiology* 94: 173-187.

Glass, M.L. and R. Soncini. 1995. Regulation of acid-base status in ectothermic vertebrates: the consequences for oxygen pressures in lung gas and arterial blood. *Brazilian Journal of Medical and Biological Research* 28: 1161-1167.

Glass, M.L., N.A. Andersen, M. Kruhøffer, E.M. Williams and N. Heisler. 1990. Combined effects on ventilation and blood gases in the carp *Cyprinus carpio* L. *Journal of Experimental Biology* 148: 1-17.

Gottllieb, G. and D.C. Jackson. 1976. Importance of pulmonary ventilation in respiratory control in the bullfrog. *American Journal of Physiology* 230: R608-R613.

Grigg, G.C. 1965. Studies on the Queensland lungfish, *Neoceratodus forsteri* (Krefft): III. Aerial respiration in relation to habits. *Australian Journal of Zoology* 13: 413-421.

Ishimatsu, A. and Y. Itazawa. 1983. Blood oxygen levels and acid-base status following air exposure in an air-breathing fish, *Channa argus*: the role of air ventilation. *Comparative Biochemistry and Physiolology* A 74: 787-793.

Jackson, D.C. 1978. Respiratory control and CO_2 conductance: temperature effects in a turtle and a frog. *Respiration Physiology* 33: 103-114.

Johansen, K. and C. Lenfant. 1967. Respiratory function in the South American lungfish, *Lepidosiren paradoxa* (Fitz.). *Journal of Experimental Biology* 46: 205-218.

Johansen, K., C. Lenfant and G.C. Grigg. 1967. Respiratory control in the lungfish *Neoceratodus forsteri* (Krefft). *Comparative Biochemistry and Physiology* 20: 835-854.

Johansen, K., J.P. Lomholt and G.M.O. Maloiy. 1976. Importance of air and water breathing in relation to size of the African lungfish (*Protopterus amphibius* Peters). *Journal of Experimental Biology* 65: 395-399.

Lenfant, C. and K. Johansen. 1968. Respiration in the African lungfish *Protopterus aethiopicus*. I. Respiratory properties of the blood and normal patterns of breathing and gas exchange. *Journal of Experimental Biology* 49: 137-152.

Lenfant, C., K. Johansen and G. C. Grigg. 1966/67. Respiratory properties of blood and pattern of gas exchange in the lungfish *Neoceratodus forsteri* (Krefft). *Respiration Physiology* 2: 1-31.

Meyer, A. and S.I. Dolven. 1992. Molecules, fossils, and the origin of tetrapods. *Journal of Molecular Evolution* 3: 102-113.

Perry, S.F., K.M. Gilmour, B. Vulesevic, B. McNeil, S.F. Chew and Y.K. Ip. 2005. Circulating catecholamines and cardiorespiratory responses in the hypoxic lungfish (*Protopterus dolloi*): a comparison of aquatic and aerial hypoxia. *Physiological and Biochemical Zoology* 78: 325-334.

Randall, D.J., J.N. Cameron, C. Daxboeck and N. Smatresk. 1981 Aspects of bimodal gas exchange in the bowfin *Amia calva* L. (Actinopterygii; Amiiformes). *Respiration Physiology* 43: 339-328.

Reeves, R.B. 1976. Temperature induced changes in blood acid-base status: pH and PCO_2 in a binary buffer. *Journal of Applied Physiology* 40: 732-761.

Sanchez, A.P., R. Soncini, T. Wang, P. Koldkjær, E.W. Taylor and M.L. Glass. 2001. The differential cardio-respiratory responses to ambient hypoxia and systemic hypoxaemia in the South American lungfish, *Lepidosiren paradoxa. Comparative Biochemistry and Physiology* A30: 677-687.

Sawaya, P. 1946. Sobre a biologia de alguns peixes de respiração aérea (*Lepidosiren paradoxa* Fitzinger e *Arapaima gigas* Cuvier). *Boletim da Faculdade de Filosofia Ciências e Letras da. Universidade de São Paulo* 11: 255-286.

Siggaard-Andersen, O. 1974. *The Acid-base Status of the Blood.* Munksgaard, Copenhagen.

Soncini, R. and M.L. Glass. 2000. Oxygen and acid-base status related drives to gill ventilation in carp. *Journal of Fish Biology* 56: 528-541.

Tohyama, Y., T. Ichimiya, H. Kasama-Yoshida, Y. Cao, M. Hasegawa, H. Kojima, Y. Tamai and T. Kurihara. 2000. Phylogenetic relation of lungfish indicated by amino-acid sequence of myelin DM20. *Molecular Brain Research* 80: 256-259.

Ultsch, G.R 1996. Gas exchange, hypercarbia and acid-base balance, paleoecology, and the evolutionary transition from water breathing to air-breathing among vertebrates. *Palaeogeography, Palaeoclimatology, Palaeoecology* 123: 1-27.

Ultsch, G.R. and D.C. Jackson. 1996. pH and temperature in ectothermic vertebrates. *Bulletin of the Alabama Museum of Natural History* 18: 1-41.

Wang, T., A.S. Abe and M.L. Glass. 1998. Temperature effects on lung and blood gases in *Bufo paracnemis:* Consequences of bimodal gas exchange. *Respiration Physiology* 113: 231-238.

Yokobori, S., M.M. Hasegawa, T. Ueda, N. Okada, K. Nishikawa and K. Watanabe. 1994. Relationship among coelacanths, lungfishes, and tetrapods: a phylogenetic analysis based on mitochondrial cytochrome oxidase I gene sequences. *Journal of Molecular Evolution* 38: 602-660.

14

Transition from Water to Land in an Extant Group of Fishes: Air Breathing and the Acquisition Sequence of Adaptations for Amphibious Life in Oxudercine Gobies

Jeffrey B. Graham*, Heather J. Lee and Nicholas C. Wegner

INTRODUCTION

This paper reports laboratory studies on the physiology and morphology of fishes in the gobiid subfamily Oxudercinae, which contains the four amphibious mudskipper genera (*Scartelaos*, *Boleophthalmus*, *Periophthalmus*, and *Periophthalmodon*) and six other genera that are either less- or entirely non-amphibious (Fig. 14.1). Oxudercines are unique

Authors' address: Center for Marine Biotechnology and Biomedicine and Marine Biology Research Division, Scripps Institution of Oceanography, University of California, San Diego, La Jolla, CA 92093-0204, USA.
Corresponding author: E-mail: jgraham@ucsd.edu

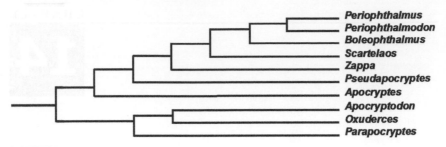

Fig. 14.1 Oxudercine cladogram. Modified from Murdy (1989).

among the vertebrates in having an assemblage of species extending along a phylogenetic continuum from aquatic to semi-terrestrial life and are unique among all fishes for the detail they provide regarding the sequence of physiological, morphological, biochemical, sensory, and behavioral character changes that occurred over the course of this transition. In this respect the radiation of the oxudericine fishes encompasses many aspects of the selection processes leading to the Paleozoic origins of both vertebrate air breathing and terrestriality (Graham and Lee, 2004).

Murdy (1989) revised the taxonomy of the Oxudercinae and built a cladogram based on 39 morphological and behavioral characters that define 10 genera and 35 species. Five of the 39 characters distinguish the Oxudercinae from other gobies (Hoese, 1984); the remaining 34 delineate the various taxa. The oxudercines include the four genera of the amphibious, air-breathing mudskipper clade, which share characters 24-26 (including, daily terrestrial excursions, lack of a membrane connection between dorsal fins 1 and 2, and the presence of a dermal cup for moistening the eyes in air [Murdy, 1989]). There are 25 mudskippers: *Scartelaos* (4 spp.), *Boleophthalmus* (5 spp.), *Periophthalmus* (13 spp. [including the recently described *Ps. spilotus* (Murdy and Takita, 1999)]), and *Periophthalmodon* (3 spp.). *Periophthalmodon* is the most terrestrial, followed by *Periophthalmus*, *Boleophthalmus*, and finally, the least amphibious, *Scartelaos* (Clayton, 1993; Kok *et al.*, 1998; Ishimatsu *et al.*, 1999; Zhang *et al.*, 2000, 2003). Some of Murdy's (1989) 39 characters were based on specializations for amphibious life. Our study quantifies additional features related to the transition from aquatic to amphibious life among the oxudercines, including gill morphology and surface area, the volume of the buccopharyngeal chamber (which, with its extensive epithelium, is the principal mudskipper air-breathing organ, ABO), the capacity of mudskippers to respire aerially and to elevate their post-exercise O_2 consumption rate (VO_2) in water and in air, and eye-lens shape.

Pioneering work by Schöttle (1931) demonstrated the correlation between a reduced gill area and dependence upon air breathing for mudskippers. In subsequent years reports on gill structure and estimates of total gill area have been made for *Boleophthalmus*, *Periophthalmus*, and *Periophthalmodon*, (Low *et al.*, 1988, 1990; Graham, 1997; Wilson *et al.*, 1999). However, gill-area measurements have not been made for *Scartelaos* and this and other comparative gill data are reported in this chapter. Recent works on *Periophthalmodon* demonstrated its large buccal chamber volume (Aguilar *et al.*, 2000) and showed that its high dependence on aerial respiration makes it an obligatory air breather that, without air access, cannot maintain its stasis blood-gas and acid-base parameters (Kok *et al.*, 1998; Ishimatsu *et al.*, 1999) or repay a post-exercise O_2 debt (Takeda *et al.*, 1999). Recent works have also amplified Schöttle's (1931) first account of the unique structure of the gill filaments of *Periophthalmodon* (Low *et al.*, 1988, 1990; Wilson *et al.*, 1999, 2000; Chew *et al.*, 2003; Randall *et al.*, 2004). We have obtained comparative data on these points for other oxudercines and review recent findings relating to the comparative respiratory physiology of the mudskippers. Studies of oxudercine lens shape tested the idea, first proposed by Karsten (1923), that the round lens of the fish eye had become oval in mudskippers to compensate for the added refractive index of the cornea in air, thereby enhancing aerial visual acuity. We compared a series of oxudercines to determine whether there is a progression in lens flatness correlating with terrestriality. A final objective of this paper is to review the recent literature dealing with oxudercine skin structure and cutaneous respiration, and burrowing.

MATERIALS AND METHODS

Specimens

Mudskippers and other oxudercines were collected by hand net in mangrove swamp habitats near Penang, Malaysia, in Cardwell and Townsville, Australia, and in Saga City, Japan. Prior to their shipment by air to Scripps Institution of Oceanography (SIO), La Jolla, CA, these specimens were maintained and studied at the School of Biological Sciences, Universiti Sains, Malaysia, the Australian Institute of Marine Science, Cape Ferguson, Queensland, Australia, and the Institute for East

China Sea Research, Nagasaki University, Tairamachi, Nagasaki, Japan. Fishes transferred to SIO were maintained in seawater (25 ppt salinity, at 25°C) aquaria and fed bloodworms or *Spirulina* algal flakes to satiation three times per week. Laboratory studies with living oxudercines were carried out in Japan (*Boleophthalmus pectinirostris, Periophthalmus modestus*), Australia (*Scartelaos histophorus*), and SIO (*Pseudapocryptes lanceolatus, S. histophorus, Ps. argentilineatus, Ps. modestus, Ps. minutus*).

Gill Structure and Surface Area

Scanning electron microscopy (SEM) was used to compare the gill structures of *Pseudapocryptes lanceolatus, Scartelaos histophorus, Boleophthalmus boddarti, Periophthalmus modestus, Ps. chrysospilos, Ps. minutus*, and *Periophthalmodon schlosseri*. SEM was also used to determine the gill dimensions and total gill surface areas for two preserved specimens of *S. histophorus* of known length and mass that had been stored in 50-70% ethyl alcohol. Gills from one side of the body were removed and prepared for high-vacuum mode SEM by critical point drying and sputter coating with gold-palladium. Standardized procedures (Gray, 1954; Hughes, 1966, 1984) were used to estimate the total gill surface area for *S. histophorus* using the equation:

$$A = N \cdot F \cdot Ln \cdot La \qquad (1)$$

where A is the total gill surface area, N is the total number of gill filaments, F is the average filament length, Ln is the mean number of lamellae per mm length on both sides of a filament, and La is the mean bilateral lamellar surface area (Hughes, 1966, 1984; Fernandes *et al.*, 1994; Graham, 1997). SEM images were used to count the total number of filaments on one side of each specimen, which was then doubled to determine N. NIH Image J software was used to view SEM gill images and measure filament lengths, lamellar abundance per length filament, and lamellar surface areas. L was estimated by measuring the length of every tenth filament and determining the mean. Ln was determined by counting the number of lamellae per unit length of filament. The surface areas of lamellae positioned at the base, middle, and tip of different filaments were then measured, averaged, and doubled to account for the dual surface. (Note: Conventional gill structure terminology has not been consistent. This report uses the terms filament and lamellae. Until recently, the terms primary lamellae [= filament] and secondary lamellae [= lamellae] were more commonly used.)

Buccal-chamber Volume

Measurements of buccopharyngeal chamber volume were made by placing specimens of *Scartelaos*, *Boleophthalmus pectinirostris*, and *Periophthalmus* in a clear plastic tube (length 20.5 cm, diameter 2 cm) covered by netting on one end and sealed with a rubber stopper on the other. A hypodermic needle (20 ga) was pushed through the stopper to the point where it just penetrated the inner surface with the needle's beveled tip emerging at the edge of the stopper's inner side and against the tube wall. The hub of the needle was removed so that PE 50 tubing could be attached to the shaft. This tubing was connected at its other end to the blunt tip of another 20-ga needle that was attached to a 3-cc syringe. A fish was placed in the tube, which was then submerged in a tank of filtered seawater (25 ppt at 25°C) that was made hypoxic by bubbling with N_2. The water entered the tube through the netted end, displacing all but a small volume of air. After being in the tube for a few minutes, the fish, in response to hypoxia, ceased water ventilation and gulped air. After the fish took and held an air breath, the excess air in the tube was gently aspirated through the needle and syringe so that the only air present was the volume contained within the fish. The syringe, PE tubing, and needle were then refilled with water from the tank in order to clear all residual air in preparation for the release of the air breath by the fish. When the fish exhaled its air-breath, the entire breath volume was quickly withdrawn into the syringe and measured before the fish could re-gulp it. Fresh air was then added to the tube and the volume estimate was repeated five to seven times for each species.

Eye-lens Axis Ratios

After body mass and total length measurement, the left and right eyes lenses were removed from preserved specimens of *Pseudapocryptes lanceolatus*, *Scartelaos histophorus*, *Boleophthalmus boddarti*, *Periophthalmus argentilineatus*, and *Periophthalmodon schlosseri*. Lens comparisons were also done on two other gobies (the frillfin, *Bathygobius ramosus*, and the long jaw mudsucker, *Gillichthys mirabilis*) and the flying fish, *Exocoetus monocirrhus*, which may have some capacity for aerial vision. (These specimens were loaned to us for study by the SIO Marine Vertebrates Collection.)

Cuts through the secondary spectacle and cornea exposed the lens, which, while still in place, was marked on its corneal and dorsal sides to preserve knowledge of lens orientation in the eye. The lenses were then removed, set into the well of a glass dissecting slide, and photographed with a Polaroid digital microscope camera (Model DMC 1) mounted on a Zeiss binocular dissecting microscope that had been calibrated with an ocular micrometer. Using NIH Image J software, the vertical (dorsal-ventral axis) and horizontal (corneal-retinal axis) diameters were measured for each lens and the ratio of the dorsal-ventral axis to corneal-retinal axis was determined. To verify that a perfect roundness ratio (vertical: horizontal = 1.0) could be determined, measurements were also made on steel ball bearings that near the size range (1.5 and 2.5 mm diameter) of the lenses being studied.

Data Analysis and Statistics

Pairs of means were compared with a Student's t-test. Groups of means were compared by single-factor analyses of variance (ANOVAs) and the Tukey-Kramer procedure was used to make multiple comparisons of pairs of means. A 5% level of significance was used in all cases.

RESULTS

Gill Structure

Figures 14.2–14.6 compare features of the gill structure of *Pseudapocryptes*, *Scartelaos*, *Boleophthalmus*, *Periophthalmus*, and *Periophthalmodon* to show morphological trends related to increased adaptations for amphibious life. The gills of *Pseudapocryptes*, *Scartelaos*, and *Boleophthalmus* have basic similarities in arch size and in the density and form of their gill rakers (Fig. 14.2). The filaments also have a comparable thickness, are straight, and lack branching (Figs. 14.2 and 14.3). The lamellae of *Pseudapocryptes*, *Scartelaos*, and *Boleophthalmus* are distributed along both sides of the filament, evenly spaced, and positioned perpendicular to the filament axis (Figs. 14.2–14.4), the optimal position for maximizing counter-current gas exchange during aquatic ventilation. The lamellae in these three genera have a similar epithelial surface (Fig. 14.4) and are of comparable thickness (Fig. 14.3); the latter implying that the diffusion distances from blood to water are also similar.

The gill arches of *Periophthalmus* and *Periophthalmodon* have a smaller relative thickness and much less prominent gill rakers than in *Pseudapocryptes*, *Scartelaos*, and *Boleophthalmus* (Fig. 14.2). *Periophthalmus* and *Periophthalmodon* also have much shorter filaments, that are usually

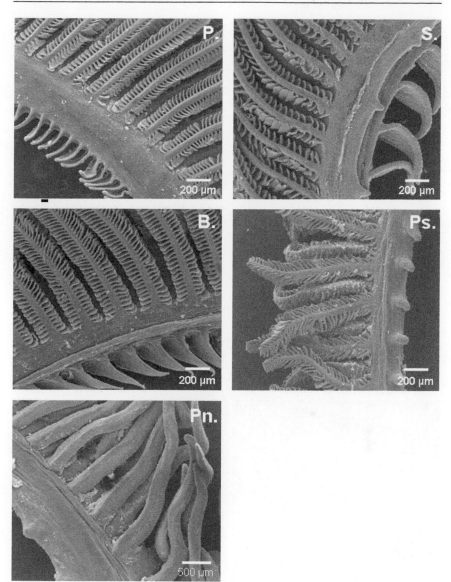

Fig. 14.2 SEM images of the gill-arch filament regions in: **P.** *Pseudapocryptes lanceolatus* (105 mm TL, 5.0 g, 150×), **S.** *Scartelaos histophorus* (140 mm, 6.5 g, 150×), **B.** *Boleophthalmus boddarti* (100 mm, 7.0 g, 150×), **Ps.** *Periophthalmus modestus* (71 mm, 3.0 g, 150×), and **Pn.** *Periophthalmodon schlosseri* (185 mm, 54.0 g, 70×).

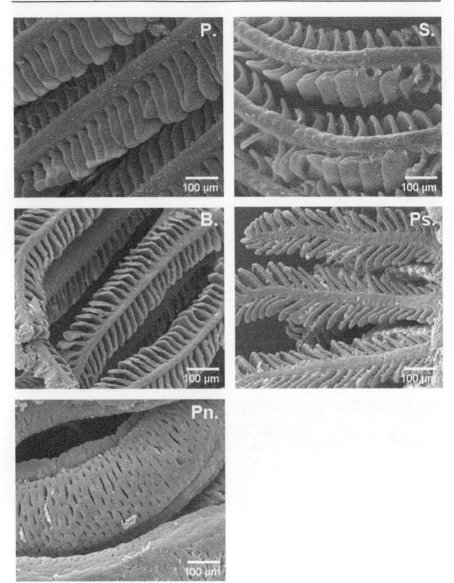

Fig. 14.3 SEM images showing the filaments and lamellae in: **P.** *Pseudapocryptes lanceolatus*, **S.** *Scartelaos histophorus*, **B.** *Boleophthalmus boddarti*, **Ps.** *Periophthalmus modestus*, and **Pn.** *Periophthalmodon schlosseri*. All images were taken at 400×. (See Fig. 14.2 for body size data.)

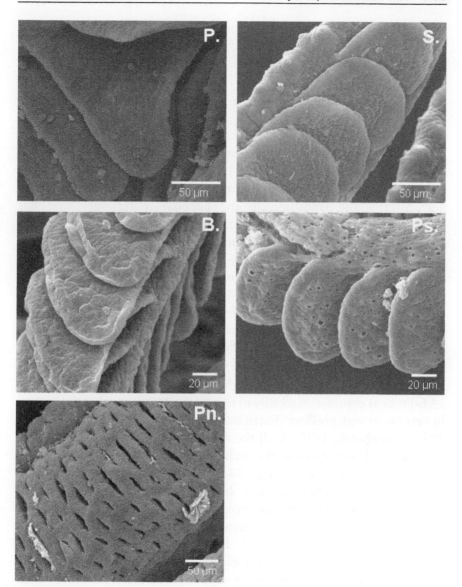

Fig. 14.4 SEM images showing the lamellar detail in: **P.** *Pseudapocryptes lanceolatus* (1200×), **S.** *Scartelaos histophorus* (1200×), **B.** *Boleophthalmus boddarti* (1500×), **Ps.** *Periophthalmus modestus* (1500×), and **Pn.** *Periophthalmodon schlosseri* (800×). (See Fig. 14.2 for body size data.)

twisted (Figs. 14.2 and 14.5) and most of which emerge from the arch in alternating series (i.e., alternating right and left hemibranchs) (Fig. 14.3). The *Periophthalmodon schlosseri* that we studied (from Sumatra) had relatively few instances of branching filaments, which is different from the high frequency of branching noted by Low *et al.* (1988) for specimens from Singapore.

The lamellae of *Periophthalmus* and *Periophthalmodon* possess morphological traits for increased terrestriality that are not present in the less amphibious genera. Figure 14.6, which compares the lamellar structure of *Pseudapocryptes*, *Periophthalmus*, and *Periophthalmodon*, demonstrates the transition in the lamellar morphology that has accompanied the shift to a more amphibious natural history. A cross-section through the lamellae of *Pseudapocryptes* (Fig. 14.6-P.) shows the typical teleost lamellar morphology that is specialized for aquatic gas exchange. (The lamellae of *Scartelaos* and *Boleophthalmus* have this same general morphology.) Pillar cells (PC) form narrow lamellar blood channels through which red blood cells (RBC) proceed in single-file (Fig. 14.6-P.), optimizing diffusion distances and the potential for gas exchange. The water-blood barrier distance (lamellar epithelium thickness, about 0.5-1.0 μm) is at the shorter end of the normal range for other teleosts.

The lamellar structure of the highly terrestrial *Periophthalmodon* (Figs. 14.6-Pn.1- 3) contrasts markedly to that of the more aquatic oxudercines. In agreement with previous descriptions (Schöttle, 1931; Low *et al.*, 1988, 1990; Wilson *et al.*, 1999, 2000) the filaments are thick, short, and curve in different directions from the arch, while the lamellae are thick and possess interlamellar fusions (Figs. 14.4 to 14.6). While these fusions help to support the filament in air, they obstruct the interlamellar spaces, thus minimizing water flow and increasing diffusion distances (Low *et al.*, 1988). Schöttle (1931) first suggested that the gill structure of *Periophthalmodon* (considered at her time as another species of *Periophthalmus*) was the result of an elaboration of the filament epithelium covering the lamellae. This idea is supported by Fig. 14.6-Pn.3, which shows the filament covering partially lifted from over the lamellae. The epithelium that covers and nearly totally encloses the lamellae has a high density of large chloride cells that function for the regulation of sodium and chloride ions and ammonia excretion (Wilson *et al.*, 1999, 2000; Chew *et al.*, 2003; Ip *et al.*, 2004a, b; Randall *et al.*, 2004). Cross-sections through the lamellae (Figs. 14.6-Pn.1 and 2) show the spatial relationship between the lamellar blood channels and the surrounding

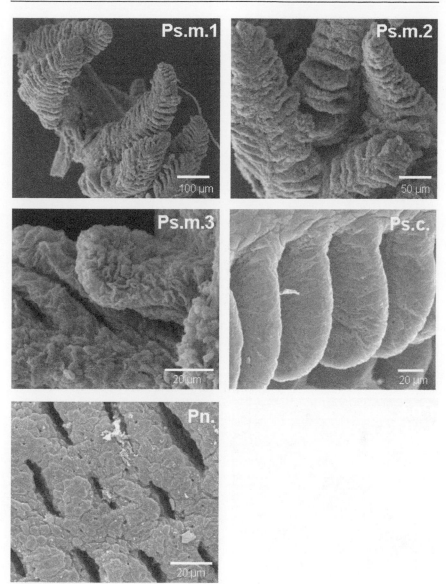

Fig. 14.5 SEM images of the gills in: **Ps.m.** *Periophthalmus minutus* (68 mm, 2.4 g, **1.** 400×, **2.** 800×, **3.** 3000×) and **Ps.c.** *Periophthalmus chrysospilos* (69 mm, 2.3 g, 1600×) (compare with *Periophthalmus modestus* [Figs. 14.2-4]). The epithelium of **Pn.** *Periophthalmodon schlosseri* (2600×) lamellae is shown for comparison. (See Fig. 14.2 for body size data.)

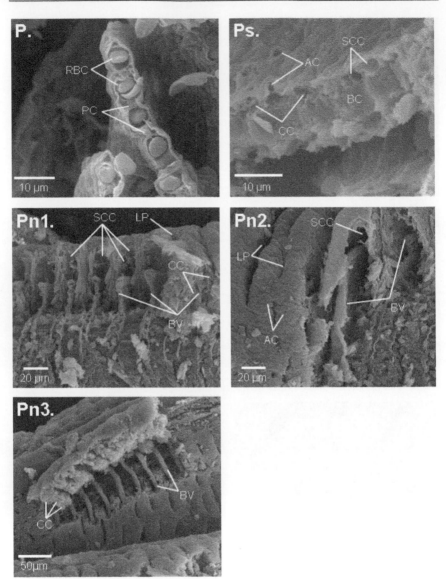

Fig. 14.6 SEM images of cross-sections through the lamellae in: **P.** *Pseudapocryptes lanceolatus* (5000×), **Ps.** *Periophthalmus modestus* (6000×); and **Pn.** *Periophthalmodon schlosseri* (**1.** 1500×, **2.** 1500×, **3.** 800×). (See Fig. 14.2 for body size data.) AC, apical crypt; BC, blood channel; BV, blood vessel; CC, chloride cell; LP, lamellar pore; RBC, red blood cell; SCC, space previously occupied by a chloride cell.

chloride cells (CC), including spaces (SCC) from which chloride cells fell during dissection. Small apical crypts (ACs) are also seen on the lamellar epithelial surface (Fig. 14.5-Pn. and 14.6-Pn.2), and are suggested as the site of water contact with the chloride cells and of ammonia excretion (Wilson *et al.*, 1999, 2000).

The lamellae of the different species of *Periophthalmus* (Figs. 14.4–14.6) appear to comprise various intermediate steps between the morphologies of the less amphibious genera and *Periophthalmodon*. Lamellae in *Ps. modestus* and *Ps. chrysospilos* (Figs. 14.4 and 14.5) more closely resemble those in *Pseudapocryptes*, *Scartelaos*, and *Boleophthalmus* (Fig. 14.4) while the lamellae of *Ps. minutus* (Fig. 14.5) appear more like those in *Periophthalmodon* (Fig. 14.4). The lamellae of *Ps. modestus* and *Ps. chrysospilos* are not significantly thicker than those of *Pseudapocryptes*, *Scartelaos*, and *Boleophthalmus* but do possess apical crypts (Figs. 14.4-Ps., 14.5-Ps.c., 14.6-Ps.) indicating the presence of chloride cells. Figure 14.6-Ps. shows a cross-section through an apical crypt of *Ps. modestus* and reveals what appear to be chloride cells on either side of the crypt. The apical crypts of *Ps. modestus* are larger than those of *Periophthalmodon* (compare Fig. 14.4-Ps. with Fig. 14.5-Pn.), but the chloride cells of are much smaller (about 5 µm in length as opposed to 15 µm) (Fig. 14.6). Although apical crypts are not easily recognized in *Ps. minutus* (Fig. 14.5-Ps.m.3), the lamellae are much thicker than in the other *Periophthalmus* species (Fig. 14.5-Ps.m.3), and fusions similar to those of *Periophthalmodon* occur near the filament tips (Fig. 14.5-Ps.m.2).

Table 14.1 shows values for the gill-area parameters (Eq. 1) determined for two specimens of *Scartelaos histophorus* together with comparative data for other oxudercines. Allowing for body size differences, most of the parameters determined for *Scartelaos* are within the range reported for the other oxudercines; the only exception is its larger number of gill filaments (N). The average filament length measured for *Scartelaos* is similar to that of a comparably sized *Pseudapocryptes* (specimen size differences preclude comparison with *Boleophthalmus*), but greater than those of *Periophthalmus* and *Periophthalmodon*. Both Ln and La are highly variable among the oxudercines, with no clear trend corresponding to the degree of specialization for terrestrial life.

The mass-specific gill areas determined for *S. histophorus* were 565.2 mm^2 g^{-1} (2.7 g specimen) and 295.4 mm^2 g^{-1} (6.2 g specimen). Figure 14.7 compares these values to those defined by the mass-specific gill area

Table 14.1   Components of the total gill surface area equation* determined for various oxudercine species (values are mean ± SD).

	Mass (g)	N	F (mm)	Ln (#/mm)	La (mm²)	A (mm²)	Mass-specific area (mm² * g⁻¹)	Reference
Pseudapocryptes lanceolatus	8.2 ± 2.6	517.7 ± 5.7	2.2 ± 0.2	44.2 ± 2.6	0.058 ± 0.022	3302.4 ± 797.2	457.6 ± 38.6	5
Scartelaos histophorus	2.7	704	1.63	40.2 ± 0.6	0.033 ± 0.006	1526.1	565.2	7
Scartelaos histophorus	6.2	728	1.93	37.3 ± 1.3	0.035 ± 0.007	1831.6	295.4	7
Boleophthalmus pectinirostris	35.2	486	4.3 ± 0.5	28.4 ± 4.0	–	3330	94 ± 30	4
B. boddaerti	53	630	3.5	35.4	0.061	4695	88.5	3
B. boddaerti **	10	–	–	31.4	0.018	1038.6	103.9	1
B. boddaerti	53	502	3.1	23.6	0.072	4587	86.5	2
Periophthalmus modestus	5.3	306	1.3 ± 0.3	47.0 ± 4.8	–	660	124 ± 44	4
Ps. chrysospilos	8.8	249	1.4	36.2	0.061	783.5	89.0	2
*Periophthalmodon schlosseri***	0.86 ± 0.31	231.3 ± 4.4	0.9 ± 0.1	55.8 ± 2.0	0.020 ± 0.004	248.4 ± 74.5	315.0 ± 20.3	6
Pn. schlosseri	8.8	317	1.2	85.6	0.025	759.8	86.3	2
Pn. schlosseri	53	354	2.6	78	0.056	4044	76.3	2

*(A = N · F · Ln · La) A = area, N = total number of filaments, F = average filament length, Ln = mean number of lamellae per mm filament, La = mean bilateral surface area of the lamellae.

** actually *B. dussumieri*, see text

*** actually *Ps. minutus*, see text

1. Hughes and Al-Kadhomiy (1986), 2. Low et al. (1990), 3. Niva et al. (1981), 4. Tamura and Moriyama (1976), 5. Yadav and Singh (1989), 6. Yadav et al. (1990) and 7. This study.

regression equations determined for *Pseudapocryptes* and the three other mudskipper genera. (Note that names of two species from the original publications have been changed in Fig. 14.7.) The *Boleophthalmus* specimens studied by Hughes and Al-Kadhomiy (1986) were obtained in the Persian Gulf, which means they were *B. dussumieri* and not *B. boddarti*. Based on both body size and gill morphology, Clayton (1993) concluded that the findings of Yadav *et al.* (1990) attributed to *Periophthalmodon schlosseri* are in fact for *Periophthalmus minutus*. We concur with this based on similarities between the gill SEM images shown by Yadav *et al.* (1990, Figs. 1-3] and those for *Ps. minutus* in this study [Fig. 14.5-Pn.m.1-3.]) The mass-specific area determinations for *S. histophorus* are positioned close to the regression line for *Pseudapocryptes* and are higher than values for comparably sized specimens of the other three mudskipper genera. It is apparent in Fig. 14.7 that the slope values for the regressions differ among genera and species. Although the data are variable, they document the trend of a smaller gill area in the more amphibious oxudercine species (Schöttle, 1931; Low *et al.*, 1990; Graham, 1997).

Buccal-chamber Volume

Figure 14.8 demonstrates the significantly greater buccal-chamber volumes for the four genera of mudskippers compared to the facultative air-breathing *Pseudapocryptes* and other non-oxudercine gobies. For *Scartelaos*, *Boleophthalmus*, and *Periophthalmus* the increase in relative buccal volume correlates directly with the trend for increased terrestriality (i.e., the mean buccal volume as a percent of mass in *Scartelaos* [14.3%] is less than *Boleophthalmus* [16.4%] and *Periophthalmus* [16.6%]). An apparent exception to this trend is the 16.0 percent of mass buccal volume determined for *Periophthalmodon schlosseri* by Aguilar *et al.* (2000). However, this difference is small, and rather than having biological significance, it may reflect differences in the method of data acquisition (Aguilar *et al.* (2000) used a pneumotach and geometric estimates based on video records). Data obtained thus far suggests there may also be species specific as well as sexually dimorphic differences in buccal volumes. For example, the buccal volume of *Ps. modestus* (17.4 ± 2.2 percent of mass) is larger than that of *Ps. argentilineatus* (13.6 ± 0.6 percent of mass). Also, the males of *Ps. modestus*, which construct burrows and have a complex egg-guarding behavior that includes the transport of air into and out of their burrows (Ishimatsu *et al.*, in review), have a

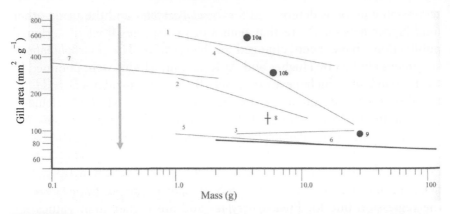

Fig. 14.7 The mass-specific gill surface areas determined for two specimens of *Scartelaos histophorus* (a: 2.7 g; b: 6.7 g) plotted (10a, b) in relation to the mass-specific gill area regressions determined for other oxudercines. The arrow indicates direction of increased terrestriality. Numbers adjacent to each regression line indicate species, the mass range of specimens studied, and the reference: **1.** *Pseudapocryptes lanceolatus*, 1 – 21 g, Yadav and Singh (1989). **2.** *Boleophthalmus boddarti*, 1 – 12 g, Niva *et al.* (1981). **3.** *B. boddarti* (*B. dussumieri*), 3.6 – 35 g, Hughes and Al-Kadhomiy (1986). **4.** *B. boddarti*, 2 – 35 g, Low *et al.* (1990). **5.** *Periophthalmus chrysospilos*, 2 – 13 g, Low *et al.* (1990). **6.** *Periophthalmodon schlosseri*, 3 – 111 g, Low *et al.* (1990). **7.** *Pn. schlosseri* (*Ps. minutus*) 0.075 – 2.3 g, Yadav *et al.* (1990). Numbers adjacent to point values identify mass-specific estimates for: **8.** *Ps. modestus* (mean mass 5.3 g, n = 11), Tomura and Moriyama (1976). **9.** *B. chinensis* (*B. boddarti*) (35.2 g, n = 15), Tomura and Moriyama (1976). **10a.** *S. histophorus* (2.7 g), this study. **10b.** *S. histophorus* (6.2 g), this study. (Names in parentheses are likely the correct species names based on geographic distribution data and morphology, [see text and Murdy, 1989; Clayton, 1993].)

significantly larger buccal volume (18.3 ± 2.3 percent of mass) than do females (15.9 ± 1.3 percent of mass). The buccal chamber volumes determined for the four mudskipper genera are three to four times greater than those of the non-oxudercine gobies, *Pseudogobius*, *Mugilogobius*, *Favonigobius*, *Chlamydogobius*, and *Arenigobius*, which were all less than 4.4 percent of mass (Fig. 14.8).

Lens-axis Ratios

Figure 14.9 compares the vertical : horizontal (cornea to retina) lens diameter ratios determined for species used in this study with ratios measured for other species, including some non-oxudercine gobies, and, to verify precision, two sizes of ball bearings. Among the mudskippers, lens flattening progresses with the extent of specialization for terrestrial life (*Scartelaos* = 1.03 < *Boleophthalmus* = 1.07 < *Periophthalmus* = 1.08

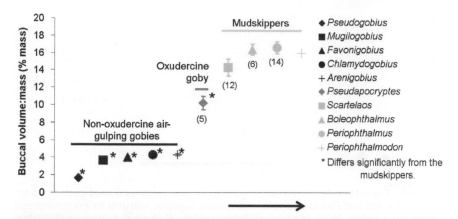

Fig. 14.8 Buccal chamber volumes, expressed as percent of body mass, for the four mudskipper genera, *Pseudapocryptes*, and five non-oxudercine air-gulping gobies. Data for *Pseudapocryptes*, *Scartelaos*, *Boleophthalmus*, and *Periophthalmus* were determined in this study, data for *Periophthalmodon* are from Aguilar *et al.* (2000), and *Pseudogobius*, *Mugilogobius*, *Favonigobius*, *Chlamydogobius*, and *Arenigobius* are from Gee and Gee (1991). (Sample size is noted in parentheses when available. Mean ± S.E.M., * denotes the buccal chamber volume is significantly different than mudskipper volumes.) Oxudercines (right half of Fig.) are in order of their increased specialization for amphibious life (arrow).

< *Periophthalmodon* = 1.12), and the difference in the lens flatness between *Periophthalmodon* and *Scartelaos* is significant. However, the non-amphibious oxudercine *Pseudapocryptes* has a significantly flatter lens (1.10) than does *Scartelaos*, *Boleophthalmus*, and *Periophthalmus*. For the other gobies investigated, *Gillichthys mirabilis* has a perfectly round eye lens (1.0) while that of *Bathygobius ramosus* is slightly flatter (1.05). *Exocoetus monocirrhus*, the flying fish, has a lens that is barely flattened (1.01). Overall, no significant differences between right and left lenses were found for any species (*n* = 93 lenses) and the ball bearings were found to be perfectly round (ratio 1.0 ± 0.0005, mean ± SE, Fig. 14.9).

DISCUSSION

The aquatic-to-amphibious continuum existing for species within the goby subfamily Oxudercinae distinguishes this group from all other fishes. The new information contained in this report adds further perspective to this continuum. We obtained data for *Pseudapocryptes lanceolatus* which, while occurring within the mudskipper clade (Fig. 14.1), is a non-amphibious intertidal mudflat species that lives in burrows, is a facultative air breather, and is capable of surviving exposure to air (Hora, 1935a, b).

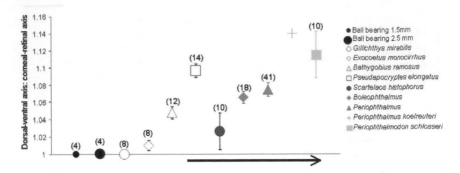

Fig. 14.9 Diameter ratios for the dorsal-ventral: corneal-retina lens axes determined for the mudskippers and *Pseudapocryptes* in this study, together with data for *Periophthalmus koelreuteri* (Karsten, 1923), two non-oxudercine gobies (*Bathygobius ramosus, Gillichthys mirabilis*), and the flying fish (*Exocoetus monocirrhus*). Ratios for ball bearings (1.5 mm and 2.5 mm diameter) are also shown. Data are Mean ± S.E.M.

We also obtained data for the four genera of mudskippers which, listed in order of increasing amphibious capability, are: *Scartelaos, Boleophthalmus, Periophthalmus, Periophthalmodon*.

A focus of our study was *Scartelaos*, about which relatively little is known. Based on its ecology and behavior, *S. histophorus* appears far less specialized for life out of water than the other mudskippers (Milward, 1974; Zhang *et al.*, 2000; Lee and Graham, 2002; Lee *et al.*, 2005). Our field observations indicate that *S. histophorus* occurs lower in the intertidal zone than other species of mudskippers. It also does not venture far from water, being confined to soft mud flat habitats often having little vertical relief and with small areas of standing water in close proximity to the opening of its burrows (Lee and Graham, 2002; Lee *et al.*, 2005). *Boleophthalmus* also occurs on mudflats, but appears less dependent upon water than *Scartelaos* (Zhang *et al.*, 2000). *Boleophthalmus* is amphibious during low tide and, like *Scartelaos*, retreats into burrows during high tide or when pursued (Low *et al.*, 1988, 1990; Zhang *et al.*, 2000).

In contrast, species of *Periophthalmus* occur at higher levels in the mudflat and even in rocky intertidal areas. Some species migrate with the tides (i.e., staying at the water's edge or ascending the branches of mangroves as the water level rises) and, when chased, do not characteristically retreat to a burrow or into water. Rather, their escape entails rapid 'skipping' over the water surface or ground and taking cover under a bush or in a hole. When lying in shallow water *Periophthalmus* species typically emerge their heads and anterior body. During the breeding season the males of some *Periophthalmus* (e.g., *Ps. modestus* and *Ps. chrysospilos*) dig burrows wherein, after inducing females to spawn, they guard and tend fertilized eggs (Ishimatsu *et al.*, in review). Other species, for example *Ps. novaeguineaensis* and *Ps. minutus*, have littoral zone burrows in which they spend considerable time and they do not migrate vertically with the tides.

Periophthalmodon similarly occurs in the high intertidal zone, however, unlike *Periophthalmus*, *Periophthalmodon* does not usually move very far away from the water and is most often near its burrow that, while containing water that is hypoxic, hypercapnic, and high in ammonia concentration, is used for refuge and for breeding (Ishimatsu *et al.*, 1998a, b; Zhang *et al.*, 2003; Randall *et al.*, 2004). *Periophthalmodon* is also an obligate air breather and when it swims along the shoreline of mangrove channels, it retains air in its buccopharyngeal chamber (Kok *et al.*, 1998; Ishimatsu *et al.*, 1999); this makes its head sufficiently buoyant to remain partially above the water surface.

Gill Structure

In 1931, Elfride Schöttle published a comprehensive comparative study of the respiratory specializations of mudskippers and other gobies. She estimated total gill surface area and relative buccopharyngeal chamber size and also obtained respiration data for fish in air and water. Schöttle was the first to document the inverse trend between gill surface area and the extent of amphibious life, suggesting a compensating respiratory function for the well-developed vascular epithelium lining the mudskipper's buccopharyngeal chamber. Schöttle was also the first to describe the unique gill structure of *Periophthalmodon schlosseri* (classified at that time as *Periophthalmus schlosseri*).

Schöttle (1931) obtained data for several species of *Periophthalmus* and two species of *Boleophthalmus*, but did not report data for any other oxudercines. With the passage of time, many workers contributed data for other mudskippers (reviewed in Clayton, 1993; Graham, 1997). However, because mudskippers have such an extensive geographic distribution and their taxonomy was poorly known, the comparative literature has been confounded by uncertainty as to which species were actually studied. With the appearance of Murdy's (1989) systematic analysis of mudskipper taxonomy it became possible to sort out species identifications and relate specific results to behavior and ecology, to examine oxudercine phylogeny

(Fig. 14.1), and to erect hypotheses about the sequence of character acquisitions relating to amphibious life.

The findings of this study show that the gill structure of *S. histophorus* is more suited for aquatic respiration. The gills of *Scartelaos* are similar to those of both *Pseudapocryptes* and *Boleophthalmus*, having long and straight filaments and un-fused lamellae that lie in parallel with branchial water flow to form an effective gas-exchanging sieve. By contrast, the gills of both *Periophthalmus* and *Periophthalmodon* have undergone modifications that compensate for greater aerial exposure, principally through reduction of the length and, in some cases, the number of filaments. The filaments of *Periophthalmus* are relatively short and twisted. While this interferes with lamellar alignment for branchial water flow, the open space between the filaments ensures lamellar contact with air during air breathing, the most common respiratory mode for *Periophthalmus*.

The gills of *Periophthalmodon* also feature short filaments (which may be branched in some populations), having relatively thick support rods and lamellar fusions (Low *et al.*, 1988). These fusions, which add support to the filament in air, largely occlude the interlamellar channels and thus limit intra-lamellar water flow and aquatic-respiration efficiency (Low *et al.*, 1988), requiring an obligatory dependence upon air breathing (Kok *et al.*, 1998; Ishimatsu *et al.*, 1999; Aguilar *et al.*, 2000; Randall *et al.*, 2004).

Based on comparisons of the lamellae of mudskippers and other gobies, Schöttle (1931) suggested a mechanism for structural change involving the progression of the gland rich filamental epithelium onto the lamellae. She reported different degrees of this progression among species of *Periophthalmus* and regarded *Pn. schlosseri* (as *Ps. schlosseri*) to be the most derived condition. Although little was known about chloride and their accessory cells at the time of Schöttle's work, her drawings depict gland-like structures invested into the lamellar walls of various *Periophthalmus* species. These drawings are somewhat like the cells seen in the lamellar walls of *Ps. modestus* (Fig. 14.6-Ps.) and in the lamellae-covering epithelium of *Periophthalmodon* (Fig. 14.6-Pn.1–3). Our SEM findings also correlated this structural change with the presence of apical crypts (i.e., the opening portals for chloride cells) on the lamellae of *Ps. modestus* (Figs. 14.3-Ps. and 14.4-Ps.), *Ps. chrysospilos* (Fig. 14.5-Ps.c.), and *Pn. schlosseri* (Figs. 14.5-Pn. and 14.6 Pn. 2).

In most fishes mitochondrial rich chloride cells occur mainly on the filaments. However, as reported by Lin and Sung (2003), chloride cells also occur on the lamellae of many air-breathing fishes and their number can be increased by acclimation to reduced salinity, as demonstrated for *Ps. modestus* (as *Ps. cantonensis*) and other species. Although apical crypts were found on the lamellae of *Ps. modestus* and *Ps. chrysospilos*, we did not see them on the lamellae of *Pseudapocryptes*, *Scartelaos*, *Boleophthalmus*, or *Ps. minutus*. However, we did note other structural differences characterizing the lamellae of *Ps. minutus*; its lamellae are thicker, have a highly uneven epithelial surface, and, near the tip of the filaments there is lamellar fusion somewhat similar to that in *Periophthalmodon* (Fig. 14.5-Ps.m.2 and 3, also see Yadav *et al.*, 1990). These structural differences may relate to natural history; *Ps. minutus* spends a good deal of its time within a burrow located in the lower part of the tidal range. The upper reaches of this burrow are dry and there may or may not be water in the lower parts. This differs from *Ps. modestus* and *Ps. chrysospilos*, which spend their time in more open areas and make regular contact with the water in their burrows.

The highly modified filament and lamellar structure of *Periophthalmodon* are best viewed as gill specializations for stagnant water conditions within the burrow that effectively supplant the organ's primary respiratory function (Randall *et al.*, 1999, 2004; Wilson *et al.*, 1999, 2000; Chew *et al.*, 2003; Ip *et al.*, 2004a, b). *Periophthalmodon* burrows occur along mangrove channels high in the intertidal zone where they are not adequately flushed by tidal flow. In addition to being hypoxic and hypercapnic, the burrow water can also have a high ammonia concentration and variable pH. The capacity of *Periophthalmodon* to reside in this water is attributable to the high density of mitochondrial-rich chloride cells in its gills. This complex arrangement enables the fish to excrete ammonia against a concentration gradient in the burrow water (Randall *et al.*, 2004). The lamellae of *Periophthalmodon* (Fig. 14.6) are thus deeply embedded within the epithelial matrix of the chloride cells and, while interlamellar pores provide water access, water turnover is minimal and the diffusion distance to the lamellar blood channels is large. This structure is therefore not suited for respiration, which must be done by the buccopharyngeal ABO.

Gill Surface Area

The mass-specific gill surface area of *S. histophorus* shows a good fit, between *Pseudapocryptes* and other mudskippers, along the progression of declining gill area with the degree of specialization for terrestrial life (Fig. 14.7). However, while our data fit the trend it is emphasized that the literature on mudskipper gill area is characterized by considerable variation in all of the parameters contained in Eq. 1, as well as gill-scaling relationships (Graham, 1997). Specifically, the scaling of mass-specific gill-area among the various mudskippers (Fig. 14.7) shows a high variability, with mass exponents ranging from –0.519 to +0.050 (Hughes and Al-Kadhomiy, 1986; Low et al., 1990; Yadav et al., 1990). These are inconsistent with findings for other air-breathing fishes, in which the mass exponents for mass-specific gill area are generally less than –0.150 (Graham, 1997).

An obvious question to ask, based on our SEM findings for the gills of *Periophthalmodon* is, does a gill-area estimate for this fish have any significance for respiration? The long-standing hypothesis, illustrated by Fig. 14.7, is that gill area reduces with the progressive development of an amphibious life history. Several works have reported gill areas for *Pn. schlosseri*. While Schöttle (1931) is often cited for having made area measurements for *Periophthalmodon* (as *Ps. schlosseri*), there are no data of this kind in her treatise. Low et al. (1988, 1990) found no significant differences between the gill areas of *Pn. schlosseri* and *Ps. chrysospilos* (Table 14.1). Thus, there is no evidence to support the contention that *Periophthalmodon* has less gill area than *Periophthalmus*. Moreover, the gill morphology seen in Fig. 14.6 indicates that the lamellae of *Periophthalmodon* have a thick epithelial cover. Thus, estimates that did not correct for this thick lamellar covering would over estimate total gill area.

In addition, the expected trend of a smaller area may have little biological relevance for *Periophthalmodon*, which appears to make minimal use of its gills to supply its total O_2 requirement. Rather, the lamellar blood channels appear to be supplying nutrients and O_2 to the surrounding matrix of mitochondria-rich chloride cells that are actively transporting Na^+, Cl^- and NH_4^+ into the interlamellar water space (Wilson et al., 1999).

Buccal-chamber Volume

Schöttle (1931) perceived the functional importance of a relative buccopharyngeal chamber volume measurement, as she approximated this from the ratio of opercular length to body length and consistently found higher values for mudskippers compared to other gobies. The mass-specific buccal chamber volumes measured for oxudercines in this study are three to four times greater than those measured for other air-gulping gobies (Gee and Gee, 1991). This may reflect a greater importance for oxudercines, of the regular use of the mouth and pharyngeal chamber in air breathing, for digging burrows, and for the transport of air into their burrows (Ishimatsu *et al.*, 1998a, b; Lee and Graham, 2002; Lee *et al.*, 2005). The volume differences between *Pseudapocryptes* and the mudskippers may, moreover, reflect the latter's additional requirements for burrow construction (requiring biting and moving chunks of mud) and in territorial display (buccal inflation). More comparative data are needed to test for specific differences in buccal chamber volumes among mudskipper species differing in body size, burrow architecture, and territorial defence behavior. It is important, for example, to determine whether species of *Periophthalmus* exhibiting marked differences in body size (e.g., *Ps. argentilineatus* vs *Ps. gracilis*) have comparable buccal chamber volumes. Also, additional tests with *Periphthalmodon* using the methods described in this study will refine the relationship between buccal-chamber volume and amphibious capacity.

Oval Eye Lens

Nearly all aquatic animals, including fishes (Walls, 1942; Graham, 1971; Fernald, 1990), amphibians (Mathis *et al.*, 1988), penguins (Sivak, 1980), and mammals (Sivak *et al.*, 1989), have round lenses. Because the cornea of the eye has the same refractive index as water, aquatic animals rely on the curvature of the round lens to focus an image (Graham, 1971; Sivak, 1990). In contrast, an oval lens permits accommodation in air by compensating for the added refraction of the cornea (Graham, 1971; Sivak, 1990). Karsten (1923) was the first to document an oval lens in the mudskipper *Ps. koelreuteri* (*Ps. kalolo*) (Fig. 14.9). We confirmed this result and have shown that all of the mudskippers and *Pseudapocryptes* have at least a slightly flat lens, and that there is a progression from near-round to increasingly oval among the four mudskipper genera. It was expected that non-oxudercine gobies would have perfectly round lenses, as does *Gillichthys mirabilis*. However, *Bathygobius ramosus* has an oval lens. This species occurs in high intertidal pools and may be occasionally trapped out of water. Why it has an oval lens is not known; this feature would result in hyperopia (i.e., farsightedness) in water. Finally, the flying fish *Exocoetus monocirrhus*, which presumably uses aerial vision to look for predators in water prior to landing, also has round lenses and thus does not gain aerial visual acuity by this compensatory refraction method.

Recent Findings on the Respiratory and Amphibious Adaptations of the Oxudercinae

This discussion's final objective is to review recent literature having relevance to the water-land transition in the Oxudercinae.

Skin Structure and Respiration

Skin specializations for amphibious life include adaptations for respiration, for desiccation resistance, and for possible protection from solar radiation and substrate abrasion (Suzuki, 1992; Zhang *et al.*, 2000, 2003; Park, 2002). Recent studies by Zhang *et al.* (2000, 2003) provide a concise comparative overview of the skin specializations of mudskippers. *Scartelaos* and *Boleophthalmus* have a well-vascularized epidermis and dermal bulges, which enhance cutaneous respiration by pushing up the vascularized epidermis to form respiratory papilla. The skin of *Boleophthalmus* appears more specialized for cutaneous respiration because of larger dermal bulges that cover a greater extent of the body (principally the head and dorsal body surfaces) than in *Scartelaos*. In both *Scartelaos* and *Boleophthalmus* respiratory areas of the skin occur mainly on the dorsal surface and head, the areas commonly exposed to air. Mucous cells, important for preventing desiccation and for cushioning the skin from abrasion, occur along the ventral and lateral surfaces, and those located near respiratory papillae are usually set low between bulges to prevent mucus from reducing gas diffusion (Zhang *et al.*, 2000).

Zhang *et al.* (2003) report that both *Periophthalmus* and *Periopthalmodon* have epidermal capillaries but lack the dermal bulges of *Scartelaos* and *Boleophthalmus*. The greatest capillary densities occur on the head and dorsal surfaces of both genera but these workers noted specific differences in this distribution pattern. Although Zhang *et al.* (2003) reported a slime-like substance on the skin of *Periophthalmus*, they did not find any mucous cells in its skin. *Periophthalmus* is, however, similar to *Scartelaos* and *Boleophthalmus* in having a thick middle epidermal layer (the layer between the epidermis and basal cells), which may have a role in ultra-violet light filtration or possibly thermal insulation (Suzuki, 1992; Park, 2002).

Rates of cutaneous gas exchange ranging from 36 to 49% of total VO_2 in water and from 43 to 77% of total VO_2 in air have been documented for both *Periophthalmus* and *Boleophthalmus* (Teal and Carey, 1967; Tamura *et al.*, 1976). There are no cutaneous respiration measurements for *Scartelaos* or *Periophthalmodon*. Based on epidermal capillary densities, Zhang *et al.* (2003) concluded that the capacity for aerial cutaneous respiration would be highest in *Periophthalmodon* and lowest in *Scartelaos*.

However, Randall *et al.* (2004) report that the skin of *Pn. schlosseri* is rich in cholesterol and saturated fatty acids and that these decrease membrane fluidity and the permeability to gases, including ammonia, while the fish is in water. Whether this impermeability applies to all regions of the skin (i.e., the head and dorsal body would nearly always be out of water) or if it affects aerial cutaneous gas transfer is unknown.

Post-exercise Aquatic VO₂

Takeda *et al.* (1999) tested the capacity of *Pn. schlosseri* to elevate its post-exercise VO_2 in air and water. After chasing and gently prodding a *Pn. schlosseri* in a small basin for 2 min, these workers found that, when placed in a respirometer containing air, the fish would elevate its VO_2 above routine until its post-exercise O_2 debt was repaid. But, post-exercise fish placed in a flow-through aquatic respirometer without access to air could not elevate their post-exercise VO_2 until an air phase was added to the chamber. This finding for *Pn. schlosseri* is consistent with its obligatory air breather status (Ishimatsu *et al.*, 1999; Takeda *et al.*, 1999). However, comparable tests with *S. histophorus* and *Ps. modestus* in this laboratory show these species can sustain high levels of post-exercise VO_2 in water (Lee and Graham, unpublished observations). This finding for *Scartelaos* reflects its greater dependence upon water. In the case of *Periophthalmus*, which is much more amphibious, this finding indicates that this group's progression to a greater dependence upon aerial respiration has not fully compromised gill function for aquatic respiration. Comparable exercise studies have not been done on species of *Boleophthalmus* or other species of *Periophthalmus*. In view of the differences in gill structures in *Ps. modestus* and *Ps. minutus*, it can be expected that their capacities for aquatic respiration will differ.

Burrowing

Figure 14.10 compares the above- and below-ground burrow structure of *Pseudapocryptes* and *Boleophthalmus*, *Periophthalmus*, and *Periophthalmodon*.

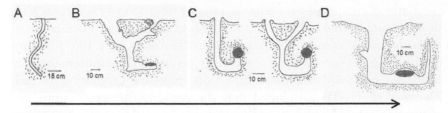

Fig. 14.10 Comparisons of oxudercine burrow shape. Shaded areas indicate portion of burrow in which air can be stored. Arrow direction indicates increasingly amphibious genera. A) *Pseudapocryptes* Modified from Swennen *et al.* (1995). B) *Boleophthalmus* Modified from Clayton and Vaughan (1986). C) *Periophthalmus* Modified from Kobayashi *et al.* (1971). D) *Periophthalmodon* Modified from Ishimatsu *et al.* (1998a).

The burrow of *Pseudapocryptes* consists of straight or slightly curved passages and there are no specialized areas that would function for air storage (Swennen *et al.*, 1995). The mudskipper burrows all have chambers with varying degrees of specialization for air storage (e.g., having an upturned or domed portion not connected to the surface) (Kobayashi *et al.*, 1971; Clayton and Vaughan, 1986; Ishimatsu *et al.*, 1998a; Lee, 2005).

Air phases have been found in the burrows of all four mudskipper genera, as well as that of one aquatic oxudercine, *Oxuderces* (Ishimatsu *et al.*, 1998a, b; Lee and Graham 2002; Lee *et al.*, 2005). *Apocryptodon* makes shallow U-shaped burrows on mudflats. During low tide it holds air in its mouth (presumably the result of burrow hypoxia) but does not appear to store air in the burrow (Graham and Lee, 2004; JBG. pers. obs.). There are no other data on oxudercine burrow air contents.

Burrow air-deposition behavior has been documented for *Scartelaos* (Lee and Graham, 2002), *Periophthalmus*, and *Periophthalmodon* (Ishimatsu *et al.*, 1998a, b). This behavior involves a rapid and repeated series of actions: a mouthful of air is gulped, transported into the burrow and released; then the fish returns to the mud surface and repeats the action. In cases where a mating pair occupies a burrow, both fishes carry out air-deposition behavior (Lee *et al.*, 2005).

Burrow air may supply the O_2 requirements of developing eggs (Ishimatsu *et al.*, 1998a, b, in review) and may also provide for the respiration of adult fish that are confined to the burrows during high tide or seasonally as is the case for species living at higher latitudes (Clayton and Vaughan, 1986; Clayton, 1993; Lee *et al.*, 2005). Studies with *Scartelaos* in a model burrow demonstrate its air-deposition behavior and show that the stored air is used for breathing during high-tide burrow confinement. This work also shows that this fish can detect the quality of burrow gas and will remove gas mixes containing low O_2 by repeatedly transporting mouthfuls to the burrow opening and releasing them and then re-filling the burrow with fresh air (Lee *et al.*, 2005).

Summary

Table 14.2 compares different oxudercine genera for the respiratory, visual, and behavioral aspects of the aquatic to amphibious transition examined in this study. It shows that, in addition to refinements in aerial respiration and terrestrial locomotion, progressive development of amphibious capacity among mudskippers has been coupled with selection for features for visual acuity in air and complex behaviors associated with burrowing.

Mudskippers are distinguished from other oxudercines by their amphibious capacities; these range from less developed in *Scartelaos* and *Boleophthalmus* to more-developed in *Periophthalmus* and *Periophthalmodon*. These genera can be sorted along a continuum for their capacity for aerial respiration, from moderately high to an obligatory dependence on aerial O_2. Some mudskippers can respire perfectly well in normoxic water and others will drown without air access.

The gill morphology of oxudercines tracks their aquatic-amphibious transition, from gills of *Scartelaos*, which are very similar to those of other teleosts, to the gills of *Periophthalmodon*, which evolved into an organ for ion regulation and active nitrogen excretion. The gill-area data for *S. histophorus* reported here completes documentation of the inverse relationship for gill area and terrestriality among mudskippers showing, as expected, that *Periophthalmus* and *Periophthalmodon* have the smallest areas. However, considerable inter- and intra-specific variability in the area data and in estimates of the scaling of total gill area with body mass preclude rigorous statistical comparisons of the different mudskipper genera and species (Graham, 1997). In addition, based on the structure revealed by SEM (Fig. 14.6), it is uncertain that the "low gill area" determined for *Pn. schlosseri* is either an accurate measurement of the actual area, or a physiologically important finding for this species, in view of the organ's diminished respiratory function and its remarkable specializations for the active transport of ions and NH_4^+.

Table 14.2. Oxudercine adaptations for amphibious life.

	Oxuderces	Pseudapocryptes	Scartelaos	Boleophthalmus	Periophthalmus	Periophthalmodon
GILLS						
Complete gill sieve for aquatic ventilation	yes	yes	yes	yes	no	no
Short filaments	no	no	no	no	yes	yes
Lamellar chloride cells	—[1]	no	no	no	yes[2]	yes
Interlamellar fusions	—[1]	no	no	no	no[2]	yes
Gill area rank[3]	—[1]	1	2	3	4	4
NH_4^+ active transport	—[1]	—[1]	—[1]	—[1]	—[1]	yes
BUCCOPHARYNGEAL CHAMBER						
Vascularized	—[1]	yes	yes	yes	yes	yes
Volume rank[3]	—[1]	4	3	2	1[2]	1
SKIN						
Dermal bulges and respiratory papillae	—[1]	—[1]	yes	yes	no	no
Epidermal capillaries	—[1]	yes	yes	yes	yes[4]	yes[4]
Mucous cells	yes	yes	yes	yes	no	yes[2]
RESPIRATION						
Obligatory air breather	—[1]	no	no	no	no[4]	yes[4]
Can repay O_2 debt in water	—[1]	—[1]	yes	—[1]	yes[4]	yes
BURROWING						
Air chamber present	yes	no	yes	yes	yes	yes
Air deposition behavior	yes	no	yes	yes	yes	yes
Monitor air chamber PO_2	—[1]	—[1]	yes	—[1]	—[1]	—[1]
FLATTENED EYE LENS						
(= aerial visual acuity)	—[1]	yes	yes	yes	yes	yes

[1]No data; [2]Interspecific differences; [3]Highest = 1; [4]Interspecific differences likely

The loss of gill area in mudskippers reflects this organ's reduced effectiveness for air breathing, which has been compensated by development of the vascular buccopharyngeal epithelium. All four mudskipper genera have buccal chamber volumes that are much larger than other oxudercines and other non-oxudercine air-gulping gobies (Fig. 14.8).

Cutaneous respiration is an important adjuvant in mudskipper respiration and all four genera have epidermal capillary beds on their heads and dorsal body surfaces. Differences in the skin structure and in the density of epidermal capillaries suggest that the more amphibious adapted mudskippers are more proficient skin breathers, however, there are no cutaneous respiratory data for either *Scartelaos* or *Periophthalmodon* and, recent reports (Randall *et al.*, 2004) suggest that *Periophthalmodon* may have a reduced skin permeability to protect against the intrusion of burrow water NH_3.

In addition to air breathing, the large buccal chambers of mudskippers have importance in behaviors such as territorial displays and burrow building. For some species burrows serve as subterranean refuges and as nests for developing eggs. Because the burrow water is typically hypoxic, this requires that, during low tide, mudskippers repeatedly gulp air and transport it into specially constructed burrow air chambers to provide an O_2 source for developing eggs, or for the adult fish during high tide or other periods of burrow confinement (Clayton, 1993; Ishimatsu *et al.*, 1998; Lee *et al.*, 2005). Burrow air-storage and associated behavior have been documented in all four mudskipper genera. Also, *Oxuderces*, the most basal oxudercine genus (Fig. 14.1) also holds air in its burrows. Because most oxudercines dig burrows and gulp air for facultative air breathing, the capacity to store air in burrows may have co-evolved as a consequence of living in littoral mudflat habitats prone to periodic hypoxia (Lee *et al.*, 2005).

Recent experiments with an artificial burrow system demonstrate that *Scartelaos* can sense the O_2 content of its burrow air and regulate this by removing old and adding new air (Lee *et al.*, 2005).

CONCLUSIONS

Species of the gobiid subfamily Oxudercinae offer insight into the sequence of character acquisition leading from aquatic to terrestrial life in the early vertebrates. Relative to the other oxudercines, the four genera of mudskippers (*Scartelaos*, *Boleophthalmus*, *Periophthalmus*, and *Periophthalmodon*) are much more specialized for amphibious air breathing and life out of water. In addition to air breathing, the progressive development of amphibious capacities by mudskippers has led to selection for specializations not present in other oxudercines. Further, there are graded differences among the mudskipper genera and among species within each genus that correlate with the range of their amphibious capabilities. *Scartelaos* is more confined to water than the other three genera; next is *Boleophthalmus*, followed by the more amphibious *Periophthalmus*, and finally *Periophthalmodon*. Differences among these genera relating to greater levels of terrestriality include: 1) Modifications in gill structure and reductions in surface area, culminating in the nearly complete loss of gill respiratory function in *Periophthalmodon*, 2) Novel epidermal structures favoring aerial cutaneous respiration while minimizing desiccation and skin permeability to ammonia, 3) Reliance upon continued air access for maintenance of normal metabolic rate and acid-base homeostasis, and for repayment of a post-exercise O_2 debt, 4) A large buccal chamber volume to body mass ratio, 5) Behavioral maintenance of the burrow-water micro-environment for pH, ammonia concentration, and air phase O_2, both for egg incubation and possibly for respiration during burrow confinement and 6) An oval eye lens for increased visual acuity in air.

Acknowledgements

We thank the following individuals for supporting research endeavors of this laboratory by hosting us at their institutions: K.K. Huat, School of Biological Sciences, Universiti Sains Malaysia; A. Ishimatsu, Institute for East China Sea Research, Nagasaki University, Tairamachi, Nagasaki, Japan; T. Done, Australian Institute of Marine Science (AIMS), Queensland, Australia. We thank P. Dayton for aid in establishing the collaboration at AIMS and in identifying study sites, M. Sheaves for assistance with obtaining an Australian collecting permit. Field studies and collections in Australia were aided by R. Lee, J. Fenger, and J. Skene. N. Aguilar transported fish specimens from Malaysia to SIO. Ms. N. Itoki provided field assistance and aided laboratory research in Japan, and also shipped specimens to SIO. A. Ishimatsu and T. Takita, both of Nagasaki University provided specimens used in the SEM studies. H. Robbins of La Jolla, CA and Evelyn York of the SIO Unified Laboratory Facility provided technical support. We thank R. Burton, P. Hastings, O. Mathieu-Costello, R. Rosenblatt, and R. Shadwick for reviewing drafts of this chapter. Part of the research reported in this paper is from the doctoral dissertation of H. Lee. This research was supported by NSF (IBN 9604699 and IBN 0111241) and the University of California Pacific Rim Research Program (01TPRRP06 0187 and 02T PRRP4 0065) and the UCSD Academic Senate. The SEM studies were partially subsidized by the SIO Unified Laboratory Facility. Australian research was conducted under collecting permit number PXH00430H from the Queensland Fisheries Management Authority. Studies on live mudskippers at SIO were done in accordance with protocols 6-073-06 approved by the University of California, San Diego Institutional Animal Care and Use Committee.

References

Aguilar, N.M., A. Ishimatsu, K. Ogawa and K.K. Huat. 2000. Aerial ventilatory responses of the mudskipper, *Periophthalmodon schlosseri*, to altered aerial and aquatic respiratory concentrations. *Comparative Biochemistry and Physiology* A127: 285-292.

Chew, S.F., L.N. Hong, J.M. Wilson, D.J. Randall and Y.K. Ip. 2003. Alkaline environmental pH has no effect on ammonia excretion in the mudskipper *Periophthalmodon schlosseri* but inhibits ammonia excretion in the related species *Boleophthalmus boddaerti*. *Physiological and Biochemical Zoology* 76: 204-214.

Clayton, D.A. and T.C. Vaughan. 1986. Territorial acquisition in the mudskipper *Boleophthalmus boddaerti* (Pisces: Gobiidae) on the mudflats of Kuwait. *Journal of Zoology* (London) 209: 501-519.

Clayton, D.A. 1993. Mudskippers. *Oceanography and Marine Biology Annual Review* 31: 507 - 577.

Fernald, R.D. 1990. The optical system of fishes. In: *The Visual System of Fish*, R.H. Douglas and M.B.A. Djamgoz (eds.). Chapman and Hall, London, pp. 45-61.

Fernandes, M.N., F.T. Rantin, A.L. Kalinin and S.E. Moron. 1994. Comparative study of gill dimensions in three erythrinid species in relation to their respiratory function. *Canadian Journal of Zoology* 72: 160-165.

Gee, J. H. and P. A. Gee. 1991. Reactions of gobioid fishes to hypoxia: Buoyancy control and aquatic surface respiration. *Copeia* 1991: 17-28.

Graham, J.B. 1971. Aerial vision in amphibious fishes. *Fauna* 3: 14-23.

Graham, J.B. 1997. *Air-Breathing Fishes: Evolution, Diversity and Adaptation*, Academic Press, San Diego.

Graham, J.B. and H.J. Lee. 2004. Breathing air in air: In what ways might extant amphibious fish biology relate to prevailing concepts about early tetrapods, the evolution of vertebrate air breathing, and the vertebrate land transition? *Physiological and Biochemical Zoology* 77: 720-731.

Gray, I.E. 1954. Comparative study of the gill area of marine fishes. *Biological Bulletin* 107: 219-225.

Hoese, D.F. 1984. Gobioidei: Relationships. In: *Ontogeny and Systematics of Fishes.* Special Publication Number 1, American Society of Ichthyologists and Herpetologists, Lawrence, Kansas, pp. 588-591.

Hora, S.L. 1935a. Ecology and bionomics of the gobioid fishes of the Gangetic Delta. *International Congress on Zoology* 12: 841-865.

Hora, S.L. 1935b. Physiology, bionomics, and evolution of the air-breathing fishes of India. *Transactions of the National Institute of Science India* 1: 1-16.

Hughes, G.M. 1966. The dimensions of fish gills in relation to their function. *Journal of Experimental Biology* 45: 177-195.

Hughes, G.M. 1984. Measurement of gill area in fishes: Practices and problems. *Journal of the Marine Biological Association of the United Kingdom* 64: 637-655.

Hughes, G.M. and J.S.D. Munshi. 1979. Fine structure of the gills of some Indian air-breathing fishes. *Journal of Morphology* 160: 169-194.

Hughes, G.M. and N.K. Al-Kadhomiy. 1986. Gill morphometry of the mudskipper, *Boleophthalmus boddaerti. Journal of the Marine Biological Association of the United Kingdom* 66: 671-682.

Ip, Y.K., S.F. Chew and D.J. Randall. 2004a. Five tropical air-breathing fishes, six different strategies to defend against ammonia toxicity on land. *Physiological and Biochemical Zoology* 77: 768-782.

Ip, Y.K., D.J. Randall, T.K.T. Kok, C. Barzaghi, P.A. Wright, J.S. Ballantyne, J.M. Wilson and S.F. Chew. 2004b. The giant mudskipper *Periophthalmodon schlosseri* facilitates active NH_4^+ excretion by increasing acid excretion and decreasing NH_3 permeability in the skin. *Journal of Experimental Biology* 207: 787-801.

Ishimatsu, A., K.H. Khoo and T. Takita. 1998a. Deposition of air in burrows of tropical mudskippers as an adaptation to the hypoxic mudflat environment. *Science Progress* 81: 289-297.

Ishimatsu, A., Y. Hishida, T. Takita, T. Kanda, S. Oikawa, T. Takeda and K.H. Khoo. 1998b. Mudskippers store air in their burrows. *Nature* (London) 391: 237-238.

Ishimatsu, A., N.M. Aguilar, K. Ogawa, Y. Hishida, T. Takeda, S. Oikawa, T. Kanda and K.K. Huat. 1999. Arterial blood gas levels and cardiovascular function during varying environmental conditions in a mudskipper, *Periophthalmodon schlosseri. Journal of Experimental Biology* 202: 1753-1762.

Ishimatsu, A., Y. Yoshida, N. Itoki, T. Takeda, H.J. Lee and J.B. Graham. Mudskippers brood their eggs in air but submerge them for hatching. (In Review).

Karsten, H. 1923. Das auge von *Periophthalmus koelreuteri. Zeitschrift für Naturwissenschaft* 59: 115-154.

Kobayashi, T., Y. Dotsu and T. Takita. 1971. Nest and nesting behavior of the mudskipper, *Periophthalmus cantonensis* in Ariake Sound. *Contributions from the Fisheries Experimental Station, Nagasaki University* 28: 27-39.

Kok, W.K., C.B. Lim, T.J. Lam and Y.K. Ip. 1998. The mudskipper *Periophthalmodon schlosseri* respires more efficiently on land than in water and vice versa for *Boleophthalmus boddaerti. Journal of Experimental Zoology* 280: 86-90.

Lee, H.J. and J.B. Graham. 2002. Their game is mud. *Natural History* 9: 42-47.

Lee, H.J., C.A. Martinez, K.J. Hertzberg, A.L. Hamilton and J.B. Graham. 2005. Burrow air phase maintenance and respiration by the mudskipper, *Scartelaos histophorus* (Gobiidae: Oxudercinae). *Journal of Experimental Biology* 208: 169-177.

Lin, H.-C. and W.-T. Sung. 2003. The distribution of mitochondria-rich cells in the gills of air-breathing fishes. *Physiological and Biochemical Zoology* 76: 215-228.

Low, W.P., D.J.W. Lane and Y.K. Ip. 1988. A comparative study of terrestrial adaptations of the gills in three mudskippers—*Periophthalmus chrysospilos, Boleophthalmus boddaerti,* and *Periophthalmodon schlosseri. Biological Bulletin* 175: 434-438.

Low, W.P., Y.K. Ip and J.W. Lane. 1990. A comparative study of the gill morphometry in the mudskippers—*Periophthalmus chrysospilos, Boleophthalmus boddaerti* and *Periophthalmodon schlosseri. Zoological Science* 7: 29-38.

Mathis, U., F. Schaeffel and H.C. Howland. 1988. Visual optics in toads. *Journal of Comparative Physiology* A163: 201-213.

Milward, N.E. 1974. Studies on the taxonomy, ecology, and physiology of Queensland mudskippers. Unpublished Ph.D. dissertation, University of Queensland, Brisbane.

Murdy, E.O. 1989. A taxonomic revision and cladistic analysis of the oxudercine gobies (Gobiidae: Oxudercinae). *Records of the Australian Museum, Supplement* 11: 1-93.

Murdy, E.O. and T. Takita. 1999. *Periophthalmus spilotus,* a new species of mudskipper from Sumatra (Gobiidae: Oxudercinae). *Ichthyological Research* 46: 367-370.

Niva, B., J. Ojha and J.S.D. Munshi. 1981. Morphometrics of the respiratory organs of an estuarine goby, *Boleophthalmus boddaerti. Japanese Journal of Ichthyology* 27: 316-326.

Park, J.Y. 2002. Structure of the skin of an air-breathing mudskipper, *Periophthalmus magnuspinnatus. Journal of Fish Biology* 60: 1543-1550.

Randall, D.J., J.M. Wilson, K.W. Peng, T.W.K. Kok, S.S.L. Kuah, S.F. Chew, T.J. Lam and Y.K. Ip. 1999. The mudskipper, *Periophthalmodon schlosseri,* actively transports NH_4^+ against a concentration gradient. *American Journal of Physiology* 277: R1562-R1567.

Randall, D.J., Y.K. Ip, S.F. Chew and J.M. Wilson. 2004. Air breathing and ammonia excretion in the giant mudskipper, *Periophthalmodon schlosseri. Physiological and Biochemical Zoology* 77: 783-788.

Schöttle, E. 1931. Morphologie und physiologie der atmung bei wasser-, schlamm- und landlebenen Gobiiformes. *Zeitschrift für Wissenschaft Zoologie* 140: 1-114.

Sivak, J.G. 1980. Avian mechanisms for vision in air and water. *Trends in Neuroscience* 12: 314-317.

Sivak, J.G. 1990. Optical variability of the fish lens. In: *The Visual System of Fish,* R.H. Douglas and M.B.A. Djamgoz (eds.). Chapman and Hall, London, pp. 63-80.

Sivak, J.G., H.C. Howland, J. West and J. Weerheim. 1989. The eye of the hooded seal, *Cystophora cristata,* in air and water. *Journal of Comparative Physiology* A165: 771-777.

Suzuki, N. 1992. Fine structure of the epidermis of the mudskipper, *Periophthalmus modestus* (Gobiidae). *Japanese Journal of Ichthyology* 38: 379-396.

Swennen, C., N. Ruttanadakul, M. Haver, S. Piummongkol, S. Prasertosongskum, I. Intanai, W. Chaipakdi, P. Yeesin, P. Horpei and S. Detsathit. 1995. The five sympatric mudskippers (Teleostei: Gobioidea) of Pattani area, southern Thailand. *Natural History Bulletin of the Siam Society* 42: 109-129.

Takeda, T., A. Ishimatsu, S. Oikawa, T. Kanda, Y. Hishida and K.H. Khoo. 1999. Mudskipper *Periophthalmodon schlosseri* can repay oxygen debts in air but not in water. *Journal of Experimental Zoology* 284: 265-270.

Tamura, O. and T. Moriyama. 1976. On the morphological feature of the gill of amphibious and air-breathing fishes. *Bulletin of the Faculty of Fisheries, Nagasaki University* 41: 1-8.

Tamura, O., H. Morii and M. Yuzuriha. 1976. Respiration of the amphibious fishes *Periophthalmus cantonensis* and *Boleophthalmus chinensis* in water and on land. *Journal of Experimental Biology* 65: 97-107.

Teal, J.M. and F.G. Carey. 1967. Skin respiration and oxygen debt in the mudskipper *Periophthalmus sobrinus. Copeia* 1967: 677-679.

Walls, G.L. 1942. *The Vertebrate Eye and its Adaptive Radiation.* Cranbrook Institute of Science, Bloomfield Hills, Michigan.

Wilson, J.M., T.W.K. Kok, D.J. Randall, W.A. Vogl and Y.K. Ip. 1999. Fine structure of the gill epithelium of the terrestrial mudskipper, *Periophthalmodon schlosseri. Cell and Tissue Research* 298: 345-356.

Wilson, J.M., D.J. Randall, M. Donowitz, A.W. Vogl and Y.K. Ip. 2000. Immunolocalization of ion-transport protein to branchial epithelium mitochondria-rich cells in the mudskipper (*Periophthalmodon schlosseri*). *Journal of Experimental Biology* 203: 2297-2310.

Yadav, A.N. and B.R. Singh. 1989. Gross structure and dimensions of the gill in an air-breathing estuarine goby, *Pseudapocryptes lanceolatus. Japanese Journal of Ichthyology* 36: 252 - 259.

Yadav, A.N., M.S. Prasad and B.R. Singh. 1990. Gross structure of the respiratory organs and dimensions of the gill in the mudskipper, *Periophthalmodon schlosseri* (Bleeker). *Journal of Fish Biology* 37: 383-392.

Zhang, J., T. Taniguchi, T. Takita and A.B. Ali. 2000. On the epidermal structure of *Boleophthalmus* and *Scartelaos* mudskippers with reference to their adaptation to terrestrial life. *Ichthyological Research* 47: 359-366.

Zhang, J., T. Taniguchi, T. Takita and A.B. Ali. 2003. A study on the epidermal structure of *Periophthalmodon* and *Periophthalmus* mudskippers with reference to their terrestrial adaptation. *Ichthyological Research* 50: 310-317.

Respiratory Function in the South American Lungfish, *Lepidosiren paradoxa*

Mogens L. Glass[1,*], A.P. Sanchez[1], J. Amin-Naves[1], M. Bassi[1] and F.T. Rantin[2]

THE LUNGFISH

The sarcopterygian (lobe-finned) fish gave rise to the lungfish (Dipnoi), the coelacanths (Actinistia) and the land vertebrates (Tetrapoda) (Carroll, 1988). Probably, the lungfish form the sister group relative to the land vertebrates (Meyer and Dolven, 1992; Yokobori *et al.*, 1994; Zardoia *et al.*, 1998; Toyama *et al.*, 2000). The Australian lungfish (*Neoceratodus forsteri*) seems similar to the Upper Devonian *Dipterus* and depends on its gills for adequate O_2 uptake. In contrast, the Lepidosireniformes are obligatory air-breathers with well-developed true lungs and rudimentary gills. They are represented by *Protopterus* (4 species) in Africa and

Authors' addresses: [1]Department of Physiology, Faculty of Medicine of Ribeirão Preto, University of São Paulo, Avenida Bandeirantes, 3900, 14049-900 Ribeirão Preto, SP, Brazil. [2]Laboratory of Zoophysiology and Comparative Biochemistry, Department of Physiological Sciences, Federal University of São Carlos, via Washington Luis, Km 235, 13565-905 São Carlos, SP, Brazil.
Corresponding author: E-mail: mlglass@rfi.fmrp.usp.br

Lepidosiren paradoxa in South America (Johansen and L.J. Lenfant, 1967). The Austrian L.J. Fitzinger (1837) gave the animal this name when amazed by the combination of gills and a well-developed lung.

The focus of this chapter is on *Lepidosiren paradoxa*, which we have studied in some detail over the last 6 years and our purpose is to place the data within a broader context.

Lepidosiren inhabits the central part of the Amazon and Paraná-Paraguai regions (Lundberg, 1993). The latter region is characterized by vegetation-covered, shallow lakes covered with mud at the bottom. These lakes expand during the rainy season and may cover large areas. The South American lungfish feeds on mollusks and other invertebrates but may also eat vegetables (Sawaya, 1946). Like amphibians, lungfish have larval forms with external gills. Aerial respiration is initiated when a length of 2 to 3 cm is reached in *P. aethiopicus* (Greenwood, 1958). Johansen *et al.* (1976) studied the relative O_2-uptake from water and air in *P. amphibius*, weighing 3 to 250 g. Within this range, the aquatic O_2-uptake became reduced from 70% of total to only 10-15%, which reflects an increasing role of pulmonary respiration.

THE ADVANTAGE OF A LUNG OR, ALTERNATIVELY, AN AIR-BREATHING ORGAN

Water at 20°C has an O_2 capacitance coefficient ($\beta O_2 = \Delta[O_2]/\Delta PO_2$) which is only about 1/30 of the value for atmospheric air (Dejours, 1981). In other words, when equilibrated to the same PO_2, air will contains about 30 times more O_2 than an equal volume of distilled water. Moreover, the Krogh diffusion coefficient (KO_2) in water is about 2×10^5 times lower than in air, which causes O_2 gradients within the habitat and these may change on a diurnal and/or seasonal basis. To cope with these ever changing conditions, exclusively water breathing fish have an O_2-oriented regulation of gill ventilation. Teleost fish possess gill arch receptors, screening PO_2 of the inspired water and the blood (Burleson and Milsom, 1990, 1995 a and b; Soncini and Glass, 2000). A decrease of water O_2-levels stimulates gill ventilation, and this response may keep blood gases at their normocarbic level (Soncini and Glass, 2000). There are, however, alternative ways of coping with exposure to hypoxic water.

Many teleost fish possess an air-breathing organ (ABO), which would increase the chances of survival during severe aquatic hypoxia. As another advantage, an ABO may increase the aerobic scope during elevated activity and/or at higher temperature (Lomholt and Johansen, 1974, 1976). Additional aerial respiration is common in species that are exposed to temporary or seasonal hypoxia. Thus, gas exchange organs have evolved independently in teleost fish based on non-pulmonary structures such as part of the gill system, the swimbladder or part of the digestive system (Graham, 1997). Atmospheric O_2 levels could have influenced development of lungs and ABOs. Recent data indicate O_2-levels of 15% O_2 during the Silurian and Devonian periods, which was less hypoxic than stated in earlier estimates (Dejours, 1981; Dudley *et al.*, 1998).

How advanced is the lung of Lepidosireniformes? Morphological and physiological studies have addressed this question (Bassi *et al.*, in press; Moraes *et al.*, in press). The physiological diffusing capacity for O_2 (D_LO_2) can be defined by a conductance equation:

$$D_LO_2 = VO_2 \cdot (P_LO_2 - PcO_2)^{-1},$$

where $P_LO_2 - PcO_2$ represents the mean O_2 partial pressure difference across the tissue barrier, separating lung gas and pulmonary capillary blood (Bohr, 1909). This difference is often written as DPO_2 (Bassi *et al.*, in press). In *Lepidosiren*, pulmonary gas exchange accounts for more than 95% of total VO_2 at elevated temperatures (Sawaya, 1946; Johansen and Lenfant, 1967). This dependence on the lung for O_2-uptake is reflected in a diffusing capacity that is practically identical to values of amphibians (Glass *et al.*, 1981a; Bassi *et al.*, in press) (See Fig. 15.1). *Lepidosiren* and *Rana* virtually have the same physiological diffusing capacity, while values for some reptiles are somewhat higher (Glass *et al.*, 1981b; Glass, 1989). A much larger increase of diffusing capacity is related to the transition from ectothermic to endothermic metabolism. Birds and mammals have a tenfold greater D_LO_2, when compared to that of reptiles, amphibians and lungfish. Even when compared at 37°C, an ectothermic vertebrate has an O_2-uptake of only 20% of that of a mammal. This difference in metabolic strategy is expressed at all levels of metabolism, including mitochondrial density of tissues (Else and Hulbert, 1981).

Alternatively, the diffusing capacity of a lung can be evaluated by morphometric measurements, using the Fick diffusion equation applied to the lung: $DO_2 = (A/T) \times KO_2$, where A = respiratory surface area of the lung, T = the thickness of the gas to blood barrier and KO_2 = the Krogh diffusing constant for the tissue at specified temperature. The morphometric diffusing capacities confirm the relative values for the vertebrate classes. There is, however, a tendency that physiological values

Fig. 15.1 Photo of the South American lungfish, *Lepidosiren paradoxa*.

are lower than morphometric estimates (cf. Perry, 1989) which indicates that the morphometric measurement provides an upper limit.

OXYGEN-HEMOGLOBIN DISSOCIATION CURVES (ODCs)

Teleost fish are characterized by ODCs of a generally low n_{Hill} (1 to 1.5) (cf. Albers *et al.*, 1983; Soncini and Glass, 2000), which implies that the co-operation of the chains is low and that the ODC is virtually hyperbolic. Conversely, sigmoid curves (n_{Hill} high) result from a high co-operation between the chains. Most anuran amphibians have such sigmoid curves, which permits a steep gradient for unloading O_2 to systemic tissues (Boutilier *et al.*, 1986; Boutilier and Heisler, 1988; Andersen *et al.*, 2001).

In an early study on *Lepidosiren*, Johansen and Lenfant (1967) obtained ODCs at 23°C with PCO_2 = 6 or, alternatively, 27 mmHg. They reported a sigmoid ODC with n_{Hill} ~ 2, which exceeds values for teleost fish. In *Lepidosiren*, the combination of a high intrapulmonary PO_2 and a sigmoid ODC, which leads to a nearly complete HbO_2-saturation and a high PaO_2 (60 to 90 mmHg) (Bassi *et al.*, in press).

Blood from teleosts may contain several types of hemoglobin, but linear Hill-plots in *Lepidosiren* indicated that only one type of hemoglobin was present (Bassi *et al.*, in press), which is consistent with Rodewald *et al.* (1984). Both n_{Hill} and the apparent heat of oxygenation heat (DH) fall within the range for amphibians (Andersen *et al.*, 2001; Bassi *et al.*, in press).

ODCs are also available for the Australian lungfish *Neoceratodus* (Kind *et al.*, 2002) and are consistent with those for *Lepidosiren*. Data are available for $PaCO_2$ (only 3.5 mmHg), and this low $PaCO_2$ permitted a 95% saturation of the blood with a PaO_2 of only 40 mmHg (Lenfant *et al.*, 1966). The low $PaCO_2$ reflects the high dependence on aquatic respiration of the species.

Oxygen dissociation curves are available for *Protopterus aethiopicus* (Lenfant and Johansen, 1968) and are compatible with those for *Lepidosiren*. A low PaO_2, however, seems incompatible with the position of the ODCs. During aerial respiration, the mean PaO_2 was only about 30 mmHg (n = 3), the interpretation of data is difficult.

PULMONARY MECHANICS

Like in amphibians, the mechanics of pulmonary ventilation in lungfish is based on positive pressure inflation of the lung (McMahon, 1969). A ribcage is absent in amphibians, whereas lungfish have ribs that are not involved in respiratory function (Foxon and Bishop, 1968). The ancestral amphibians had solid ribs, which were probably used for sustention and not involved in lung ventilation. In lungfish, buccal movements are both involved in gill and lung ventilation. The pressure changes for lung ventilation are much larger than those for gill irrigation (Johansen and Lenfant, 1967; MacMahon, 1969). Reaching the surface, *Protopterus* closes the mouth and compresses the buccal cavity to expel water via the operculum. Then the buccal and opercular cavities expand while the mouth is open. Subsequently, the glottal sphincter muscles open and the lung gas exits from the lung. This is followed by one or more movements that force buccal gas into the lung (MacMahon, 1969; Lomholt, 1993). MacMahon (1969) emphasized that 'the air-breathing mechanism of tetrapods was powered by a buccal force-pump mechanism which evolved directly from the aquatic system'. Many similarities exist between the respiratory mechanics in lungfish and anuran amphibians. Thus, toads (*Bufo* sp.) often inflate the lungs by a series of buccal movements (Kruhøffer *et al.*, 1987; Branco *et al.*, 1992).

There is, however, a striking difference between the sizes of the tidal volume. Lungfish have a large tidal volume even when at rest (~30 ml/kg) (Lomholt 1993; Sanchez *et al.*, in press). By contrast, the tidal volumes in anuran amphibians are small (~2.5 ml/kg). This difference reflects that lungfish spend much time submerged and, therefore, have a low respiratory frequency. Conversely, toads and frogs respire at a higher frequency and a low tidal volume (cf. Branco *et al.*, 1992).

CONTROL OF PULMONARY VENTILATION

Most terrestrial vertebrates inhabit environments of constant O_2 availability. For this reason, their ventilatory responses are linked to regulation of acid-base status of the blood rather than to O_2 homeostasis (cf. Dejours, 1981). This regulatory pattern involves central chemoreceptors bi-laterally located in the ventral region of the medulla oblongata and, probably, also in a dorsal position (Wellner-Kienitz and Shams, 1998).

There is no evidence for central chemoreceptors in teleost fish. Consistently, Hedrick *et al.* (1991) applied central superfusion of mock CSF to the ventricular system of the holeost fish *Amia calva*, which failed to alter patterns of aerial ventilation. Its relative, the long nosed garpike

(*Lepisosteus osseus*) gave a different but somewhat controversial result. Wilson *et al.* (2000) found that hypercarbia increased the frequency of the air-breathing motor output from *in vitro* brain stem of this animal. Earlier, Smatresk and Cameron (1982) had earlier reported a modest increase of aerial ventilation in the garpike (*Amia calva*) exposed to hypercarbia.

As mentioned above, the terrestrial vertebrates and the Dipnoi are descendants from the lobe-finned fish Sarcopterygii, whereas the teleost fish originated from Actinopterygii. All classes of land vertebrates possess central chemoreceptors that monitor acid-base status of the blood and the cerebrospinal fluid (Hitzig and Jackson, 1978; Smatresk and Smits, 1991; Branco *et al.*, 1992; Wellner-Kienitz and Shams 1998; Ballantyne *et al.*, 2001; Milsom, 2002). This raises questions of the status of the lungfish. As expressed by Fishman *et al.* (1989): '...the sensory mechanism mediating these hypercapnic responses are unknown, and it is currently uncertain whether or not there is a structure equivalent to the central chemoreceptors in mammals.' This comment was based on frequencies of lung and gill ventilation rather than on ventilated volumes.

In tetrapods, the central chemoreceptors are involved in a negative feed-back control of acid-base status, according a relationship:

$$V_{eff}/VCO_2 = R \cdot T/P_L CO_2,$$

where V_{eff} = effective ventilation of the gas exchange regions of the lung; VCO_2 = rate of elimination of CO_2 from the lung; R = the gas constant; T = absolute temperature and $P_L CO_2$ = the CO_2 partial pressure of the lung gas.

Direct measurements of pulmonary ventilation in *Lepidosiren* were recently performed during application of hypercarbic gases and/or superfusion with mock CSF solutions. These superfusion procedures clearly indicated the presence of central chemoreceptors involved in chemical control of pulmonary ventilation (Sanchez *et al.*, 2001; Sanchez *et al.*, in press). Superfusion of mock CSF with step-wise decreasing pH increased pulmonary ventilation of *Lepidosiren*. With return to a higher pH, pulmonary ventilation returned to its original set point (Sanchez *et al.*, 2001).

Lepidosiren and *Protopterus* also possess intrapulmonary CO_2-sensitive receptors, that are stimulated by stretch and inhibited by increased CO_2 levels (Delaney *et al.*, 1983), and these were located within the gas exchange region of the lung rater than in the upper airways. In general, this type of chemoreceptor is present in the upper airways and/or in the gas exchange region of the lung of land vertebrates with one exception – the mammals (Milsom, 1995, 2002). As a hallmark of these receptors, a shift from hypercarbia to inhalation of atmospheric air causes a transient increase of ventilation, which exceeds the ventilation during the previous hypercarbic exposure. The consequent reduction of intrapulmonary CO_2 elicits a transient increase of ventilation, which initially exceeds the previous hypercarbic response levels (Milsom, 1995). This response known as 'post-hypercarbic hyperpnea' was also obtained in *Lepidosiren* (Sanchez and Glass, 2001). Similar responses occur in amphibians (Kinkead and Milsom, 1996), indicating an early origin of these intrapulmonary receptors.

Protopterus aethiopicus possess a pulmonary control that includes a mechanism, which functions very much like the classical Hering-Breuer reflex, which is particularly interesting, since the lung mechanics are based on inflation by buccal pressure (Pack *et al.*, 1990, 1992). As a further feature, surfactant is present and its composition in *Lepidosiren* and *Protopterus* is very similar to that of amphibians, whereas the composition in *Neoceratodus* is different (Orgeig and Daniels, 1995).

The information cited above indicates that the basic features of pulmonary function evolved early and before the transition to the terrestrial mode of life. Moreover, amphibians and lungfish share many very specific components of respiratory control. This suggests that these key features were present in the common ancestor of the two groups. This assumption seems more likely than the suggestion that very specific components were developed independently in amphibians and lungfish. As expressed by Johansen *et al.* (1967): '...vertebrates acquired functional lungs before they possessed a locomotor apparatus for invasion of the terrestrial environment.'

An intriguing problem is the function of gills in extant lungfish, in particular in *Lepidosiren*, which has a negligible surface (Moraes *et al.*, in press). Johansen and Lenfant (1967) reported that gill movements in *Lepidosiren* were almost imperceptible at rest. They further reported that stirring of the water caused large increases of gill movements, even when visual and acoustical disturbances were absent. Responses to aquatic hypercarbia and hypoxia were also inconclusive, possibly because of the sensitivity to agitation of the water. It was possible to measure a 30% O_2 extraction from the water exiting the gills. This value is low relative to those of teleost fish (60-90% extraction). The extraction in *Lepidosiren* could even result from a high metabolism of local cells that are not involved in gas exchange. More likely, aquatic gas exchange occurs via the skin. Relatively small O_2 losses to nearly anoxic water in adult specimens (Abe and Steffensen, 1996b) but were absent in small specimens (Lomholt and Glass, 1987).

Johansen and Lenfant (1968) also studied *Protopterus aethiopicus* and found that aquatic hypercarbia depressed gill ventilation and enhanced surface breathing. Gill movements increased with low levels of hypercarbia but were progressively depressed with further increases of CO_2 (Jesse *et al.*, 1967).

REGULATION OF ACID-BASE STATUS

In teleost fish, acid-base status is predominantly regulated by active modulation of $[HCO_3^-]pl$, whereas a regulation by gill ventilation is limited or absent (Dejours, 1981; Milsom, 2002). More than 90% of transfers of acid-base relevant ions are performed by specialized cells of the gill epithelia with a minor contribution by the kidneys (Heisler, 1984). Hypercarbia-exposed teleosts may increase $[HCO_3^-]_{pl}$ and return pHa close to the set point (Heisler, 1984; Claiborne and Heisler, 1986). In this regard, few data are available for lungfish.

As stated by Dejours (1981) there is a tendency that "the more an animal depends on pulmonary, the higher it's PCO_2." *Lepidosiren* has a high $PaCO_2$ (~ 27 mmHg), which reflects its dependence on the lung and partly prolonged dives. At 25°C, however, as much as 50% of total CO_2-elimination is aquatic (Amin-Naves *et al.*, 2004). Its habitat may turn hypercarbic, which would increase $PaCO_2$ and lower pHa. If this happens, *Lepidosiren* increases ventilation to match rising levels of CO_2 (Sanchez *et al.*, in press). As a consequence, the animal maintained $PaCO_2$ below that of the water (49 mmHg), and this was possible because the animal exhaled part of the surplus CO_2 to the normocarbic atmosphere (Sanchez *et al.*, in press). Consistent data are available for the urodela salamanders *Amphiuma means* and *Siren lacertina*, which were exposed to aquatic hypercarbia that often occurs in their habitat (Heisler *et al.*, 1982a). Some urodela amphibians achieve a partial compensation of extra-cellular pH, as in predominantly skin breathing salamander *Cryptobranchus alleganiensis* (Boutilier and Toews, 1981). It should also be pointed out that anuran amphibians exposed to hypercarbia are capable of a partial compensation of pHa (Alvarado *et al.*, 1975; Boutilier and Heisler, 1988; Toews and Stiffler, 1990; Snyder and Nestler, 1991; Busk *et al.*, 1997).

The set points for normocarbic blood gases in *Lepidosiren* (25°C) are: pHa = 7.5; PCO_2 = 22 mmHg; [HCO_3^-]pl = 24 mM; PaO_2 = 81 mmHg (Amin-Naves, 2004). These values are in agreement with data for *Protopterus aethiopicus* at the same temperature: pHa = 7.6; PCO_2 = 26 mmHg; [HCO_3^-]pl = 31.6 mM (Delaney *et al.*, 1977). These values are distinct from those of anuran amphibians. Thus, at 27°C, the toad *Bufo paracnemis* has a pHa of 7.73 and a $PaCO_2$ close to 8 mmHg (Wang *et al.*, 1998). In part, this reflects the prolonged dives that characterize lungfish. In fact, the values are close to those for the snakehead fish (*Channa argus*) that ventilates its air-breathing organ (modified part of the gill system) at low frequency (Ishimatsu and Itazawa, 1983). The pHa of *Lepidosiren* is very low when compared to teleost fish (normal range: 7.7 < pH < 8.1, 25°C; cf. Cameron, 1984; Heisler, 1984) and amphibians (normal range: 7.6 < pH < 7.9, 25°C; cf. Boutilier *et al.*, 1986; Kruhøffer *et al.*, 1987).

In a pioneering study, Johansen and Lenfant (1967) obtained some blood gas samples from *Lepidosiren* of PaO_2 ~ 35 mmHg (20°C), which is considerably lower than PaO_2 ~ 90 mmHg as measured by Amin-Naves *et al.* (2004). Johansen and Lenfant (1967) catheterized the pulmonary artery and the dorsal aorta, which might have influenced blood gases. As an alternative, it should pointed out that their animals were small (104-212 g) with non-functional traces of external gills and, therefore, the low values could depend on size of the animals.

EFFECTS OF BIMODAL RESPIRATION AND TEMPERATURE-DEPENDENT ACID-BASE STATUS

A general tendency in ectothermic vertebrates is that pHa decreases with rising temperature (Robin, 1962; Reeves, 1972; Heisler, 1984). Rahn (1966) proposed that this negative $\Delta pHa/\Delta t$ would parallel $\Delta pH/\Delta t$ of neutral water, which requires a reduction of pHa by 0.017 units/°C (temperature range 15–30°C). An alternative hypothesis was proposed by Reeves (1972), who claimed that ventilation and PCO_2 are regulated so that the fractional dissociation of peptide-linked histidine imidazole is kept constant. The pK value of imidazole changes with temperature by – 0.018 to –0.024 units/°C (temp. ~25°C), depending on ligands and steric arrangements (Edsall and Wyman, 1958). The tendency is that $\Delta pHa/\Delta t$ is lesser than predicted from the range cited above (Heisler, 1984).

Considering this discrepancy, it seems more useful to discuss how this negative $\Delta pH/\Delta t$ is achieved. In trout fish, this negative $\Delta pH/\Delta t$ is achieved by adjustments of $[HCO_3^-]$pl (Randall and Cameron, 1973), whereas PCO_2-dependent adjustments dominate in amphibians and reptiles. Reptiles adjust pulmonary ventilation to CO_2-output and non-pulmonary gas is negligible for most species. Additional CO_2 elimination by the skin and, in some cases, also the gills must be taken into account in amphibians (Jackson, 1978; Wang et al., 1998).

Temperature-dependent regulation of blood acid-base status in *Lepidosiren* is similar to that of amphibians (Amin-Naves et al., 2004). In amphibians, the lung becomes increasingly important for gas exchange as temperature rises, because skin respiration is a diffusive process with a limited capacity for gas exchange (Jackson, 1978). Likewise, many air-breathing teleost fish depend on an increased role of aerial respiration at high temperature (Lomholt and Johansen, 1974; Glass et al., 1986). As a general rule: In combined aquatic and aerial gas exchange, the tendency is that the aerial exchanger plays the major role when total VO_2 increases (Burggren et al., 1983). Amphibians and lungfish share the feature of combined aerial and aquatic respiration. Thus, amphibians control acid-base status according to the equation (Glass and Soncini, 1995):

$$PaCO_2 = VCO_{2,\ tot}/(D_{cut} + V_{eff} \cdot \beta g),$$

where $VCO_{2,tot}$ = total CO_2, i.e., the sum of cutaneous and pulmonary output; D_{cut} = CO_2-diffusing capacity of the skin; V_{eff} is the effective (= equivalent to alveolar) ventilation of the lung and βg = capacitance coefficient of gases ($\beta g = 1/RT$). For lungfish and salamanders with external gills, the equation becomes:

$$PaCO_2 = VCO_{2,\ tot}/(D_{cut} + V_{eff} \cdot \beta g + Ggill),$$

where Ggill is the conductance of the gills. In *Lepidosiren*, the gill conductance is negligible (Moraes et al., in press) and the gas exchange equation can be simplified.

Temperature-dependent acid-base regulation in *Lepidosiren* is, in fact, very similar to that of many amphibians (Jackson, 1978, 1989).

In its Pantanal region habitat the water temperatures range from 18°C to 34°C (Harder et al., 1999). At 30 to 35°C, *Lepidosiren* obtains more than 95% of total O_2-uptake by pulmonary respiration (Sawaya, 1946; Johansen and Lenfant, 1967; Amin-Naves et al., 2004), but as much as 30% of total CO_2-output is eliminated to the water (Amin-Naves et al., 2004). In part, this reflects the relative Krogh diffusion constants for CO_2 and O_2 (Dejours, 1981).

The Krogh diffusion constant for CO_2 (KCO_2) increases by only 10%/10°C. This implies that cutaneous gas exchange cannot expand to meet larger needs for O_2 at high temperature and/or increased activity (Gottlieb and Jackson, 1976; Jackson, 1978, 1989; Mackenzie and Jackson, 1978; Burggren and Moalli, 1984; Wang et al., 1998). The solution to this problem is to increase pulmonary gas exchange to cope with increased metabolic demands at higher temperature.

As metabolism augmented with temperature, a larger fraction of total CO_2 produced became eliminated by the lung of Lepidosiren (compare: Wang et al., 1998; Amin-Naves et al., 2004). As a consequence, the pulmonary gas exchange ratio (R_E) rose with temperature, which increased intra-pulmonary PCO_2 (cf. Fenn et al., 1946). Thereby, $PaCO_2$ equilibrated to this higher intrapulmonary PCO_2, and the consequent increase of $PaCO_2$ produced a negative $\Delta pHa/\Delta t$. Meanwhile, bicarbonate levels remained constant.

Not only Bufo but also many other amphibians increase lung ventilation at high temperature, while the gas conductance of the skin changes little. Consistent data have been obtained for urodela salamanders (for a review see Ultsch and Jackson, 1996). At 5°C, 90 to 100% of VCO_2 was aquatic in the Congo eel, Amphiuma means, while at 25°C only 40 to 80% of the VCO_2 was cutaneous (Guimond and Hutchinson, 1974).

Such gas exchange patterns have also been reported for teleost fish of bimodal respiration. Studying the air-breathing Amphipnous cuchia, Lomholt and Johansen (1974) reported that R_E for the air-breathing organ (ABO) increased with temperature. Data for the snakehead fish, Channa argus are consistent (Glass et al., 1986).

AESTIVATION

Aestivation is a dormancy that may occur during the dry season. In this condition, metabolism can be reduced without any decrease of ambient temperature. Aestivating amphibians and reptiles remain little active in burrows during adverse conditions such as a dry season (Abe, 1995). Some amphibians form cocoon and the reductions of metabolism can be substantial (20 to 50% of previous non-aestivating resting conditions) (Seymour, 1973; Flanigan et al., 1991). This low metabolism is accompanied by a down-regulation of cardio-vascular function (Glass et al., 1997). Aestivation in lungfish has attracted special attention, due to the impressive survival of *Protopterus* within a cocoon for years (cf. Lomholt, 1993). *Protopterus* has been studied in detail (Smith, 1935; DeLaney *et al.*, 1974, 1977), whereas data for *Lepidosiren* are few.

The dry season of the Pantanal region starts in April, and the *Lepidosiren* may assume a U-shape posture, approaching its tail to the head, a posture that also has been observed in *Protopterus* (DeLaney *et al.*, 1974). *Lepidosiren* burrows into the clay without forming a cocoon. This is accompanied by a significant but transient reduction of heart rate, while air breath became more frequent (Harder *et al.*, 1999). There is a lack of information on what happens over longer periods, but it seems that aestivation in *Lepidosiren* is a less dramatic than in *Protopterus*. Abe and Steffensen (1996a) reported a 30% reduction of total VO_2 after 4 months of aestivation at 25°C. Non-pulmonary VO_2 was reduced considerable more than pulmonary uptake by the lung.

COMPARISON TO *PROTOPTERUS* sp.

Protopterus is represented by four species *P. aethiopicus*, *P. amphibius*, *P. annectens* and *P. dolloi*. They have a robust appearance and are less eel-like than *Lepidosiren*. Moreover, *Protopterus* possess much longer appendages. Over many years' information accumulated on the respiratory and cardiovascular characteristics of *Protopterus*. Delaney *et al.* (1974, 1977) reported on aestivation in *Protopterus aethiopicus*. This species inhabits shallow waters that dry during the hot season. *Protopterus* burrows into the mud, while the animal assumes a U-shaped posture with the mouth positioned to breathe air. In contrast to *Lepidosiren*, a cocoon develops to protect the animal. The cardiac frequency falls from 25 beats/min in water to 3 beats/min when in the cocoon, while O_2-uptake became reduced to half of the value in water.

There are examples of aestivation for 6 years in *Protopterus amphibius* (Lomholt, 1993). After that time, pulmonary ventilation had diminished from 5 to 1 mlBTPS·kg^{-1}·min^{-1}. Concomitantly, O_2-uptake fell from 0.70 to 0.05 mlSTPD kg^{-1}·min^{-1}. Concurrently, the peak values for expired PCO_2 and PO_2 were around 40 and 120 mmHg, which is not very far from the values of active fish in water. Data are also available for shorter periods of aestivation, which seems more relevant to the ecology of the animal. Such measurements were obtained for dry season torpor in *Protopterus aethiopicus* and indicate an active modulation of $[HCO_3^-]$pl. The waterproof cocoon accumulates CO_2 to produce a severe respiratory acidosis that is partially compensated by an increased $[HCO_3^-]_{pl}$ (DeLaney et al., 1977). The elevated $[HCO_3^-]_{pl}$ could, however, result from dehydration rather than active modulation (Lomholt, 1993).

In *Lepidosiren* the gill system that accounts for less than 0.1% of the total respiratory surface (Moraes et al., in press). This indicates that the skin, rather than the gill system, is involved in aquatic gas exchange. *Protopterus* seems to possess a more complex circulatory control, although the two genera have many anatomic components in common (Szidon et al., 1969; Laurent et al., 1978; Fishman et al., 1989). In *Protopterus*, the left side of the heart receives pulmonary venous blood that preferentially enters the ventral part of the bulbus. This part supplies the three anterior arches. Of these the first arch is equipped with gill system that supplies the brain. The arches II and III have no gill system and fuse to form the dorsal aorta.

The right-side part of the heart receives systemic venous return that preferentially enters the dorsal part of the bulbus, which supplies the arches III, IV, V and VI, carrying pulmonary arterial blood to the lung. A muscular and highly innervated ductus arteriosus provides a potential shunt to connect the pulmonary artery and the dorsal aorta. A detailed description is outside the scope of this text. While the system seems very sophisticated and advanced, a challenge is to evaluate its function under various environmental conditions. In amphibians, microsphere techniques has been applied to assess blood shunts and relative perfusion of the skin and the lung under conditions of differential O_2-availability in the gas phase and in the water Boutilier et al. (1986). Similar measurements in lungfish might be useful.

Little is known about circulatory function in *Lepidosiren*, but it is important that PaO_2 of this animal is high and exceeds the values for the better known amphibians, in particular at lower temperature. Different from the reptiles, the lung gas PO_2 of the anuran amphibians and *Lepidosiren* is relatively constant with temperature, which can be explained based on bimodal bimodal respiration (Wang et al., 1998; Amin-Naves et al., 2004; Bassi et al., in press). In amphibians and reptiles, however, large central vascular shunts limit values for PaO_2. In this situation, PaO_2 becomes determined by the magnitude of central systemic venous admixture rather than by exchange at the lung (Johansen and Ditadi, 1966; Wood and Hicks, 1985; Wood and Glass, 1991; Glass and Soncini, 1995). Given a shunt, PaO_2 is predicted to increase with rising temperature. As explained by Wood (1982), the arterial point on the ODC is right-shifted with increase of temperature. When temperature increases, the arterial point approaches the lung gas values at high temperature. Oppositely large gas-to-blood gradients (DPO$_2$ ~ 50 mmHg) develop at low temperature (Glass and Soncini, 1995).

Apparently, *Lepidosiren* has a constant PaO$_2$, indicating absence of any important shunt, which is consistent with high values for PaO$_2$ and O$_2$-hemoglobin saturation (Bassi *et al.*, in press). Unfortunately, no data are available for Lepidosireniformes in this context.

COMPARISON TO *NEOCERATODUS*

Consistent with the focus of this chapter the Australian lungfish will only be discussed shortly. *Neocaratodus forsteri* Krefft grow much larger and heavier than the Lepidosireniformes and, moreover, look much like ancestral forms such as the upper Devonian *Dipterus* (Carroll, 1988). There is no evidence of aestivation in this animal, which inhabits river systems in south-east Queensland. As a special feature, it has electroreceptors to locate prey. To our knowledge such systems have not been reported for lepidosirenid lungfish (Watt *et al.*, 1999). Recently, true enamel tooth covering was detected in juvenile lungfish. Once again this emphasizes the similarities between the lungfish and the tetrapod vertebrates (Satchell *et al.*, 2000).

Johansen *et al.* (1967) studied respiratory function in *Neoceratodus* and reported that more than 1 h could pass between air breaths (temp. 18°C). When surfacing, the animal initially sucked air into the oral cavity, while the pneumatic duct was closed. Then the duct opened while the lung forced gas into the lung by compression of the buccal pump, which reminds of lung ventilation in *Protopterus* (MacMahon, 1969).

Neoceratodus doubled gill ventilation when exposed to a decrease of water PO$_2$ from 125 mmHg to 25 mmHg (Johansen *et al.*, 1967). By contrast, neither *Lepidosiren* nor *Protopterus* showed such responses (Johansen and Lenfant, 1968; Sanchez *et al.*, 2001b) which, as pointed out by Fishman *et al.* (1989), indicates that the reduced gills of Lepidosireniformes is not equipped with O$_2$-receptors. By contrast, in teleost fish, such receptors are located in the gill system (cf. Burleson and Milsom, 1995a, b). As mentioned above, the diving time is long in *Neoceratodus*, but it surfaces in normoxic water (Johansen *et al.*, 1967; Fritsche *et al.*, 1993). Nevertheless, its PaCO$_2$ is about 3.5 mmHg, which superposes the range for exclusively water-breathing fish (Lenfant et al., 1966) of the value for other lungfish, which indicates a correspondingly low dependence on the lung for gas exchange in normoxic water (Johansen et al., 1967). Pulmonary ventilation, however, increases with spontaneous activity (Grigg, 1965).

COMPARISON TO AMPHIBIANS AND REPTILES

As mentioned above, amphibians and lungfish share many features of respiratory control. Both groups possess central pH/CO_2 receptors to monitor acid-base status. In addition, intrapulmonary CO_2-sensitive stretch receptors have been identified in lungfish (DeLaney et al., 1983). This general type of receptor is present in land vertebrates with the exception of mammals (Milsom, 2002). Likewise, in tetrapods and lungfish (Sanchez et al., 2001b), the dominant specific O_2 stimulus is partial pressure rather than O_2-content. This combination of central and peripheral receptors characterizes anuran amphibians (Wang et al., 1994; Branco and Glass, 1995). In amphibians, O_2-receptors output has been recorded from the aortic and carotid nerves and from the carotid labyrinth (Ishii et al., 1985; van Vliet and West, 1992). Moreover, peripheral CO_2/pH receptors are present in Bufo (Branco et al., 1993).

The Lepidosireniformes and anuran amphibians are distinctly different as to set points for blood gases. Thus, the lungfish have a much lower pHa and $PaCO_2$ and $[HCO_3^-]$pl values are much higher than in Bufo and Rana. In spite of this, the temperature-dependent acid-base regulation follows very similar patterns. Both groups eliminate substantial amounts of CO_2 via the skin and/or gills, which is the basis a constant lung gas PO_2 constant over a wide range of temperatures (Glass and Soncini, 1995). Such non-pulmonary exchanges are negligible in most reptiles and, consequently, lung PO_2 decreases with higher temperature (Jackson, 1978 1989). In reptiles, the negative $\Delta pHa/\Delta t$ is obtained by a reduction of the ratio of ventilation to pulmonary CO_2 output. The lung of reptiles is non-alveolar, but the equation is still valid for a 'faveolar' lung (Wang et al., 1998). The alveolar ventilation equation states:

$$V_A/VCO_2 = RT/P_ACO_2,$$

where V_A = effective ventilation of the lung; VCO_2 = pulmonary CO_2-output; R = gas constant; T = absolute temperature and P_ACO_2 = the PCO_2 within the pulmonary gas exchange region and R_E = the pulmonary gas exchange ratio. In reptiles, the relationship (V_A/VCO_2) decreases with

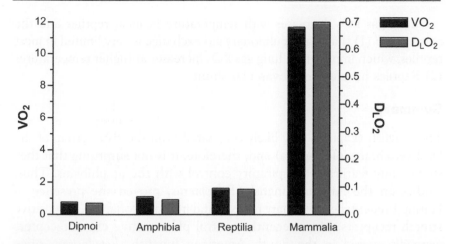

Fig. 15.2 Oygen uptakes (VO_2) and diffusing capacities (D_LO_2) for various vertebrate groups. (1). Dipnoi: *Lepidosiren paradoxa* (W = 0.58 kg; 35°C). Adapted from: Bassi *et al.* (2005). (2). Amphibia: *Rana catesbeiana* (W = 0.36 kg; 30°C). Adapted from: Glass *et al.* (1981a). (3). Reptilia: *Varanus exanthematicus* (2.2 kg; 35°C). Adapted from: Glass *et al.* (1981b). (4) Mammalia: Combined small mammals (0.5 kg; 37°C). Adapted from: Takezawa *et al.* (1980).

rising temperature, leading to an increasing PCO_2. As evident from a simplified version of the relationship:

$$P_AO_2 = P_IO_2 - P_ACO_2/R_E$$

Evidently an increasing PCO_2 reduces lung gas PO_2 (see Fenn *et al.*, 1946 for the complete equation).

Few data exist on effects of temperature on PaO_2 in on teleost fish, but carp, trout and pacu *Piaractus mesopotamicus* have a constant PaO_2 over a wide range of temperatures. Furthermore, the absolute values seem species-specific, e.g., high PaO_2 in active species (trout and pacu) and low PaO_2 in the hypoxia-resistant carp. This reflects that acid-base regulation in teleost fish depends on $[HCO_3^-]pl$ rather than on a $PaCO_2$-oriented regulation by gill ventilation. As a further aspect, PaO_2 in carp and pacu increased considerably with application of hyperoxia, indicating absence of significant blood shunts within the gill system (Glass and Soncini, 1999)

Lung gas PO_2 is also constant with temperature in *Lepidosiren* and in *Bufo*, which, as explained above, is a consequence of bimodal gas exchange. There is, however, a striking difference: *Lepidosiren* has a constant PaO_2, while that of *Bufo* decreases at low temperature due to central shunt (Fig. 15.3 A, B).

Lung gas PO_2 decreases with temperature in most reptiles and the reasons are: (1) that extra-pulmonary gas exchange is very limited in most reptiles, which implies that lung gas PO_2 increases at higher temperatures (2) Reptiles have a central vascular shunt.

Summary

The lungfish (Dipnoi) very likely originated from the sister group of the land vertebrates (Tetrapoda) and, therefore, it is not surprising that they share many features of respiratory control with the amphibians. Thus, studies on the African lungfish, *Protopterus*, proved the presence of Hering-Breuer type pulmonary reflexes and intrapulmonary CO_2-sensitive stretch receptors. More recently, central pH-sensitive chemoreceptors were discovered in the South American lungfish, *Lepidosiren*, since

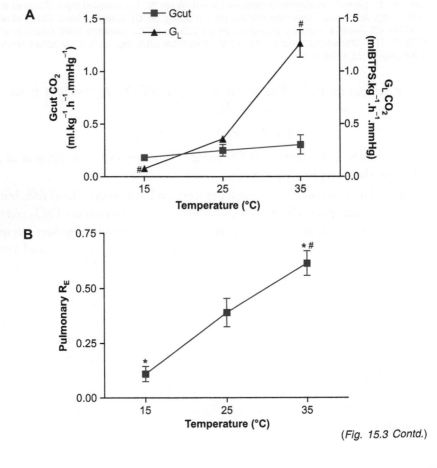

(*Fig. 15.3 Contd.*)

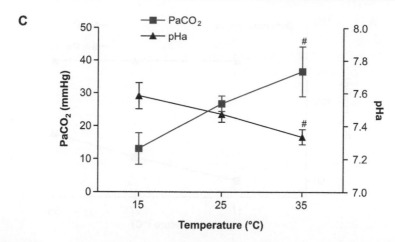

Fig. 15.3 **(A)** Upper panel: the relative CO_2-conductance of pulmonary ventilation (G_LCO_2) to that of the skin ($GcutCO_2$). Increases of temperature are accompanied by large increases of the pulmonary gas conductance. In contrast, the skin conductance ($GcutCO_2$) remained virtually constant. This reflects that pulmonary ventilation can increase to meet metabolic demands, while diffusion by skin is a virtually passive process with a low Q_{10}. Mean values ± SEM; n = 5. * indicates significant differences between groups. Same symbols below.

(B) At higher temperatures, the bulk part of CO_2-output becomes eliminated by the lung. This causes an increase of the pulmonary gas exchange ratio ($R_E = VCO_2/VCO_2$), implying a higher lung gas PCO_2 at elevated temperatures.

(C) In turn, arterial PCO_2 equilibrates with the lung gas PCO_2. This leads to the negative slope for pHa versus temperature.

The figure is modified.

reductions of CSF pH increased pulmonary ventilation. In most tetrapods the specific O_2-stimulus is O_2 partial pressure rather than O_2 content and this also applies to *Lepidosiren*. Finally, transition from hypercarbia to atmospheric air, released responses that distinguish intrapulmonary CO_2-receptors. These ventilatory responses in lungfish are very similar to those of amphibians. As temperature increased in *Lepidosiren*, the lung took over a larger part of total CO_2 output. Accordingly, PCO_2 of lung gas and arterial blood increased at higher temperature leading to a negative $\Delta pHa/\Delta t$ as predicted for ectothermic vertebrates. All these features of respiratory control are very similar to those of amphibians.

Acknowledgements

This research was supported by FAPESP (Fundção de Ampora á Pesquisa do Estado de São Paulo); Proc 98/06731-5, CNPq (Conselho Nacional de

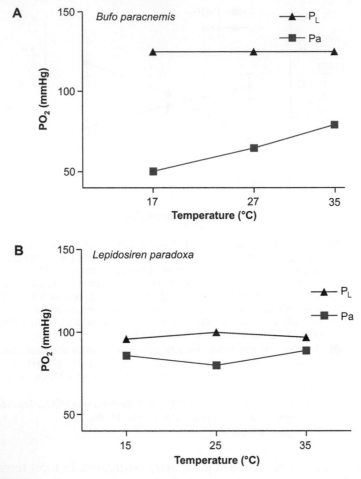

Fig. 15.4 (A). Lung gas PO_2 (P_LO_2) and arterial PO_2 (PaO_2) in the toad *Bufo paracnemis* and (B) in *Lepidosiren paradoxa*. Notice the large difference ($P_LO_2 - PaO_2$) in *Bufo*, which also increases PaO_2. By contrast, P_LO_2 and PaO_2 are constant with temperature in the lungfish and the difference between the pressures is much smaller than in the toad. See text for further explanations.

Desenvolvimento Científico o Tecnológico); Proc. 520769/93-7, FAEPA (Fundação de Apoio ao Ensino, Pesquisa e Assistência do Hospital das Clínicas da FMRP-USP).

References

Abe, A.S. 1995. Estivation in South American amphibians and reptiles. *Brazilian Journal of Medical and Biological Research* 28: 1241-1247.

Abe, A.S. and J.F. Steffensen. 1996a. Perda cutânea de oxigênio na pirambóia, *Lepidosiren paradoxa*, exposta à água hipóxica (Osteichthyes, Dipnoi). *Revista Brasileira da Biologia* 56: 211-216.

Abe, A.S. and J.F. Steffensen. 1996b. Respiração pulmonary e cutânea na pirambóia, *Lepidosiren paradoxa*, durante a atividade e a estivação (Osteichthyes, Dipnoi). *Revista Brasileira da Biologia* 56: 485-489.

Albers, C., R. Manz, D. Muster and G.M. Hughes. 1983. Effects of acclimation temperature on oxygen transport in the blood of the carp, *Cyprinus carpio*. *Respiration Physiology* 52: 165-179.

Alvarado, R.H., T.H. Diets and T.L. Mullen. 1975. Chloride transport across isolated skin of *Rana pipiens*. *American Journal of Physiology* 229: 869-876.

Amin-Naves, J., H. Giusti and M.L. Glass. 2004. Effects of acute temperature changes on aerial and aquatic gas exchange, pulmonary ventilation and blood gas status in the South American lungfish, *Lepidosiren paradoxa*. *Comparative Biochemistry and Physiology* A 138: 133-139.

Andersen, J.B., F.B. Jensen and T. Wang. 2001. Effects of temperature and oxygen availability on circulating catecholamines in the toad *Bufo marinus*. *Comparative Biochemistry and Physiology* A129: 473-486.

Ballantyne, D. and P. Scheid. 2001. Central chemosensitivity of respiration: A brief overview. *Respiration Physiology* 129: 5-12.

Bassi, M., W. Klein, M.N. Fernandes, S.F. Perry and M.L. Glass. Pulmonary oxygen diffusing capacity of the South American lungfish *Lepidosiren paradoxa*: Physiological values by the Bohr integration method. *Physiological and Biochemical Zoology*. (In Press).

Bohr, C. 1909. Über die spezifische Tätigkeit der Lungen bei der respiratorischen Gasaufnahme und ihr Verhalten zu der durch die Alveolenwand stattfindende Gasdiffusion. *Skandinavisches Arciv für. Physiologie* 22: 221-280.

Boutilier, R.G. and N. Heisler. 1988. Acid-base regulation and blood gases in the anuran amphibian, *Bufo marinus*, during environmental hypercapnia. *Journal of Experimental Biology* 134: 79-98.

Boutilier, R.G. and D.P. Toews. 1981. Respiratory, circulatory and acid-base adjustments to hypercapnia in a strictly aquatic and predominantly skin-breathing urodele, *Cryptobranchus alleganiensis*. *Respiration Physiology* 46: 177-192.

Boutilier, R.G., M.L. Glass and N. Heisler. 1986a. The relative distribution of the pulmocutaneous blood flow in *Rana catebeiana*: Effects of pulmonary or cutaneous hypoxia. *Journal of Experimental Biology* 126: 33-41.

Boutilier, R.G., M.L. Glass and N. Heisler. 1986b. Blood gases, and extracellular/intracellular acid-base status as a function of temperature in the anuran amphibians *Xenopus laevis* and *Bufo marinus*. *Journal of Experimental Biology* 130: 13-25.

Branco, L.G.S. and M.L. Glass. 1995. Ventilatory responses to carboxyhaemoglobinaemia and hypoxic hypoxia in *Bufo paracnemis*. *Journal of Experimental Biology* 198: 1417-1421.

Branco, L.G.S., M.L. Glass and A. Hoffmann. 1992. Central chemoreceptor drive to breathing in unanesthetized toads, *Bufo paracnemis*. *Respiration Physiology* 87: 195-204.

Branco, L.G.S., M.L. Glass, T. Wang and A. Hoffmann. 1993. Temperature and central chemoreceptor drive to ventilation in toad (*Bufo paracnemis*). *Respiration Physiology* 93: 337-346.

Burggren, W.W. and R. Moalli. 1984. Active regulation of cutaneous gas exchange in amphibians experimental evidence and a revised model for skin respiration. *Respiration Physiology* 55: 379-392.

Burggren, W.W., M.E. Feder and A.W. Pinder. 1983. Temperature and the balance between aerial and aquatic respiration in larvae of *Rana berlandieri* and *Rana catesbeiana*. *Physiological Zoology* 56: 263-273.

Burleson, M.L. and W.K. Milsom. 1990. Propanol inhibits O_2-sensitive chemoreceptor activity in trout gills. *American Journal of Physiology* 27: R1089-R1091.

Burleson, M.L. and W.K. Milsom. 1995a. Cardio-ventilatory control in rainbow trout: I. Pharmacology of branchial oxygen-sensitive chemoreceptors. *Respiration Physiology* 100: 231-238.

Burleson, M.L. and W.K. Milsom. 1995b. Cardio-ventilatory control in rainbow trout: II. Reflex effects of exogenous neurochemicals. *Respiration Physiology* 101: 289-299.

Busk, M., E.H. Larsen and F.B. Jensen. 1997. Acid-base regulation in tadpoles of *Rana catesbeiana*, exposed to environmental hypercapnia. *Journal of Experimental Biology* 200: 2507-2512.

Cameron, J.N. 1984. Acid-base status of fish at different temperatures. *American Journal of Physiology* 246: R452-R459.

Carroll, R.L. 1988. *Vertebrate Palaeontology and Evolution*. 1st edition. Freeman and Company, New York.

Claiborne, J.B., Jr. and N. Heisler. 1986. Acid-base regulation and ion transfers in the carp (*Cyprinus carpio*): pH compensation during graded long and short-term environmental hypercapnia, and the effect of bicarbonate infusion. *Journal of Experimental Biology* 126: 41-61.

Dejours, P. 1981. *Principles of Comparative Respiratory Physiology*. 2nd ed. Elsevier/North Holland, Amsterdam.

DeLaney, R.G., S. Lahiri and A.P. Fishman. 1974. Aestivation of the African lungfish *Protopterus aethiopicus*: Cardiovascular and pulmonary function. *Journal of Experimental Physiology* 6: 111-128.

DeLaney, R.G., S. Lahiri, R. Hamilton and A.P. Fishman. 1977. Acid-base balance and plasma composition in the aestivating lungfish (*Protopterus*). *American Journal of Physiology* 232: R10-R17.

DeLaney, R.G., P. Laurent, R. Galante, A.I. Pack and A.P. Fishman. 1983. Pulmonary mechanoreceptors in the dipnoi lungfish *Protopterus* and *Lepidosiren*. *American Journal of Physiology* 244: R418-R428.

Dudley, R. 1998. Atmospheric oxygen, giant Palaeozoic insects and the evolution of aerial locomotor performance. *Journal of Experimental Biology* 201: 1043-1050.

Edsall, J.T. and J. Wyman. 1958. *Biophysical Chemistry*. Academic Press, New York.

Else, P.L. and A.J. Hulbert. 1981. Comparison of the "mammal machine" and the "reptile machine": Energy production. *American Journal of Physiology* 240: R3-R9.

Feder, M.E. and W.W. Burggren. 1985. Cutaneous gas exchange in vertebrates: design, patterns, control and implications. *Biological Reviews* 60: 1-45.

Fenn, W.O., H. Rahn and A.B. Otis. 1946. A theoretical study of alveolar air at altitude. *American Journal of Physiology* 146, 637-653.

Fishman, A.P., R.G. DeLaney and P. Laurent. 1985. Circulatory adaptation to bimodal respiration in the dipnoan lungfish. *Journal of Applied Physiology* 59: 285-294.

Fishman, A.P., R.J. Galante and A.I. Pack. 1989. Diving Physiology: Lungfish. In: *Comparative Respiratory Physiology: Current Concepts*, S.C. Wood (ed.). Marcel Dekker Inc. New York. pp. 645-677.

Fitzinger, L.J. 1837. Vorläufiger Bericht über eine höchst interessante Entdeckung Dr. Natterers in Brasilien. *Isis* 1837: 379-380.

Flanigan, J.E., P.C. Withers and M. Guppy. 1991. *In vitro* metabolic depression of tissues from the aestivating frog *Neobatrachus pelobatoides*. *Journal of Experimental Biology* 161: 273-283.

Foxon, G.E.H. and I.R. Bishop. 1968. The mechanism of breathing in the South American lungfish *Lepidosiren paradoxa*; radiological study. *Journal of Zoology* (London) 154: 263-272.

Fritsche, R., M. Axelsson, C.E. Franklin, G.C. Grigg, S. Holmgren and S. Nilsson. 1993. Respiratory and cardiovascular responses to hypoxia in the Australian lungfish. *Respiration Physiology* 94: 173-187.

Glass, M.L. 1989. Diffusing capacity. In: *Comparative Respiratory Physiology: Current Concepts*, S.C. Wood (ed.). Marcel Dekker Inc., New York, pp. 417-439.

Glass, M.L. and R. Soncini. 1995. Regulation of acid-base status in ectothermic vertebrates: The consequences for oxygen pressures in lung gas and blood. *Brazilian Journal of Medical and Biological Research* 28: 1161-1167.

Glass, M.L. and R. Soncini. 1999. Physiological shunts in gills of teleost fish: Assessment of the evidence. In: *Biology of Tropical Fishes*, A.L. Val and V.M.A. Val. (eds.). Instituto Nacional de Pesquisas da Amazônia (INPA), Manaus, Brazil, pp. 333-341.

Glass, M.L., W.W. Burggren and K. Johansen. 1981a. Pulmonary diffusing capacity of the bullfrog (*Rana catesbeiana*). *Acta Physiologica Scandinavica* 113: 485-490.

Glass, M.L., K. Johansen and A.S. Abe. 1981b. Pulmonary diffusing capacity in reptiles (relations to temperature and O_2-uptake). *Journal of Comparative Physiology* 142: 509-514.

Glass, M.L., A. Ishimatsu and K. Johansen. 1986. Responses of aerial ventilation to hypoxia and hypercapnia in *Channa argus*, an air-breathing fish. *Journal of Comparative Physiology* B 156: 425-430.

Glass, M.L., N.A. Andersen, M. Kruhøffer, E.M. Williams and N. Heisler. 1990. Combined effects of environmental PO_2 and temperature on ventilation and blood gases in the carp *Cyprinus carpio* L. *Journal of Experimental Biology* 148: 1-17.

Glass, M.L., M.S. Fernandes, R. Soncini, H. Glass and J. Wasser. 1997. Effects of dry season dormancy on oxygen uptake, heart rate, and blood pressures in the toad *Bufo paracnemis*. *Journal of Experimental Zoology* 279: 330-336.

Gottlieb, G. and D.C. Jackson. 1976. Importance of pulmonary ventilation in respiratory control in the bullfrog. *American Journal of Physiology* 230: 608-613.

Graham, J.B. 1997. *Air-breathing Fishes: Evolution, Diversity and Adaptation*. Academic Press, San Diego.

Greenwood, P.H. 1958. Reproduction in the East African lungfish *Protopterus aethiopicus* Heckel. *Proceedings of the Zoological Society of London* 130: 547-567.

Grigg, G.C. 1965. Studies on the Queensland lungfish *Neoceratodus forsteri* (Krefft). III. Aerial respiration in relation to habits. *Australian Journal of Zoology* 13: 413-421.

Guimond, R.W. and V.H. Hutchinson. 1974. Aerial and aquatic respiration in the Congo eel, *Amphiuma means* (Garden). *Respiration Physiology* 20, 147-159.

Harder, V., R.H.S. Souza, W. Severi, F.T. Rantin and C.R. Bridges. 1999. The South American lungfish – adaptations to an extreme habitat. In: *Biology of Tropical Fishes*, A.L. Val and V.M.F. Almeida-Val (eds.). Instituto Nacional de Pesquisas da Amazônia (INPA), Manaus, pp. 99-110.

Hedrick, M.S., D.R. Burleson, D.R. Jones and W.K. Milsom. 1991. An examination of central chemosensitivity in an air-breathing fish (*Amia calva*). *Journal of Experimental Biology* 155: 165-174.

Heisler, N. 1982. Intracellular and extracellular acid-base regulation in the tropical freshwater teleost fish *Synbranchus marmoratus* in response to the transition from water breathing to air breathing. *Journal of Experimental Biology* 99: 9-28.

Heisler, N. 1984. Acid-base regulation in fishes. In: *Fish Physiology*, W.S. Hoar and D.J. Randall (eds.). Academic Press, Orlando, Vol. 10A, pp. 315-401.

Heisler, N. 1986. *Acid-Base Regulation in Animals*. Elsevier, Amsterdam.

Heisler, N., G. Forcht, G.R. Ultsch and J.F. Anderson. 1982. Acid-base regulation to environmental hypercapnia in two aquatic salamanders, *Siren lacertina* and *Amphiuma means*. *Respiration Physiology* 49: 141-158.

Hitzig, B.M. and D.C. Jackson. 1978. Central chemical control of ventilation in the unanaesthetized turtle. *American Journal of Physiology* 235: R257-R264.

Ishii, K., K. Ishii and T. Kusanabe. 1985. Chemo- and baroreceptor innervation of the aortic trunk of the toad *Bufo vulgaris*. *Respiration Physiology* 60: 365-375.

Ishimatsu, A. and T. Tazawa. 1983. Differences in blood oxygen levels in the outflow vessels of the heart of an air-breathing fish, *Channa argus*: Do separate blood stream exist in the telesostean heart? *Journal of Comparative Physiology* 149: 435-440.

Jackson, D.C. 1978. Respiratory control and CO_2 conductance: temperature effects in a turtle and a frog. *Respiration Physiology* 33: 103-114.

Jackson, D.C. 1989. Control of breathing: effects of temperature. In: *Comparative Respiratory Physiology: Current Concepts*, S.C. Wood (ed.). Marcel Dekker Inc., New York, pp. 621-637.

Jesse, M.J., C. Shub and A.P. Fishman. 1967. Lung ventilation of the African lungfish. *Respiration Physiology* 3: 267-287.

Johansen, K. and A.S.F. Ditadi. 1966. Double circulation in the giant toad *Bufo paracnemis*. *Physiological Zoology* 39: 140-150.

Johansen, K. and C. Lenfant. 1967. Respiratory function in the South American lungfish, *Lepidosiren paradoxa*. *Journal of Experimental Biology* 46: 305-218.

Johansen, K. and C. Lenfant. 1968. Respiration in the African Lungfish *Protopterus aethiopicus*: II Control of breathing. *Journal of Experimental Biology* 49: 453-468.

Johansen, K., J.P. Lomholt and G.M. Maloiy. 1976. Importance of air and water breathing in relation to size of the African lungfish *Protopterus amphibius* Peters. *Journal of Experimental Biology* 65: 395-399.

Kind, P.K., G.C. Grigg and D.T. Booth. 2002. Physiological responses to prolonged aquatic hypoxia in the Queensland lungfish, *Neoceratodus forsteri. Respiratory Physiology and Neurology* 132: 179-190.

Kinkhead, R. and W.K. Milsom. 1996. CO_2-sensitive olfactory and pulmonary receptor modulation of episodic breathing in bullfrogs. *American Journal of Physiology* 270: R134-R144.

Kruhøffer, M., M.L. Glass, A.S. Abe and K. Johansen. 1987. Control of breathing in an amphibian *Bufo paracnemis*: Effects of temperature and hypoxia. *Respiration Physiology* 69: 267-275.

Laurent, P., R.G. DeLaney and A.P. Fishman. 1978. The vasculature of the gills in the aquatic and aestivating lungfish (*Protopterus aethiopicus*). *Journal of Morphology* 156: 173-208.

Lenfant, C. and K. Johansen. 1968. Respiration in the African lungfish *Protopterus aethiopicus*. I. Respiratory properties of blood and normal patterns of breathing and gas exchange. *Journal of Experimental Biology* 49: 137-452.

Lenfant, C., K. Johansen and G.C. Grigg. 1966. Respiratory proporties of blood and pattern of gas exchange in the lungfish *Neoceratodus forsteri* (Krefft). *Respiration Physiology* 2: 1-21.

Lomholt, J.P. 1993. Breathing in the aestivating African lungfish, *Protopterus amphibius*. In: *Advances in Fish Research*, B.R. Singh (ed.). Narendra Publishing House, Delhi Vol. 1, pp. 17-34.

Lomholt, J.P. and M.L. Glass. 1987. Gas exchange of air-breathing fishes in near-anoxic water. *Acta Physiologica Scandinavica* 129: 45 A.

Lomholt, J.P. and K. Johansen. 1974. Control of breathing in *Amphipnous cuchia*, an amphibious fish. *Respiration Physiology* 21: 325-340.

Lomholt, J.P. and K. Johansen. 1976. Gas exchange in the amphibious fish, *Amphipnous cuchia. Journal of Comparative Physiology* 107: 141-157.

Lundberg, J.G. 1993. African fresh water fish clades and continental drift: problem with a paradigm. In: *Biological Relationships between Africa and South America*, Goldblatt (ed.) Yale University Press, New Haven.

Mackenzie, J.A. and D.C. Jackson. 1978. The effect of temperature on cutaneous CO_2, loss and conductance in the bullfrog. *Respiration Physiology* 32: 313-323.

McMahon, B.R. 1969. A functional analysis of the aquatic and aerial respiratory movements of an African lungfish, with reference to the evolution of the lung ventilation mechanism in vetebrates. *Journal of Experimental Biology* 51: 407-430.

Meyer, A. and S.I. Dolven. 1992. Molecules, fossils, and the origin of tetrapods. *Journal of Molecular Evolution* 35: 102-113.

Milsom, W.K. 1995. The role of CO_2/pH chemoreceptors in ventilatory control. *Brazilian Journal of Medical and Biological Research* 28: 1147-1160.

Milsom, W.K. 2002. Phylogeny of CO_2/H+ chemoreception in vertebrates. *Respiratory Physiology and Neurobiology* 131: 29-41.

Milsom, W.K., D.R. Jones and G.R.J. Gabbott. 1981. On chemoreceptor control of ventilatory responses to CO_2 in unanesthetized ducks. *Journal of Applied Physiology* 50: 1121-1128.

Moraes, M.F.P.G., S. Höller, O.P.F. Costa, M.L. Glass, M.N. Fernandes and S.F. Perry. Morphomtric comparison of the respiratory organs of the South American lungfish *Lepisosiren paradoxa* (Dipnoi). *Physiological and Biochemical Zoology* 78: 546-559.

Orgeig, S. and C.B. Daniels. 1995. The evolutionary significance of pulmonary surfactant in lungfish (*Dipnoi*). *American Journal of Respiratory and Molecular Biology* 13: 161-166.

Pack, A.J., R.J. Galante and A.P. Fishman. 1990. Control of the interbreath interval in the African lungfish. *American Journal of Physiology* 259: R139-R146.

Pack, A.J., R.J. Galante and A.P. Fishman. 1992. Role of lung inflation in control of air breath duration in African lungfish (*Protopterus annectens*). *American Journal of Physiology* 262: R879-R884.

Perry, S.F. 1989. Structure and function of the reptilian respiratory system. In: *Comparative Respiratory Physiology: Current Concepts*, S.C. Wood (ed.). Marcel Dekker Inc., New York, pp. 193-237.

Perry, S.F., R.J.A. Wilson, C. Straus, M.B. Harris and J. Remmers. 2001. Which came first, the lung or the breath? *Comparative Biochemistry and Physiology* A129: 37-47.

Rahn, H. 1966. Aquatic gas exchange: theory. *Respiration Physiology* 1: 1-12.

Randall, D.J. and J.N. Cameron. 1973. Respiratory control of arterial pH as temperature changes in rainbow trout *Salmo gairdneri*. *American Journal of Physiology* 225: 997-1002.

Reeves, R.B. 1972. An imidazole alphastat hypothesis for vertebrate acid-base regulation: tissue carbon dioxide content and body temperature in bullfrogs. *Respiration Physiology* 14: 219-236.

Robin, E.D. 1962. Relationship between temperature and plasma pH and carbon dioxide tension in the turtle. *Nature* (London) 195: 249-251.

Rodewald, K., A. Stangl and G. Braunitzer. 1984. Primary structure, biochemical and physiological aspects of hemoglobin from the South American lungfish (*Lepidosiren paradoxa*, Dipnoi). *Hoppe Seylers Zeitschrift für Physiologische Chemie* 365: 639-649.

Sanchez, A.P. and M.L. Glass. 2001. Effects of environmental hypercapnia on pulmonary ventilation of the South American lungfish. *Journal of Fish Biology* 58: 1181-1189.

Sanchez, A.P., A. Hoffman, F.T. Rantin and M.L. Glass. 2001a. The relationship between pH of the cerebro-spinal fluid and pulmonary ventilation of the South American lungfish, *Lepidosiren paradoxa*. *Journal of Experimental Zoology* 290: 421-425.

Sanchez, A.P, R. Soncini, T. Wang, P. Koldkjaer, E.W. Taylor and M.L. Glass. 2001b. The differential cardio-respiratory responses to ambient hypoxia and systemic hypoxaemia in the South Amercian lungfish, *Lepidosiren paradoxa*. *Comparative Biochemistry and Physiology* A130: 677-687.

Sanchez, A.P., H. Giusti, M. Bassi and M.L. Glass. Acid-base regulation in the South American lungfish, *Lepidosiren paradoxa*: effects of prolonged hypercarbia on blood gases and pulmonary ventilation. *Biochemical and Physiological Zoology*. (In Press).

Satchell, P.G., C.F. Shuler and T.G. Diekwisch. 2000. True enamel covering in teeth of the Australian lungfish *Neoceratodus forsteri*. *Cell and Tissue Research* 299(1): 2737.

Sawaya, P. 1946. Sobre a biologia de alguns peixes de respiração aérea (*Lepidosiren paradoxa* Fitzinger e *Arapaima gigas* Cuvier). *Boletim da Faculdade de Filosofia Ciências e Letras da Universidade de*. São Paulo 11: 255-286.

Seymour, R.S. 1973. Gas exchange in spadefoot toads beneath the ground. *Copeia* 1973: 452-460.

Smatresk, N.J. and A.W. Smits. 1991. Effects of central and peripheral chemoreceptor stimulation on ventilation in the marine toad, *Bufo marinus*.

Smatresk, N.J. and J.N. Cameron. 1982. Respiration and acid-base physiology of the spotted gar, a bimodal breather. II. Responses to temperature change and hypercapnia. *Journal of Experimental Biology* 96: 281-293.

Smith, H.M. 1935. The metabolism of a lungfish. I. General considerations of the fasting metabolism in an active fish. *Journal of Cellular Comparative Physiology* 6: 43-67.

Snyder, G.K. and J.R. Nestler. 1991. Intracellular pH in the toad *Bufo marinus* following hypercapnia. *Journal of Experimental Biology* 161: 415-422.

Soncini, R. and M.L. Glass. 2000. Oxygen and acid-base related drives to gill ventilation in carp. *Journal of Fish Biology* 56: 528-541.

Szidon, J.P., S. Lahiri, M. Lev and A.P. Fishman. 1969. Heart rate of circulation of the African lungfish. *Circulatory Research* 25: 23-38.

Toews, D.P. and D.F. Stiffler. 1990. Compensation of progressive hypercapnia in the toad (*Bufo marinus*) and the bullfrog (*Rana catesbeiana*). *Journal of Experimental Biology* 148: 293-302.

Toyama, Y., T. Ichimiya, H. Kasama-Yoshida, Y. Cao, M. Hasegava, H. Kojima, Y. Tamai and T. Kurihari. 2000. Phylogenetic relation of of lungfish indicated by the amino acid sequence of myelinDM20. *Molecular Brain Research* 8: 256-259.

Ultsch, G.R and D.C. Jackson. 1996. pH and temperature in ectothermic vertebrates. *Bulletin of the Alabama Museum of Natural History* 18: 1-41.

Van Vliet, B.N. and N.H. West. 1992. Functional characteristics of arterial chemoreceptors in an amphibian (*Bufo marinus*). *Respiration Physiology* 88: 113-127.

Wang, T., L.G.S. Branco and M.L. Glass. 1994. Ventilatory responses to hypoxia in the toad (*Bufo paracnemis* Lutz) before and after reduction of HbO_2 concentration. *Journal of Experimental Biology* 186: 1-8.

Wang, T., A.S. Abe and M.L. Glass. 1998. Temperature effects on lung and blood gases in the toad *Bufo paracnemis*: The consequences of bimodal gas exchange. *Respiration Physiology* 113: 231-238.

Watt, M., C.S. Evans and J.M. Joss. 1999. Use of electroreception during foraging by the Australian lungfish. *Animal Behavior* 58: 1039-1045.

Wellner-Kienitz, M.-C. and H. Shams. 1998. CO_2-sensitive neurons in organotypic cultures of the fetal rat medulla. *Respiration Physiology* 111: 137-151.

Wilson, R.J., M.B. Harris, J.E. Remmers and S.F. Perry. 2000. Evolution of air-breathing and central CO_2/H^+ respiratory chemosensitivity: new insights from an old fish? *Journal of Experimental Biology* 203: 3505-3512.

Wood, S.C. and J.W. Hicks. 1985. Oxygen homeostasis in vertebrates with cardiovascular shunts. In: *Cardiovascular Shunts, Phylogenetic, Ontogenetic and Clinical Aspects*, K. Johansen and W.W. Burggren (eds.). Munksgaard, Copenhagen, pp. 354-372.

Wood, S.C. 1982. Effect of O_2 affinity on arterial PO_2 in animals with central vascular shunts. *Journal of Applied Physiology* 53: R1360-R1364.

Wood, S.C. and M.L. Glass. 1991. Respiration and thermoregulation in amphibians and reptiles. In: *Physiological Strategies for Gas Exchange and Metabolism*, A.J. Woakes, M.K. Grieshaber and C.R. Bridges (eds.). Society for Experimental Biology Seminar Series 41. Cambridge University Press, Cambridge, pp. 107-124.

Yokabori, S., M. Hasegawa, T. Ueda, N. Okada, K. Nishikawa and K. Watanabe. 1994. Relationship among coelacanths, lungfishes, and tetrapods: A phylogenetic analysis bases on mitochondrial cytochrome oxidase I gene sequences. *Journal of Molecular Evolution* 38: 602-609.

Zardoya, R., Y. Cao, M. Hasegava and A. Meyer. 1998. Searching for the closest living relative(s) of tetrapods through evolutionary analyses of mitochandrial ansd nuclear data. *Journal of Molecular Biology and Evolution* 15: 506-517.

Respiration in Infectious and Non-infectious Gill Diseases

Mark D. Powell

INTRODUCTION

Gill diseases represent one of the most prolific and economically important groups of diseases and disorders of fish. Although it is impossible to put an absolute figure on the economic importance of gill diseases their widespread nature, aetiological diversity and indiscriminate species susceptibility make them of sufficient significant standing to warrant discussion in terms of respiration and the environment of fishes. The majority of published works on fish physiology and ecophysiology discuss the physiology of apparently healthy fish. Unfortunately most of the discussion of fish pathology focuses upon the aetiological agents of disease and the host patho-physiology is often discussed in terms of speculation or extrapolation from the healthy condition.

The paucity of research in the area of fish disease physiology can be attributed to the fact that the techniques used to study respiratory

Author's address: School of Aquaculture, Tasmanian Aquaculture and Fisheries Institute, University of Tasmania, Locked Bag 1370 Launceston 7250 Tasmania Australia. E-mail: Mark.Powell@utas.edu.au

physiology in healthy fishes are difficult to transfer to disease compromised fish without extensive mortality. This questions whether the gill disease or stress of the surgical procedure killed the fish. The reasons for this are embedded within the fact that the diseases, even those of the gill affect more than just structural anatomy or individual physiological systems but rather represent a more systemic effect on the fish. Surgical techniques such as cannulation and masking that are useful for sampling blood for gas measurements (cannulation such as that described by Soivio *et al.*, 1975) or measuring ventilatory volume or flow (such as the techniques used by Davis and Cameron, 1971 and Smith and Jones, 1982) place a large stress upon the fish which when compromised make post-surgical recovery challenging.

FACTORS AFFECTING RESPIRATION

The physiology of ventilation, gas exchange and transport in fishes has been widely studied in recent decades. The structure of the gill, the main gas exchange organ in most fish species, intuitively plays a critical role in fish respiration. The countercurrent blood and water flow through or across a gill, respectively, that comprises a thin epithelium of large surface area allows the diffusional transfer of respiratory gases (oxygen and carbon dioxide) into and out of the blood down a partial pressure gradient according the Fick principle of diffusion (for discussion, see reviews by Perry and MacDonald, 1993; Gilmour, 1998).

$$MO_2 = \frac{kO_2 \cdot A \cdot (\Delta PO_2)}{d} \qquad MCO_2 = \frac{kCO_2 \cdot A \cdot (\Delta PCO_2)}{d}$$

where MO_2 and MCO_2 are the rates of oxygen consumption and carbon dioxide excretion respectively, k is the diffusivity constant for the gas, A is the functional surface area of the gill and d the blood-water diffusion distance. ΔPO_2 and ΔPCO_2 are the partial pressure gradients across the gill maintained by the afferent and efferent gas partial pressures of blood in the gill:

$$\Delta PO_2 = \frac{1}{2}(P_IO_2 + P_EO_2) - \frac{1}{2}(P_aO_2 + P_vO_2)$$

$$\Delta PCO_2 = \frac{1}{2}(P_ICO_2 + P_ECO_2) - \frac{1}{2}(P_aCO_2 + P_vCO_2)$$

where $P_I$gas and $P_E$gas are the partial pressures of respiratory gases in the inspired and expired water, $P_a$gas and $P_v$gas are the partial pressures or respiratory gases in the efferent and afferent blood of the gills respectively (after Perry and MacDonald, 1993; Gilmour, 1998).

Therefore changes in any of the parameters affecting gill diffusion would potentially lead to respiratory compromise.

In addition to changes in structural anatomy of gills that may occur in response to disease, the flow of blood through the gills will affect the rate of gas exchange, in some cases more significantly than structural changes. This is because:

$$MO_2 = Q \cdot \alpha \ (P_aO_2 - P_vO_2) \quad MCO_2 = Q \cdot \beta \ (P_aCO_2 - P_vCO_2)$$

where Q is the cardiac output, α and β are the solubility coefficients for oxygen and carbon dioxide in plasma, respectively. Thus changes in cardiac output would have a significant effect on the rates of gas transfer across the gill. Since cardiac output is a direct function of cardiac performance and is affected by and affects vascular resistance, it would be reasonable to assume that any diseases that affect blood flow would impact on gas transfer regardless of the effect on gill morphology.

CLASSIFICATION OF DISEASES

Gill diseases can be classified into broad categories as being infectious or non-infectious. With infectious gill diseases the aetiological agent is a pathogen (viral, bacterial or parasitic). Non-infectious gill diseases are not associated with a pathogen, but arise as a consequence of either physical injury, chemical insult or nutritional deficiency (Ferguson, 1989).

Infectious gill diseases can be further arbitrarily classified as being necrotic or proliferative in character. With necrotic diseases, the gill lamellae and filaments are eroded by action of bacteria or by grazing parasites. It is apparent that with necrotic diseases, erosion of the gill epithelium results primarily in the reduction in gill surface area. Conversely proliferative diseases result in a hyperplasia of the pavement cells, chloride cells or mucous cells of the lamellar and filamental epithelium. Alternatively other cell types or parasites may proliferate (see below). This increase in cell numbers and often cell layers results in a thickening of the blood-water diffusion distance and or restriction of blood flow through parts of the gill potentially leading to a localized redistribution of blood flow.

PROLIFERATIVE GILL DISEASES

Pavement Epithelial Cell Proliferation

The process by which gill epithelial (pavement) cell proliferation occurs is not known. Similarly it is not known if cell proliferation (hyperplasia) of the pavement cells themselves occur or whether hyperplastic responses in gills is due to proliferation of other filamental epithelial cells (e.g., filament basal cells). Nevertheless, epithelial cell hyperplasia is a common inflammatory response to gill injury particularly to bacterial and parasitic pathogens and to a limited extent, physical trauma (for example see Figs. 16.1 and 16.2). Although thickening of the blood-water diffusion barrier would lead to a reduction in the rate of oxygen (and carbon dioxide) transfer across the gill, studies have shown that this may not occur in every case. The reason for this being that not all the gill surface is uniformly affected by the hyperplastic response. For example, with diseases like bacterial gill disease (BGD, caused by *Flavobacterium branchiophilum*), although there is often extensive hyperplasia of the filamental and lamellar epithelium, blood PO_2 values of BGD affected brook trout (*Salvelinus fontinalis*) and rainbow trout (*Oncorhynchus mykiss*) are relatively unchanged under resting conditions (Byrne *et al.*, 1991, 1995). This is a similar situation, with amoebic gill disease (AGD) of marine Atlantic salmon (*Salmo salar*). Amoebic gill disease is caused by a reaction of the gill to the amphizoic parasite *Neoparamoeba* sp. (*N. pemaquidensis* and the newly identified *N. branchiphila*, Dykova *et al.*, 2005). Despite numerous multifocal hyperplastic lesions that occur across the filamental and lamellar epithelium (up to 80% of filaments may be affected), there are large parts of the gill unaffected. The result being that blood PO_2 of resting fish, although slightly depressed, remains close to that of unaffected animals (Powell *et al.*, 2000; Powell and Nowak, 2003). However, under hypoxic challenge, the PaO_2 of diseased animals equates with that of healthy animals suggesting that fish are capable of compensating for the effect of localized reduced gill surface area and increase diffusion distance as affected by AGD (Powell *et al.*, 2000). AGD affected fish, however, exhibit a respiratory acidosis (Powell *et al.*, 2000; Powell and Nowak, 2003; Leef *et al.*, 2005a) that is the first physiological sign to develop even though blood PO_2 effects are not seen (Leef *et al.*, 2005a). Under a hypoxic challenge however, P_aCO_2 levels remain elevated relative to unaffected animals. This occurs as a consequence of the diffusion limited nature of CO_2 excretion across the teleost gill (see explanation below).

It is unclear as to whether gill epithelial hyperplasia is as a direct result of increases in the number of pavement cells *per se* or a proliferation of the underlying basal cell layer. Indicators of cell proliferation are rarely

Fig. 16.1 Examples of proliferative responses to gill parasitic infections. A. Amoebic gill disease in Atlantic salmon (*Salmo salar*) showing focal hyperplasia of the gill epithelium (asterisk) (H & E, Bar = 100 μm). B. *Neoparamoeba pemaquidensis* (arrow), the causative agent of amoebic gill disease in Atlantic salmon (*S. salar*) (H & E, Bar = 50 μm). C. Bacterial gill disease in Arctic charr (*Salvelinus alpinus*) showing prolific filamental and lamellar epithelial hyperplasia (asterisk) (H & E, Bar = 100 μm). D. Filamentous bacteria associated with the lamellae of Arctic charr (*Salvelinus aplinus*) gills exhibiting signs of bacterial gill disease (H & E, Bar = 50 μm). E. Hyperplastic filamental epithelium in response to infection of the gills of rainbow trout (*Oncorhynchus mykiss*) with *Loma salmonae* (H & E, Bar = 100 μm). *Loma salmonae* xenomas and localized epithelial response in the gills of brook trout (*Salvelinus fontinalis*) (Bar = 50 μm).

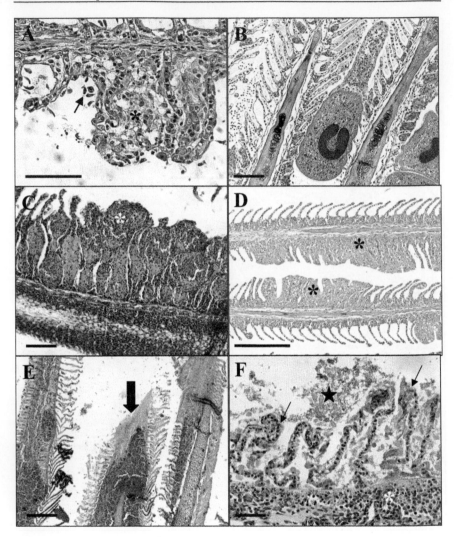

Fig. 16.2 A. Focal hyperplasia (asterisk) of the filamental epithelium in response to infestation of the gills with *Icthyobodo necator* (arrow) in blackfish (*Gadopsis marmoratus*) (H&E, Bar = 50 μm). B. *Ichthyophthirius multifiliis* (with its characteristic bean-shaped nucleus) infestation of rainbow trout (*O. mykiss*) gills (H&E, Bar = 10 μm). C. Hyperplasia and inflammatory response (asterisk) to acute trauma (48 h after injury) in the gills of marine Atlantic salmon (*S. salar*) (H&E, Bar = 10 μm). D. Hyperplasia of the filamental epithelium (asterisk) of the gills of kingfish (*Seriola lalandii*) infested with the monogenean gill fluke *Zeuxapta seriolae* (H& E, Bar = 180 μm). E. Bacterial erosion of a marine Atlantic salmon (*S. salar*) gill filament (arrow) infected with *Tenacibaculum maritimum* (H&E, Bar = 300 μm). F. *T. maritimum* colonisation (star), erosion, lamellar necrosis (arrow) and inflammatory response (asterisk) in the gills of marine Atlantic salmon (*S. salar*) (H&E, Bar = 50 μm).

reported in pathology or physiology studies on gills. However, Powell *et al.* (1995) reported a doubling of mitotic figures in the base of the filament epithelium in fish that showed hypertrophy and hyperplasia of chloride cells in response to chemical insult. More recently Adams and Nowak (2003) showed convincingly that there was an increased expression of proliferating cell nuclear antigen (PCNA) in hyperplastic tissue and the underlying basal cell layer of the filament epithelium in salmon gills affected by amoebic gill disease (AGD). However, a number of other cell types including leukocytes and eosinophilic granule cells also showed PCNA positive staining, suggesting that proliferation of cells other than filament epithelium also occurs in AGD affected gills (Adams and Nowak, 2003).

Mucus Cell Hyperplasia

Mucus cell hyperplasia is a common and non-specific response to many infectious and non-infectious gill diseases (for example Ellis and Wootten, 1978; Speare *et al.*, 1991; Ferguson *et al.*, 1992; Powell *et al.*, 1998; Tumbol *et al.*, 2001; Woo *et al.*, 2002; Roberts and Powell, 2003). The production of mucus serves to protect the gill from potentially pathogenic bacteria and parasites by providing a non-stable environment over the surface of the gill. It is a relatively viscous medium that varies in viscosity between fish species and environment (Roberts and Powell, 2005). In response to gill infections, mucus becomes more prolific and less viscous, thereby providing a less stable microenvironment on the gill, presumably to aid the flushing of the gill epithelium of offending pathogens (Roberts and Powell, 2005). Another role for mucus is that of ionic and acid-base regulation of the microenvironment of the gill surface that has been reviewed by Randall and Wright (1989). Mucus is polyanionic and so would serve to trap cations at the gill surface while allowing anions to pass through (Scott, 1989; Shephard 1989, 1992, 1994). Indeed ionic losses of chloride tend to be greater than those for sodium when fish are challenged with gill irritants such as chlorine (Powell and Perry, 1998).

Mucus has long since been considered a barrier to gas transport across the gill. Ultsch and Gros (1979) showed that oxygen diffusion across carp (*Cyprinus carpio*) mucus was retarded *in vitro*. *In vivo*, the thickness of the mucous layer comprises two principal components, the cellular glycocalyx (a mucopolysaccharide extracellular component of glycosylated proteins in epithelial cell membranes) and the secreted mucus itself produced and secreted from mucous cells. The gill epithelial cell glycocalyx varies with cell type with ionically active cells having a thicker glycocalyx than secretory cells (Powell et al., 1994). The mucous coat can vary in thickness depending upon environment and disease condition. The thickness and density of the mucous layer is probably not as significant a hindrance to oxygen transfer as suggested by Ultsch and Gros (1979), which has never been demonstrated in vivo. However, rainbow trout with a mucous cell hyperplasia were able to maintain blood oxygen levels even at low water oxygen tensions compared with unaffected fish (Powell et al., 1998). However, Powell and Perry (1999) did demonstrate that CO_2 excretion across the gills was affected by chemical irritation (vis-à-vis mucus secretion). This confirmed the suggestions by Cameron and Polhemus (1974) and Malte and Weber (1985) that CO_2 excretion across the gill was primarily diffusion limited whereas oxygen uptake was primarily perfusion limited (Daxboeck et al., 1982; Randall and Daxboeck, 1984; Malte and Weber, 1985). The reasons for this lay primarily in the fact that oxygen transfer is driven by cardiac output (Q) and diffusion of oxygen across the gill membrane into the red blood cells is limited to only a minor extent by the mucous layer as shown by (Powell and Perry, 1996, 1997a, 1999). In addition, qualitative changes on oxygen binding to haemoglobin can increase the efficiency of transport (see review by Gilmour, 1998). Carbon dioxide excretion, on the other hand, involves the rate-limiting step of HCO_3^- entering the red blood cell and then being catalyzed to molecular CO_2 by carbonic anhydrase before diffusion of CO_2 can take place (Perry, 1986; Perry and McDonald, 1993). The result is that a relative increase in the mucous barrier thickness has a more magnified effect on CO_2 excretion than O_2 uptake since CO_2 excretion can only occur when haemoglobin oxygenation is occurring. If the latter is slowed at all by increases in the gill diffusion distance, then the effect on CO_2 excretion is magnified. However, if mucous secretion is extreme (induced in response to acute intoxication such as exposure to chlorine (Bass and Heath, 1977; Bass et al., 1977), or other environmental pollutants and metals (see review by Fernandes and Mazon, 2003), then oxygen uptake will become affected. However, under such conditions, other structural responses caused by ionic dysfunction or cardiovascular collapse may be having a more significant effect on gas transfer than the mucous layer itself.

Chloride/Mitochondrial Rich Cell Hyperplasia

Chloride cells and mitochondrial rich cells in the gills (as they are also known) are involved in ionoregulation and acid-base regulation and their physiological role have been extensively reviewed (Perry, 1997, 1998). Ionoregulatory and acid-base dysfunction as a consequence of non-infectious insult can result in a hyperplastic even changes in the chloride cell population of the gill (Fernandes and Mazon, 2003). The end result of a chloride cell hyperplasia is an increase the blood-water diffusion thickness of the gill. Chloride cell hyperplasia as has been demonstrated using growth hormone and cortisol injection (Bindon et al., 1994a,b) and using an environmental stimulus such as artificial soft water (Greco et al., 1995, 1996) or saline ground water poor in specific ions such as potassium (Partridge and Creeper, 2004). In both hormonally and environmentally stimulated chloride cell hyperplasia, when fish were exposed to progressive hypoxia, there was compensation to maintain arterial PO_2 up to a point at which oxygen transfer became diffusion limited (Bindon et al., 1994; Greco et al., 1996). Carbon dioxide excretion, on the other hand, was compromised even at water PO_2 equivalent to normoxia (Bindon et al., 1994; Greco et al., 1996). This suggests that increases in chloride cell hyperplasia can have a significant effect on the respiratory physiology of the gill (as reviewed by Perry, 1998) in particular CO_2 excretion as discussed above.

Chloride cell hyperplasia is not often reported as a consequence of infectious diseases and is most likely to occur only when an ionoregulatory or osmotic stress occurs as a consequence of infection. Where parasitic infections such as Ichthyophthirius multifiliis (Ich) infestations of trout gills (Fig. 16.2) or skin of channel catfish (Ictalurus punctatus, Ewing et al., 1994) or goldfish (Carassius auratus, Tumbol et al., 2001) occur, the mature trophont emerges from under the epithelium and grazes on branchial or cutaneous epithelial cells. Consequently an acute ionic disturbance occurs with a lowering of blood plasma chloride concentrations (Ewing et al., 1994). This results in a progressive hyperplasia of gill chloride cells (Ewing et al., 1994; Tumbol et al., 2001) corresponding to a decrease in whole body net ionic losses (Tumbol et al., 2001). A similar recovery of plasma ion levels and reduction in whole body ionic net efflux is also seen in rainbow trout following infection with the microsporean, Loma salmonae (Fig. 16.1) (Powell et al., 2006). However, as with Ich, the consequences of any chloride cell hyperplasia on respiration in parasitized fish has not been investigated to date.

OTHER CELL TYPE PROLIFERATION

X-cell Disease

Although X-cell disease has been reported in more than 20 species including pleuronectids and gadids (as reviewed by Davison, 1998) the origins of the so-called X-cells remains an enigma. The etiology of X-cell disease (Fig. 16.3) remains ambiguous with the X-cells reported as being cancerous (Alpers *et al.*, 1977), virally transformed epidermal cells (Peters *et al.*, 1983) and unicellular parasites (Daimant and McVicar, 1990; Khattra *et al.*, 2000). Although reported in many species of Antarctic fishes including *Trematomus bernachii*, *T. hansonii* and *Chionodraco hammatus* (Davison, 1998) it also forms skin lesions in *Notothenia squamifrons* (Bucke and Everson, 1992). The most intensively studied cases of X-cell disease have been those of the Antarctic notothenid fish *Pagothenia borchgrevinki* ('borchs') which seem particularly susceptible with estimates of more than 15% of the population potentially affected within some local areas (Davison and Franklin, 2003). The disease is characterized by a hyperplasia of granulated 'X-cells' in the filament epithelium and progressive occlusion of the lamellar blood vessels and lacunae of the gills (Fig. 16.3) (Franklin *et al.*, 1993). This infiltration leads to a reduction in the functional surface are of the gill for gas exchange (Fig. 16.3). The study of *P. borchgrevinki* with X-cell disease has been limited by the relatively small size of the fish. It has not been possible to obtain reliable arterial blood gas samples. However, it has been shown that X-cell affected fish are restricted in their ability to take up oxygen and the occlusion of parts of the gill (Franklin *et al.*, 1993) does limit their metabolic scope for aerobic activity (Davison *et al.*, 1990). Similarly, through measurements of ventral aortic pressure and cardiac output, it has been shown that with X-cell disease, total peripheral vascular resistance is elevated (Davison and Franklin, 2003). X-cell affected borchs also appear to have elevated CO_2 concentrations in the blood and a consequential decrease in blood pH indicating the presence of a respiratory acidosis compared with unaffected fish (Table 16.1) clearly suggesting that branchial CO_2 excretion is also impaired.

Necrotic Gill Diseases

Necrotic diseases include those that cause either focal or extensive necrotic damage to the respiratory epithelium. For example, with flexibacteriosis (caused by the freshwater bacterium *Flavobacterium*

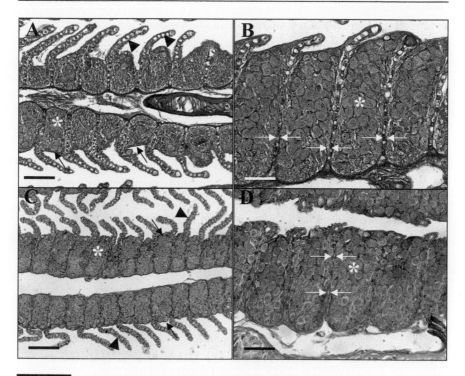

Fig. 16.3 X-cell gill disease in Antarctic fish. A. Progressive infiltration of the filamental epithlium of *Trematomus bernachii* with X-cells (asterisk) (Mallory's, Bar = 100 μm). B. X-cell infiltration of the inter-lamellar spaces in the gills of *T. bernachii* (asterisk) and constriction of lamellar lacunae (arrows) (Mallory's, Bar = 50 μm). C. Progressive X-cell infiltration (asterisk) of the hyperplastic filamental epithelium (arrow) *of Pagothenia borchgrevinki*. Normal lamellar structure indicated by arrowhead (H&E, Bar = 100 μm). D. Complete infiltration of the inter-lamellar space with X-cells (asterisk) and constriction of gill blood lacunae (arrows) in *P. borchgrevinki* (H&E, 50 μm).

Table 16.1 Comparison of mean ($\pm$ SE) mass, standard length caudal blood pH and total CO_2 concentration (TCO_2) and haematocrit (Hct) from non-diseased and X-cell gill diseased *P. borchgrevinki* (Powell, unpublished). Asterisk indicates statistical difference between groups.

	Non diseased (n = 7)	X-cell disease (n = 2)
Mass (g)	78.9 (11.9)	139.0 (62.2)
St. Length (mm)	169.1 (23.3)	210.5 (19.1)
pH	8.03 (0.05)	7.86 (0.04)*
TCO_2 (mM)	5.24 (0.68)	6.21 (0.03)
Hct (%)	14.0 (2.2)	12.7 (0.7)

columnare or marine bacterium *Tenacibaculum maritimum*) colonization of the gill filaments leads to progressive erosion of lamellae and filaments (Fig. 16.2). Hughes and Nyholm (1979) reported increases in ventilation rate in fish with missing gill filaments and arches assuming that the fish had a reduced gill surface area and therefore ventilation was increased to facilitate oxygen uptake although no measurements were made with respect to blood gases. Recent studies using an experimentally induced gill infection with *T. maritimum*, a particularly aggressive necrotizing infection of marine fish (Powell *et al.*, 2004) has shown some interesting results. Even severe gill infection with *T. maritimum*, resulting in erosion of lamellae and parts of the gill filament has relatively little effect on respiratory blood gases and the saturation of haemoglobin (Powell *et al.*, 2005). With this disease, fish appear to succumb to osmoregulatory dysfunction before respiratory compromise (Powell *et al.*, 2004, 2005). Any respiratory effects are probably compensated for by the increased perfusion of the gill (as has been seen histologically, Powell *et al.*, 2005) such that respiration is defended even during acute osmoregulatory dysfunction.

OTHER TYPES OF CHANGES IN GILL MORPHOLOGY

Lamellar Epithelial Lifting

The separation of the epithelium from the basal membrane of the pillar cells is a commonly referred to pathological sign (Fig. 16.4). Indeed, Mallet (1986) refers to it as a characteristic pathology of chemical intoxication and it often appears as a consequence of post-mortem deterioration of the gill architecture (Speare and Ferguson, 1989; Munday and Jaisankar, 1998). The consequence of such a pathological change would logically be that the blood-water diffusion distance is significantly increased and therefore the rates of gas exchange across the gill compromised. Although acute changes in the ionic permeability of the gill associated with chemical intoxication may occur, these usually are reflected in changes in the morphology of the ion regulating or pavement epithelial cells (Fernandes and Mazon, 2003) and reliance upon epithelial lifting as a pathognemonic sign is unreliable. However, these are often seen as more extensive responses whereas epithelial lifting is either focal or only appears to affect certain parts of the gills with no clearly defined pattern (pers. obs.). Epithelial lifting also occurs in response to acute hypoxia and in response to handling and anaesthesia (Fig. 16.4). Epithelial lifting in seahorse gills

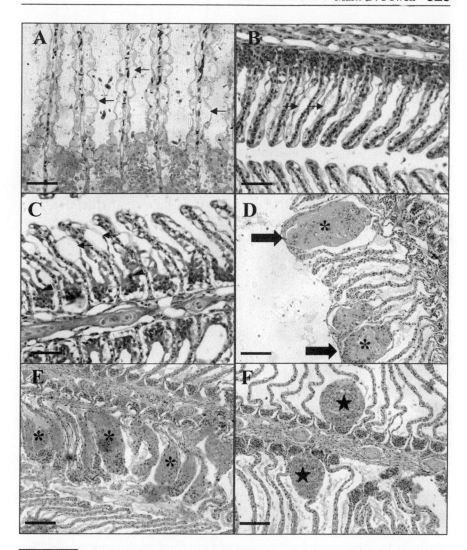

Fig. 16.4   A. Intercellular oedema (arrow) in the gills of hyperoxia acclimated rainbow trout (*O. mykiss*) (0.5 μm section, toluidine blue, Bar = 100 μm). B. Epithelial separation (arrow) and lifting in the gills of marine Atlantic salmon (*S. salar*) (H&E, Bar = 100 μm). C. Epithelial separation, intercellular oedema (arrow) and epithelial cell necrosis (arrowhead) in the gills of marine Atlantic salmon (*S. salar*) exposed to hydrogen peroxide (H&E, Bar = 100 μm). D and E. Lamellar aneurysms (asterisks) of unknown aetiology with evidence of lamellar synctia (arrows) in marine Atlantic salmon (*S. salar*) (H&E, Bar = 100 μm). Telangiectasis-like aneurysms (asterisk) in the gills of marine Atlantic salmon (*S. salar*) killed by cerebral percussion (H&E, Bar = 100 μm).

occurs following exposure to elevated environmental nitrite and ammonium concentrations, however, there was no association between epithelial lifting and the pre-mortem ammonium or nitrite exposure level (Adams, 1999). A more likely explanation of epithelial lifting is that pre-mortem, stress effects (driven primarily by acute releases of adrenaline and noradrenaline) cause an elevation of the intrabranchial pressure (Booth, 1979). Consequently, oedema forms between the structural pillar cells and the overlaying pavement cells, interpreted as the lifting of the epithelium. Since not all the gill filament is uniformly perfused (Booth, 1978) localized changes in gill intrabranchial pressure may account for the lack of uniformity of epithelial lifting in gills. It is important, therefore, to interpret the pathological changes carefully when making assumptions with regard to the physiological implications of that change.

Aneurysms

The development of aneurysms has been reported in response to a number of chemical insults (see reviews by Mallet ,1986; Fernandes and Mazon, 2003) and is rarely reported for pathogens. Aneurysms form as the blood pools in the gill lacunae (Fig. 16.4). This may result from occlusion of the efferent blood vessels. A more acute type of gill aneurysm is referred to as telangiectasis (rupture of capillaries, Fig. 16.4), whereby the pillar cells of the gill lamellae rupture, perhaps in response to acute high pressure and blood appears to pool in the expanded blood space. Telangiectasis is a common artefact of euthanasia when fish are killed by a cranial blow and as a consequence of acute abrasive branchial trauma (Fig. 16.4, Powell et al., 2004, 2005). The physiological effects of these structural abnormalities have not been characterized although it would be reasonable to surmise that where blood pools localized haemostasis may occur and thus functional gas exchange would be compromised. The relatively localized and focal nature of branchial aneurysms means that there is probably little measurable effect on respiration in most fish species, because perfusion of non-affected areas of the gill is possible due to the labile distribution of blood flow through the gills (Booth, 1978; Stenslokken et al., 1999).

Hyperaemia and Congestion

Hyperaemic and congestive changes in the gill probably reflect changes in cardiovascular parameters that may ultimately be driven by acute hypoxemia. For example, following acute exposure to oxidative toxicants (such as chlorine or hydrogen peroxide), the gill epithelium quickly shows necrotic changes that are often accompanied by osmotic or ionic changes in the blood chemistry (Powell and Perry, 1997b; Powell and Harris, 2004). However, associated with these changes is a noticeable infiltration of the central venous sinus (CVS) red blood cells (for example, see Powell and Harris, 2004; Powell et al., 2004, 2005). In addition, the gills often appear congested with large numbers of red blood cells (more than are usually observed) in the gill lacunae. This congestion of the gills and presence of large numbers of cells in the CVS suggests an acute increase in gill perfusion. Although rarely reported, hyperaemia, hyperperfusion and congestion of gills occur in response to infectious disease (Adams and Nowak, 2004; Powell et al., 2005)

Despite the hyperaemia and congestion in gills exposed to oxidative toxicants, fish appear to suffer acute hypoxaemia (Powell and Perry, 1997b; Powell and Harris, 2004). Haemoglobin is readily oxidized by reactive oxygen species generated by oxidative toxicants such that methaemoglobin levels rise sharply in intoxicated fish. Although methaemoglobin cannot transport oxygen, the levels of methaemo-globinaemia that occur with lethal levels of chlorine (Grothe and Eaton, 1975) and hydrogen peroxide (Powell and Perry, 1997b) are below those at which respiratory performance are compromised (Brauner et al., 1993; Williams et al., 1997) suggesting that methaemoglobinaemia is rarely a cause of death in chemically intoxicated fish, with the exception of acute nitrite poisoning (Russo and Thurston, 1991).

Cardiovascular Responses to Gill Diseases

Although very few gill diseases have been studied for physiological effects, even fewer have investigated the effects beyond those at the level of the gill. Reference has already been made to two diseases in which cardiovascular effects of the 'gill disease' have been studied: X-cell disease in *Pagothenia borchgrevinki* and amoebic gill disease in salmonids. In the latter case, it has been demonstrated that the disease results in a systemic hypertension in clinical cases on fish farms (Powell et al., 2002) but presents as increases in systemic vascular resistance and reduced cardiac output in acute laboratory infections (Table 16.2, Leef et al., 2005b). Interestingly, although other salmonid species can become affected by AGD and the characteristic focal hyperplastic gill lesions develop (Fig. 16.1), rainbow trout (*Oncorhynchus mykiss*) do not appear to be affected

Table 16.2 Comparative effects of amoebic gill disease on mean ($\pm$ SE) dorsal aortic pressure (DAP, cmH$_2$O), cardiac output (Q, mL.m^{-1} kg^{-1}) and systemic vascular resistance (R$_s$, cmH$_2$O mL^{-1} min^{-1} kg^{-1}) 6 h following surgery for three species of salmonid affected by amoebic gill disease (% lesioned filaments). Asterisk indicates "significant difference" from control values (Modified from Leef et al., 2005b).

	Control (unaffected)			AGD affected			% lesioned filaments
	DAP	Q	R$_s$	DAP	Q	R$_s$	
Alt. salmon	32.0 (2.6)	19.0 (1.5)	1.86 (0.20)	34.4 (2.2)	12.6 (1.8)*	3.43 (0.50)*	52.5 (11.6)
Brown trout	41.7 (2.5)	22.0 (4.7)	2.26 (0.53)	35.8 (5.8)	20.1 (8.4)	1.91 (0.70)	53.4 (4.6)
Rainbow trout	32.2 (4.0)	21.5 (2.6)	1.49 (0.06)	31.4 (4.9)	27.6 (6.7)	1.69 (0.58)	24.0 (4.9)

in terms of the cardiovascular system compared with Atlantic salmon or brown trout (*Salmo trutta*). Brown trout similarly show less severe effects than Atlantic salmon that are the most susceptible (Table 16.2, Leef *et al.*, 2005b). It remains to be seen whether systemic hypertension is a common feature of all gill diseases. However, with hypertension seen in different diseases such as those of Antarctic fishes (Davison and Franklin, 2003) and AGD in salmon (Powell *et al.*, 2002; Leef *et al.*, 2005b), it would be reasonable to surmise that other gill diseases may present cardiovascular effects. Whether these effects are as a direct effect of the host-parasite interaction (i.e., production of a parasite derived 'exotoxin') or as a consequence of an inflammatory response, remains to be investigated.

Conclusions and Future Directions

What we determine as a gill disease is based almost entirely upon the fact that the gills are primarily affected organ. Since the gills' primary function is that of respiration and gas exchange, we logically assume that changes in gill structure will translate into respiratory compromise, based upon physiological models. The truth is that so-called gill diseases rarely affect only the gills structurally. In reality, gill diseases also have significant impacts upon ion regulation (another primary function of the gill) as well as the cardiovascular system with which the gills are intimately associated. In fact, for the few gill diseases that have been intensively studied to date, it would appear that respiration is only mildly affected by the structural changes that have been reported but the other physiological systems with which the gills are so closely associated are more severely affected. While this at first appears to be at odds with logic, we must consider the evolution of gills. The gill is a fundamental organ for respiration and gas transfer. The vital process of oxygen uptake is a robust one that although is physically limited by diffusional changes in the gill, gill physiology is designed to compensate for diffusional limitations that occur in some infectious and non-infectious diseases. For example, the increased perfusion of the gill by blood (hyperaemia and congestion) is the end-point of a compensation process for increasing oxygen uptake. Similarly, processes such as lamellar recruitment play a vital role in increasing functional surface area for gas exchange. Qualitative changes on oxygen binding to haemoglobin enhance the efficiency by which oxygen uptake and transport occurs. The end result of these compensations is that the animal tolerates the pathological state. Only when the capacity to compensate is outstripped by the physiological demands of the disease (or the host such as in response to predation, migration or a sudden increase in oxygen demand), does respiration become compromised to the point at which mortality occurs. However, in most cases (diseases studied to date), mortality occurs before significant respiratory dysfunction. This is a reflection of the multidimensional effect of disease and that physiological systems are not affected in isolation. Therefore it must be recognized that fish that die of gill diseases don't usually die of respiratory failure! Rather it is the combination of the partial failure of multiple physiological systems that results in death.

Despite the plethora of studies that have examined the physiology of gills and the functional role of the gill in fishes from numerous environments, there remain very few studies that have comprehensively examined the effects of disease on gill function. The limited studies have been reviewed here. The difficulties in obtaining reliable data from fish that are physiologically compromised and having standardized methods for determining the extent of disease are the biggest challenges that face research in this area. However, the few disease models that have been extensively studied to date, have been revealing. Hopefully the future will see a continuation of research into the pathophysiology of fishes and an ever increase of our understanding of pathological processes in fishes.

Acknowledgments

I would like to thank M. Leef for use of unpublished data, to Dr W. Davison for gill sections of X-cell disease and to B. Mansell for the picture of *Zeuxapta seriolae* infestation of kingfish gills. The Australian Research Council is acknowledged for the Large Grant and Discovery programs, as well as the Fisheries Research and Development Corporation and the Cooperative Research Centre for sustainable finfish aquaculture (Aquafin CRC) for financial research for support Dr Powell's research.

References

Adams, M.B. 1999. The effects of ammonia and nitrite on the respiratory physiology and morphology of the juvenile big bellied seahorse *Hippocampus abdominalis*. B. App. Sc. Honours thesis, University of Tasmania, Launceston.

Adams, M.B. and B.F. Nowak. 2003. Amoebic gill disease: sequential pathology in cultured Atlantic salmon, *Salmo salar* L. *Journal of Fish Diseases* 26: 601-614.

Alpers, C.E., B.B. McCain, M. Myers, S.R. Wellings, M. Poore, J. Bagshaw and C.J. Dawe. 1977. Pathologic anatomy of pseudobranch tumours in Pacific cod, *Gadus macrocephalus*. *Journal of the National Cancer Institute* 59: 277-298.

Bass, M.L. and A.G. Heath. 1977. Cardiovascular and respiratory changes in rainbow trout *Salmo gairdneri*, exposed intermittently to chlorine. *Water Research* 11: 497-502.

Bass, M.L., C.R. Berry and A.G. Heath. 1977. Histopathological effects of intermittent chlorine exposure on bluegill (*Lepomis macrochirus*) and rainbow trout (*Salmo gairdneri*). *Water Research* 11: 731-735.

Bindon, S.D., K.M. Gilmour, J.C. Fenwick and S.F. Perry. 1994a. The effects of branchial chloride cell proliferation on respiratory function in the rainbow trout *Oncorhynchus mykiss*. *Journal of Experimental Biology* 197: 47-63.

Bindon, S.D., J.C. Fenwick and S.F. Perry. 1994b. Branchial chloride cell proliferation in the rainbow trout, *Oncorhynchus mykiss*: implications for gas transfer. *Canadian Journal of Zoology* 72: 1395-1402.

Booth, J.L. 1978. The distribution of blood flow in gills of fish: application of a new technique to rainbow trout (*Salmo gairdneri*). *Journal of Experimental Biology* 73: 119-129.

Booth, J.L. 1979. The effects of oxygen supply, epinephrine and acetylcholine on the distribution of blood flow in trout gills. *Journal of Experimental Biology* 83: 31-39.

Brauner, C., A.L. Val and D.J. Randall. 1993. The effects of graded methaemoglobin levels on the swimming performance of chinook salmon (*Oncorhynchus tshawytscha*). *Journal of Experimental Biology* 185: 121-135.

Bucke, D. and I. Everson. 1992. "X-cell" lesions in *Notothenia* (*Lepidonotothen*) *squamifrons* Gunther. *Bulletin of the European Association of Fish Pathologists* 12: 83-86.

Byrne, P., H.W. Ferguson, J. Lumsden and V.E. Ostland. 1991. Blood chemistry of bacterial gill disease in brook trout *Salvelinus fontinalis*. *Diseases of Aquatic Organisms* 10: 1-6.

Byrne, P.J., V.E. Ostland, J.S. Lumsden, D.D. MacPhee and H.W. Ferguson. 1995. Blood chemistry and acid-base balance in rainbow trout *Oncorhynchus mykiss* with experimentally-induced acute bacterial gill disease. *Fish Physiology and Biochemistry* 14: 509-518.

Cameron, J.N. and J.A. Polhemus. 1974. Theory of CO_2 exchange in trout gills. *Journal of Experimental Biology* 60: 183-194.

Daimant, A. and A.H. McVicar. 1990. Distribution of X-cell disease in common dab, *Limanda limanda* L., in the North Sea, and ultrastructural observations of previously undescribed developmental stages. *Journal of Fish Diseases* 13: 25-37.

Davis, J.C. and J.N. Cameron. 1971. Water flow and gas exchange at the gills of rainbow trout, *Salmo gairdneri*. *Journal of Experimental Biology* 54: 1-18.

Davison, W. 1998. X-cell gill disease in *Pagothenia borchgrevinki* from McMurdo Sound, Antarctica. *Polar Biology* 19: 17-23.

Davison, W and C.E. Franklin. 2003. Hypertension in *Pagothenia borchgrevinki* caused by X-cell disease. *Journal of Fish Biology* 63: 129-136.

Davison, W., C.E. Franklin and P.W. Carey. 1990. Oxygen uptake in the Antarctic teleost *Pagothenia borchgrevinki*. Limitations imposed by X-cell disease. *Fish Physiology and Biochemistry* 8: 69-77.

Daxboeck, C., P.S. Davie, S.F. Perry and D.J. Randall. 1982. Oxygen uptake in a spontaneously ventilating blood-perfused trout preparation. *Journal of Experimental Biology* 101: 35-45.

Dykova, I., B.F. Nowak, P.B.B. Crosbie, L. Fiala, H. Peckova, M.B. Adams, B. Machackova and H. Dvorakova. 2005. *Neoparamoeba branchiphila* n. sp., and related species of the genus *Neoparamoeba* Page, 1987: morphological and molecular characterization of selected strains. *Journal of Fish Diseases* 28: 49-64.

Ellis, A.E. and R. Wootten. 1978. Costiasis of Atlantic salmon, *Salmo salar* L. Smolts in seawater. *Journal of Fish Diseases* 1: 389-393.

Ewing, M.S., M.C. Black, V.S. Blazer and K.M. Kocan. 1994. Plasma chloride and gill epithelial response of channel catfish to infection with *Ichthyophthirius multifiliis*. *Journal of Aquatic Animal Health* 6: 187-196.

Ferguson, H.W. 1989. *Systemic Pathology of Fish*. Iowa State University Press, Ames, USA.

Ferguson, H.W., D. Morrison, V.E. Ostland, J. Lumsden and P. Byrne. 1992. Responses of mucus-producing cells in gill disease of rainbow trout (*Oncorhynchus mykiss*). *Journal of Comparative Pathology* 106: 255-265.

Fernandes, M.N. and A. de F. Mazon. 2003. Environmental pollution and fish gill morphology. In: *Fish Adaptations*, A.L. Val and B.G. Kapoor (eds.). Science Publishers Inc., Enfield, (NH), USA, pp. 203-231.

Franklin, C.E., J.C. McKenzie, W. Davison and P.W. Carey. 1993. X-cell gill disease obliterates the lamellar blood supply in the Antarctic teleost, *Pagothenia borchgrevinki* (Boulenger, 1902). *Journal of Fish Diseases* 16: 249-254.

Gilmour, K.M. 1998. Gas exchange. In: *The Physiology of Fishes*, D.H. Evans (ed.). CRC Press, Boca Raton, 2nd edition, pp. 101-127.

Greco, A.M., J.C. Fenwick and S.F. Perry. 1996. The effects of softwater acclimation on gill structure in the rainbow trout *Oncorhynchus mykiss*. *Cell and Tissue Research* 285: 75-82.

Greco, A.M., K.M. Gilmour, J.C. Fenwick and S.F. Perry. 1995. The effects of softwater acclimation on respiratory gas transfer in the rainbow trout *Oncorhynchus mykiss*. *Journal of Experimental Biology* 198: 2557-2567.

Grothe, E.R. and J.W. Eaton. 1975. Chlorine-induced mortality in fish. *Transactions of the American Fisheries Society* 104: 800-802.

Hughes, G.M. and K. Nyholm. 1979. Ventilation in rainbow trout (*Salmo gairdneri*, Richardson) with damaged gills. *Journal of Fish Biology* 14: 285-288.

Khattra, J.S., S.J. Gresoviac, M.L. Kent, M.S. Myers, R.P. Hedrick and R.H. Devlin. 2000. Molecular detection and phylogenetic placement of a microsporidian from English sole (*Pleuronectes vetulus*) affected by X-cell pseudotumours. *Journal of Parasitology* 86: 867-871.

Leef, M.J., J.O. Harris and M.D. Powell. 2005a. Respiratory pathogenesis of amoebic gill disease (AGD) in experimentally infected Atlantic salmon (*Salmo salar*). *Diseases of Aquatic Organisms* 66: 205-213.

Leef, M.J., J.O. Harris and M.D. Powell. 2005b. Cardiovascular responses of three salmonid species affected with amoebic gill disease (AGD). *Journal of Comparative Physiology* B175: 523-532.

Malte, H. and R.E. Weber. 1985. A mathematical model for gas exchange in the fish gill based on non-linear blood gas equilibrium curves. *Respiration Physiology* 62: 359-374.

Munday, B.L. and C. Jaisankar. 1998. Postmortem changes in the gills of rainbow trout (*Oncorhynchus mykiss*) in freshwater and seawater. *Bulletin of the European Association of Fish Pathologists* 18: 127-131.

Partridge, G.J. and J. Creeper. 2004. Skeletal myopathy in juvenile barramundi, *Lates calcarifer* (Bloch), cultured in potassium-deficient saline ground water. *Journal of Fish Diseases* 27: 523-530.

Perry, S.F. 1986. Carbon dioxide excretion in fish. *Canadian Journal of Zoology* 64: 565-572.

Perry, S.F. 1997. The chloride cell: structure and function in freshwater fishes. *Annual Reviews in Physiology* 59: 325-347.

Perry, S.F. 1998. Relationships between branchial chloride cells and gas transfer in freshwater fish. *Comparative Physiology and Biochemistry* A119: 9-16.

Perry, S.F. and D.G. MacDonald. 1993. Gas exchange. In: *The Physiology of Fishes*, D.H. Evans (ed.) CRC Press, Boca Raton, pp. 251-278.

Peters, N., W. Schmidt, H. Kranz and H.F. Stich. 1983. Nuclear inclusion in the X-cells of skin papillomas on Pacific flatfish. *Journal of Fish Diseases* 6: 533-536.

Powell, M.D. and J.O. Harris. 2004. Influence of oxygen on the toxicity of chloramine-T to Atlantic salmon (*Salmo salar* L.) smolts in fresh and salt water. *Journal of Aquatic Animal Health* 16: 83-92.

Powell, M.D. and S.F. Perry. 1996. Respiratory and acid-base disturbances in rainbow trout (*Oncorhynchus mykiss*) blood during exposure to chloramine-T, paratoluenesulphonamide and hypochlorite. *Canadian Journal of Fisheries and Aquatic Sciences* 53: 701-708.

Powell, M.D. and S.F. Perry. 1997a. Respiratory and acid-base disturbances in rainbow trout blood during exposure to chloramine-T under hypoxia and hyperoxia. *Journal of Fish Biology* 50: 418-428.

Powell, M.D. and S.F. Perry. 1997b. Respiratory and acid-base pathophysiology of hydrogen peroxide in rainbow trout (*Oncorhynchus mykiss* Walbaum). *Aquatic Toxicology* 27: 99-112.

Powell, M.D. and S.F. Perry. 1998. Acid-base and ionic fluxes in rainbow trout (*Oncorhynchus mykiss*) during exposure to chloramine-T. *Aquatic Toxicology* 43: 13-24.

Powell, M.D. and S.F. Perry. 1999. Cardio-respiratory effects of chloramine-T exposure in rainbow trout. *Experimental Biology Online* 4:5.

Powell, M.D., D.J. Speare and G.M. Wright. 1994. Comparative ultrastructural morphology of lamellar epithelial, chloride and mucous cell glycocalyx of the rainbow trout (*Oncorhynchus mykiss*) gill. *Journal of Fish Biology* 44: 725-730.

Powell, M.D., D.J. Speare and J.A. Becker. 2006. Whole body net ion fluxes, plasma electrolyte concentrations and haematology during a *Loma salmonae* infection in juvenile rainbow trout, *Oncorhynchus mykiss* (Walbaum). *Journal of Fish Diseases* 29: 727-735.

Powell, M.D., G.M. Wright and D.J. Speare. 1995. Morphological changes in rainbow trout (*Oncorhynchus mykiss*) gill epithelia following repeated intermittent exposure to chloramine-T. *Canadian Journal of Zoology* 73: 154-165.

Powell, M.D., F. Haman, G.M. Wright and S.F. Perry. 1998. Respiratory responses to graded hypoxia of rainbow trout (*Oncorhynchus mykiss*) following repeated intermittent exposure to chloramine-T. *Aquaculture* 165: 27-39.

Powell, M.D., D. Fisk and B.F. Nowak. 2000. Effects of graded hypoxia on Atlantic salmon infected with amoebic gill disease. *Journal of Fish Biology* 57: 1047-1057.

Powell, M.D., M.E. Forster and B.F. Nowak. 2002. Apparent vascular hypertension associated with Amoebic Gill Disease in cultured Atlantic salmon (*Salmo salar*) in Tasmania. *Bulletin of the European Association of Fish Pathologists* 22: 328-333.

Powell, M.D., J. Carson and R. van Gelderen. 2004. Experimental induction of gill disease in Atlantic salmon *Salmo salar* smolts with *Tenacibaculum maritimum*. *Diseases of Aquatic Organisms* 61: 179-185.

Powell, M.D., J.O. Harris, J. Carson and J.V. Hill. 2005. Effects of gill abrasion and experimental infection with *Tenacibaculum maritimum* on the respiratory physiology of Atlantic salmon *Salmo salar* affected by amoebic gill disease. *Diseases of Aquatic Organisms* 63: 169-174.

Randall, D.J. and C. Daxboeck. 1984. Oxygen and carbon dioxide transfer across fish gills. In: *Fish Physiology*, W.S. Hoar and D.J. Randall (eds.). Academic Press, New York, Vol. 10A, pp. 263-314.

Randall, D.J. and P.W. Wright. 1989. The interaction between carbon dioxide and ammonia excretion and water pH in fish. *Canadian Journal of Zoology* 67: 2936-2942.

Roberts, S.D. and M.D. Powell. 2003. Comparative ionic flux and gill mucous cell histochemistry: effects of salinity and disease status in Atlantic salmon (*Salmo salar* L.). *Comparative Biochemistry and Physiology* A134: 525-537.

Roberts, S.D. and M.D. Powell. 2005. The viscosity and glycoprotein biochemistry of salmonid mucus varies with species, salinity and the presence of amoebic gill disease. *Journal of Comparative Physiology* B175: 1-11.

Russo, R.C. and R.V. Thurston. 1991. Toxicity of ammonia, nitrite and nitrate to fishes. In: *Aquaculture and Water Quality*, D.E. Brune and J.R. Tomasso (eds.). World Aquaculture Society, Baton Rouge. USA, pp. 58-89.

Scott, J.E. 1989. Ion binding patterns of affinity depending upon type of acid groups. In: *Mucus and Related Topics*, E. Chantler and N.A. Ratcliffe (eds.). The Company of Biologists Ltd., Cambridge, UK, pp. 111-115.

Shephard, K.L. 1989. The effects of mucus and mucilagenous materials on ion-distributions at epithelial surfaces. In: *Mucus and Related Topics*, E. Chantler and N.A. Ratcliffe (eds.). The Company of Biologists Ltd., Cambridge, UK, pp. 123-130.

Shephard, K.L. 1992. Studies on the fish gill microclimate: interaction between gill tissue, mucus and water quality. *Environmental Biology of Fishes* 34: 409-420.

Shephard, K.L. 1994. Functions for fish mucus. *Reviews in Fish Biology and Fisheries* 4: 401-429.

Smith, F.M. and D.R. Jones. 1982. The effect of changes in blood oxygen-carrying capacity on ventilation volume in the rainbow trout (*Salmo gairdneri*). *Journal of Experimental Biology* 97: 325-334.

Soivio, A., K. Nyholm and K. Westman. 1975. A technique for repeated sampling of the blood of individual resting fish. *Journal of Experimental Biology* 62: 207-217.

Speare, D.J. and H.W. Ferguson. 1989. Fixation artefacts in rainbow trout (*Salmo gairdneri*) gills: a morphometric evaluation. *Canadian Journal of Fisheries and Aquatic Sciences* 46: 780-785.

Speare, D.J., H.W. Ferguson, F.W.M. Beamish, J.A. Yager and S. Yamashiro. 1991. Pathology of bacterial gill disease: sequential development of lesions during natural outbreaks. *Journal of Fish Diseases* 14: 21-32.

Stenslokken, K-O., L. Sundin and G.E. Nilsson. 1999. Cardiovascular and gill microcirculatory effects of endothelin-1 in Atlantic cod: evidence for pillar cell contraction. *Journal of Experimental Biology* 202: 1151-1157.

Tumbol, R.A., M.D. Powell and B.F. Nowak. 2001. Ionic effects of *Ichthyophthirius multifiliis* infection in goldfish *Carassius auratus*. *Journal of Aquatic Animal Health* 13: 20-26.

Williams, E.M., J.A. Nelson and N. Heisler. 1997. Cardio-respiratory function in carp exposed to environmental nitrite. *Journal of Fish Biology* 50: 137-149.

Woo, P.T.K., D.W. Bruno and L.S. Lim. 2002. *Diseases and Disorders of Finfish in Cage Culture*. CABI Publishing, Wallingford, UK.

Control of the Heart in Fish

Edwin W. Taylor[1,*], Cleo Leite[2], Hamish Campbell[1],
Itsara Intanai[3] and Tobias Wang[4]

INTRODUCTION

The single circulatory system of all fish consists of a four-chambered heart (sinus venosus, atrium, ventricle and bulbous arteriosus), in series with the branchial and systemic vascular beds (Randall, 1968; Farrell and Jones, 1992). The matching of rates of water and blood flow over the functional countercurrent at the gills, according to their relative capacities for oxygen, is essential for effective respiratory gas exchange (Piiper and Scheid, 1977) and must be capable of rapid adjustment to varying metabolic rates. Cardiac output, and its components stroke volume and heart rate, are robust indicators of metabolism (Farrell and Jones, 1992), and the changes in heart rate with altered metabolic

Authors' addresses: [1]School of Biosciences, University of Birmingham, Edgbaston, Birmingham B15 2TT, UK.
[2]Department of Physiological Sciences, Federal University of São Carlos, 13565-905 São Carlos, SP, Brazil.
[3]Prince of Songkla University, Thailand.
[4]Department of Zoophysiology, Aarhus University, 8000 Aarhus, Denmark.
E-mail: tobias.wang@biol.au.dk
*_Corresponding author:_ E-mail: E.W.TAYLOR@bham.ac.uk

demand requires that the pacemaker activity is controlled. Fine control of heart rate includes beat-to-beat modulation by the respiratory cycle that can be manifested as one-to-one cardiorespiratory synchrony (Taylor, 1992).

The autonomic nervous system is the main short-term modulator of heart rate in fish. In all fish except cyclostomes, the heart receives inhibitory innervation via the vagus nerve. In most teleosts there is evidence for adrenergic, excitatory control via sympathetic innervation of the heart. Here we review innervation of the heart in the major taxonomic groups of fish and we then focus on the generation of cardio-respiratory synchrony. Variations in inhibitory vagal tone imposed by activity in cardiac vagal preganglionic neurones (CVPN) within the medulla oblongata are the predominant factors generating cardiorespiratory synchrony. We review the location of CVPN and the putative roles of their inputs from chemoreceptors and mechanoreceptors in determining their activity.

PARASYMPATHETIC INNERVATION OF THE HEART

Cyclostomes

The heart of myxinoids is aneural, that is, it is not innervated by the vagus or the sympathetic nervous system (Green, 1902). Consistent with the lack of innervation, the isolated heart of myxinoids is insensitive to exogenously applied acetylcholine. The heart of the lamprey (although similarly devoid of a sympathetic supply) is innervated by the vagus (Augustinsson et al., 1956).

The main effect of vagal stimulation in lampetroids is an *acceleration* of the heart, a response unique amongst vertebrates, which is accompanied by a decreased force of contraction (Falck et al., 1966). Acetylcholine induces an acceleration of the heart and nicotinic cholinoceptor agonists, such as nicotine, have a similar effect (Augustinsson et al., 1956; Falck et al., 1966). The excitatory effect of vagal stimulation or nicotinic agonists can be blocked by nicotinic cholinoceptor antagonists such as tubocurarine and hexamethonium (Augustinsson et al., 1956; Falck et al., 1966). Thus, in contrast to other vertebrates, the lamprey heart has nicotinic rather than muscarinic receptors on their hearts.

Elasmobranch Fish

Phylogenetically, the elasmobranchs are the earliest vertebrates with a well developed autonomic nervous system that is clearly differentiated into parasympathetic and sympathetic components (Taylor, 1992). They are also the earliest group known to have an inhibitory vagal innervation of the heart. In the dogfish, *Scyliorhinus canicula*, the vagus nerve divides to form, at its proximal end, branchial branches 1, 2, 3 and 4 which contain skeletomotor fibres innervating the intrinsic respiratory muscles of gill arches 2, 3, 4 and 5 respectively (Fig. 17.1). The first gill arch is innervated by the glossopharyngeal (IXth cranial) nerve. The vagus also sends, on each side of the fish, two branches to the heart. The visceral cardiac arises close to the origin of the visceral branch of the vagus, the branchial cardiac from the post-trematic, fourth branchial branch of the vagus (Taylor, 1992). The two cardiac vagi pass down the ductus Cuveri and then break up into an interwoven plexus on the sinus venosus, terminating at the junction with the atrium (Young, 1933). The sino-atrial node is thought to be the site of the pacemaker in elasmobranch fishes (Satchell, 1971).

Continuous stimulation of the vagus nerve, as well as application of acetylcholine, has an inhibitory effect on heart rate. Variations in the degree of cholinergic vagal tonus on the heart, in the absence of an adrenergic innervation (see below), serves as an important mode of nervous cardioregulation in elasmobranchs (Butler and Taylor, 1971; Taylor *et al.*, 1977). There is a normoxic vagal tone on the heart, which is augmented by increases in ambient temperature (Butler and Taylor, 1975; Taylor *et al.*, 1977; Taylor, 1992). Application of acetylcholine exerts a negative chronotropic effect on the elasmobranch heart, which is antagonized by atropine, implying that the effect is mediated by muscarinic cholinoceptors, as in the higher vertebrates (Taylor, 1992; Taylor *et al.*, 1999).

Retrograde intra-axonal transport of HRP for the identification of vagal preganglionic neurons showed that the vagal motor column of the dogfish, *Scyliorhinus canicula* extends over 5 mm in the hindbrain from 2 mm caudal to 3 mm rostral of obex (Withington-Wray *et al.*, 1986). Most of the vagal motoneurons were found in a rostromedial division of the vagal motor column, identified as the dorsal vagal motor nucleus (DVN). A clearly distinguishable lateral group of cells was identified which had a rostrocaudal extent of approximately 1 mm, rostrally from obex. This population of motoneurons contributed axons solely to the branchial cardiac branch of the vagus innervating the heart (Barrett and Taylor, 1985b) and comprised 8% of the total population of vagal motoneurons. However, the cells in this lateral division supply 60% of the efferent axons running in the branchial cardiac nerve, with the other 40% supplied by cells in the rostromedial division. When the medial cells contributing efferent axons to the heart via the visceral cardiac branches are taken into account, then the lateral cells are found to supply 45% of vagal efferent output to the heart. Thus cardiac vagal preganglionic neurones (CVPN) providing axons to the branchial cardiac nerve are found rostromedially in the elasmobranch equivalent of the DVN, and solely comprise the lateral division of the vagal motor column, possibly the primitive equivalent of the nucleus ambiguous (NA) of mammals. It is thought that this dual location of CVPN has important functional implications (see below).

Teleost Fish

The vagus is cardioinhibitory in teleosts. Variation in vagal tone affects heart rate and a vagal, inhibitory, resting tonus has been demonstrated in some species (e.g., Cameron, 1979). The level of resting vagal tone decreases with increasing temperature in trout (cf., elasmobranchs) with adrenergic mechanisms taking over the cardioacceleratory function at higher temperatures (Wood et al., 1979). Although the negative inotropic influence of the vagi does not reach the ventricle, cardiac output is greatly affected by the inotropic control of the atrium, which directly regulates the filling of the ventricle (Jones and Randall, 1978; Johansen and Burggren, 1980). This differs from the situation described for mammals. The negative chronotropic effects that can be elicited by vagal stimulation or infusion of ACh are completely abolished by atropine, which shows that the effects are mediated by stimulation of muscarinic receptors.

Application of HRP to the whole vagus nerve and to selected branches of the vagus in two species of teleost, the cod (Gadus morhua) and trout (Oncorhynchus mykiss) revealed that vagal preganglionic neurons were located over a distance of 2.8 mm in the ipsilateral hindbrain from 1.2 mm caudal to 1.6 mm rostral to obex, and that approximately 11% of these neurons were located ventrolaterally, while the others were found in a dorsomedial location clustered close to the edge of the 4th ventricle (Withington-Wray et al., 1987). The lateral group of vagal motoneurons were divided into two groups: a caudal group extending for approximately 1 mm (from 0.75 mm caudal to 0.25 mm rostral of obex) and a more rostral group which extended for approximately 0.75 mm

(0.75-1.5 mm rostral of obex). When HRP was applied to the cardiac branch of the vagus, labelled CVPN were found in the caudal lateral division as well as in the dorsomedial part of the vagal motor column. The application of HRP to one of the branchial branches of the vagus also labelled both lateral and dorsomedial cells, this time with the lateral cells located in the more rostral group of cells. We have speculated that these cells may provide vasomotor input to the branchial circulation (Taylor, 1992) as, in contrast to elasmobranchs, the branchial branches in teleosts have both a vasomotor and skeletomotor function. The vagus contains vasomotor fibres, which have been shown to innervate sphincters at the base of the efferent filament arteries (Nilsson, 1984). In a more recent study, the fluorescent tracer True Blue was applied to the cardiac branch of the vagus in the pacu, *Piaractus mesopotamicus* (E.W. Taylor, C.A.C. Leite and F.T. Rantin, unpublished observation). Cardiac vagal preganglionic neurones (CVPN) were located over a distance of about 3.0 mm, from 1.5 mm rostral of obex to 1.5 mm caudal of obex, having an overlapping distribution with somatic motor neurones supplying respiratory muscles via the branchial branches of the vagus and glossopharyngeal nerves. These CVPN were distributed in separate nuclei with two medial groups of cells in the DVN, a ventral group containing about 60% of cell bodies and a dorsal group containing about 40% of cell bodies. In addition there were a small number of cell bodies scattered laterally outside of the DVN, constituting only about 2% on the total number of CVPN. This relative paucity of CVPN outside of the DVN may be offset by the division into 2 separate groups of CVPN in the DVN, a novel observation in fish, though Taylor *et al.* (2001) noted a similar separation of CVPN in the DVN of the lizard *Uromastyx aegyptius microlepis*. Its possible functional significance is not yet known.

Air-breathing Fish

Following application of HRP to the second and third branchial branches of the vagus nerve in the bowfin, *Amia calva*, retrogradely labelled cell bodies were found in the DVN over a rostro-caudal distance of 4 mm either side of obex (Taylor *et al.*, 1996). There was a sequential topography in the distribution of the cell bodies innervating branchial nerves 2 and 3, as described in dogfish and cod (Taylor, 1992). In addition, motor cell bodies were located in lateral locations outside the DVN as described in teleosts (e.g., cod) and all tetrapod vertebrates (Taylor, 1992, 1994). In the bowfin, some of these lateral cells were of an unusual appearance with large cell bodies and thick, branching dendrites.

Application of HRP to the nerve supplying the glottis and ABO revealed cell bodies in a ventrolateral location in the brainstem and the ventral horn of the anterior spinal cord over a rostro-caudal distance of 5.3 mm, predominantly caudal of obex. From their location it was possible to identify them as cell bodies which typically supply axons to the hypobranchial nerve (i.e., occipital and anterior spinal nerves). Consequently, it is apparent that the glottis and ABO are innervated by nerves of the hypobranchial complex, which provides nerves to elements of musculature normally associated with feeding movements in water-breathing fish. Given that feeding-type movements are implicated in air-breathing in bowfin (Liem, 1989; and see above), the observed hypobranchial innervation of the glottis and ABO may imply nervous coordination of air-gulping and glottal opening, which would ensure effective ventilation of the swimbladder. Indeed, even in mammals, there are functional similarities and a close connection between the central nervous mechanisms controlling breathing and those controlling swallowing (Jean, 1993). An additional group of stained cell bodies were identified as VPN in the DVN. These probably provide efferent axons to smooth muscle in the swimbladder wall, comparable with the vagal efferents controlling reflex broncho-constriction in the mammalian lung. An afferent vagal supply to the ABO would seem axiomatic but was not revealed by this study (Taylor et al., 1996).

SYMPATHETIC INNERVATION OF THE HEART

Cyclostomes, elasmobranchs, dipnoans and some teleosts (particularly pleuronectids) seem to lack adrenergic innervation of the heart (Laurent et al., 1983). However, cardiac cells containing adrenaline or noradrenaline are found in all vertebrates, including lampreys (reviewed by Taylor et al., 1999) and in those vertebrates without a sympathetic innervation of the heart the effect of circulating or locally released catecholamines is excitatory.

Cyclostomes

The heart of cyclostomes is not innervated by the sympathetic nervous system, but the heart contains large quantities of adrenaline and noradrenaline stored in chromaffin cells. Although not as pronounced as the effects of acetylcholine, injected catecholamines have marked cardiostimulatory effects in the intact animal (Nilsson and Holmgren,

1994). Infusion of β-agonists and tyramine also stimulate the lampetroid heart, and their effects are blocked by propranalol, suggesting an effect via β-adrenoceptors in lampetroids as in the higher vertebrates (Augustinsson *et al.*, 1956; Falck *et al.*, 1966). Infusion of the beta-adrenoreceptor antagonist sotalol causes a marked reduction of heart rate, suggesting a tonic release of catecholamines from the chromaffin stores within the heart of the normal animal (Nilsson and Holmgren, 1994). Spinal autonomic nerve fibres, containing adrenergic elements, innervate some blood vessels in lampreys and both catecholamines and acetylcholine increase vascular resistance in *Myxine* (Axelsson *et al.*, 1990).

Elasmobranch Fish

The sympathetic nervous system of elasmobranchs does not extend into the head. The most anterior pair of paravertebral ganglia is situated within the posterior cardinal sinuses. This condition is unique amongst vertebrates, but it is not clear whether it is primary or the result of a secondary loss (Young, 1950). As a result, there is no direct sympathetic innervation of the heart or the branchial circulation in elasmobranchs, and there is no evidence for branchial vasomotor control, other than by circulating catecholamines (Butler *et al.*, 1978), which exert a tonic influence on the cardiovascular system (Short *et al.*, 1977). In dogfish, circulating catecholamines are important for maintaining and increasing heart rate (Randall and Taylor, 1991) and have been shown to modulate vagal control of the heart (Agnisola *et al.*, 2003). In addition, an adrenergic influence on the heart may be exerted by specialized catecholamine-storing endothelial cells in the sinus venosus and atrium. These cells are innervated by cholinergic vagal fibres (Pettersson and Nilsson, 1979). The effects of adrenaline and noradrenaline on the elasmobranch heart are somewhat variable, so the possibility of a selective cardiac control via the two naturally occurring amines exists, although the mechanisms of this action remain unknown (Nilsson, 1983). The effects of adrenaline on the heart are modulated by neuropeptide Y (Xiang *et al.*, 1994).

Teleost Fish

Historically, a sympathetic cardio-acceleratory innervation had been generally assumed to be lacking in teleosts, however, the sympathetic chains extend into the head where they contact cranial nerves, forming a vagosympathetic trunk, and adrenergic fibres have been found to innervate the heart of some teleosts (e.g., trout, Gannon and Burnstock, 1969). Thus, teleosts may be considered the earliest group of vertebrates with both sympathetic and parasympathetic control of the heart (Taylor, 1992).

The positive chronotropic and inotropic effects on the teleost heart produced by adrenergic agonists and adrenergic nerves are mediated via β-adrenoceptor mechanisms associated with the pacemaker and the myocardial cells (Gannon and Burnstock, 1969; Holmgren, 1977; Cameron and Brown, 1981). β-adrenoreceptors generally mediate positive chronotropy in teleost fish, while α-adrenoreceptors mediate negative chronotropic responses (Farrell, 1984, 1991; Nilsson, 1984). In addition, there are sympathetic adrenergic fibres innervating the peripheral circulation, which increase vascular resistance by stimulation of muscarinic and α-adrenergic receptors or decrease it by β-adrenoreceptor stimulation (Farrell et al., 1980; Randall, 1982; Nilsson, 1984).

There is adrenergic innervation of the branchial and systemic vascular beds in teleost and other actinopterygian fishes. They contain adrenergic fibres which innervate vessels in both the arterio-arterial respiratory circuit and the arterio-venous nutritive circulation, in the gill filaments; possibly controlling and directing blood flow through these alternate pathways (Morris and Nilsson, 1994). Thus the patterns of blood flow in the gills are regulated by vagal cholinergic (see above) and sympathetic adrenergic fibres

CALCULATION OF AUTONOMIC TONES

The role of the autonomic innervation in the regulation of heart rate can be quantified by calculating the autonomic tones. Most studies examining the dual mechanism affecting heart rate have been conducted by pharmacological blockade (Axelsson et al., 1987, 1988; Altimiras et al., 1997). Atropine, a muscarinic receptor antagonist introduced via cannulae blocks cholinergic post-synaptic receptors, thus reducing inhibitory parasympathetic drive resulting in a tachycardia. Alternatively, β-adrenoceptor antagonists such as propranolol or sotalol competitively inhibit endogenous catecholamines and result in a bradycardia. The sequential infusion of each drug will result in changes to the heart beat

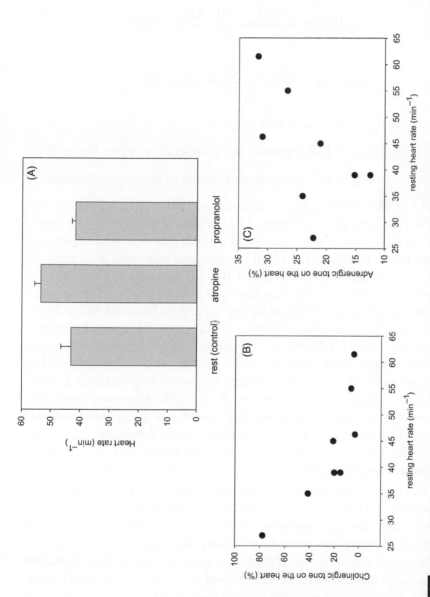

Fig. 17.1 *Platichthys flesus* A (top trace) Mean heart rates in normal fish (rest, control) and following intra-arterial injection of atropine then propranolol. B and C (traces at bottom of figure) scattergrams of cholinergic or adrenergic tones plotted against heart rate.

interval. The following equations can be used for the quantification of cholinergic and adrenergic tonus, if atropine is infused prior to β-adrenoceptor antagonists (Altimiras *et al.*, 1997):

$$\mathrm{Chol}(\%) = \frac{(R\text{-}R)_{cont} - (R\text{-}R)_{musc}}{(R\text{-}R)_0} \times 100$$

$$\mathrm{Adr}(\%) = \frac{(R\text{-}R)_0 - (R\text{-}R)_{musc}}{(R\text{-}R)_0} \times 100$$

where: $(R\text{-}R)_{cont}$ = control R-R interval

$(R\text{-}R)_{musc}$ = R-R interval after muscarinic receptor blockade

$(R\text{-}R)_0$ = R-R interval after complete autonimic blockade

An example of the autonomic regulation of heart rate is shown for the flatfish *Platichthys flesus* in Figure 17.1. In these experiments, the caudal artery had been previously cannulated with PE 50 containing heparinized Ringer for measurements of blood pressure and heart rate 24 h after the operation to allow for recovery of normal autonomic control. It is, nevertheless, quite possible that longer recover times would be required. Average heart rate is presented for resting and undisturbed fish in Figure 17.1A, and it is clear that inhibition of cholinergic receptors by infusion of atropine caused a significant rise in heart rate, whereas subsequent inhibition of the β-adrenergic receptors with propranolol caused a decline in heart rate. In this case, the average cholinergic tone was 23.3 ± 7.8%, while the adrenergic tone was 41.5 ± 1.2%. However, when the autonomic tones are expressed relative to heart rate at the individual level (Fig. 17.1 B and C), it is clear that cholinergic tone dominates at low heart rates, whereas adrenergic tone seems to dominate at high heart rates. A similar relationship has been shown for other fish (Altimiras *et al.*, 1997).

Removal of vagal tonus by surgical intervention, and the comparison of vagotomized resting and exercised fish, with intact fish, can also provide information on the relative neural (cholinergic) and humoral (adrenergic) action upon the heart (Campbell *et al.*, 2004). Comparison of calculated relative tonus obtained by surgical intervention (Campbell *et al.*, 2004) and pharmacological blockade (Axelsson *et al.*, 1987) in the sculpin (*Myoxcephalus scorpius*) showed that pharmacological studies can underestimate cholinergic tonus. This is due to the fish withdrawing vagal tonus on the heart due to an increased stress response arising from cannulation.

REFLEX CONTROL OF THE HEART

Chemoreceptors

There is no clear evidence of a role for central O_2 chemoreceptors in the control of cardiorespiratory function in fish (Randall, 1990; Randall and Taylor, 1991; Burleson *et al.*, 1992; Taylor, 1992). However, fish typically respond to ambient hypoxia with a reflex bradycardia and increased ventilatory effort and many studies have identified putative peripheral locations for chemosensory reflexogenic sites. These include: in the elasmobranch fish *Scyliorhinus*, the orobranchial cavity, innervated by cranial nerves V, VII, IX and X in the dogfish (Butler *et al.*, 1977) and possible venous receptors (Barrett and Taylor, 1984); while in teleosts studies have implicated the first gill arch innervated by cranial nerves IX and X and the pseudobranch, innervated by cranial nerves VII and IX. These studies are generally based on indirect investigation, using denervation or ablation of the reflexogenic region and subsequent analyses of the cardiorespiratory reflex to hypoxia and/or hypoxemia, or NaCN injections. Recent studies have identified externally oriented receptors on the gills of trout that elicit cardiorespiratory responses to CO_2 (Perry and Reid, 2002).

The recording of afferent vagal activity from branchial respiratory branches elucidates important information about the way the chemoreception works in response to variation on oxygen levels (Milsom and Burleson, 1986). These studies revealed an exponential increase in afferent activity with a progressive decrease in O_2 availability. This resembles the recorded responses of the mammalian carotid body, similarly innervated by the IXth cranial nerve, leading to the suggestion that the receptor would resemble the mammalian O_2-receptor, the glomus cell. In the Atlantic cod, (*Gadus morhua*, Fritsche and Nilsson, 1989), trout (*Salmo gairdneri*, Smith and Jones, 1978), coho salmon (*Oncorhynchus kisutch*, Smith and Davie, 1984) and traíra (*Hoplias malabaricus*, Sundin *et al.*, 1999), denervation of the first gill arch completely abolished the hypoxic bradycardia. This has not been the case in all fish species, however. O_2-sensitive receptors involved in producing a reflex bradycardia have been found on the first three pairs of gill arches in the catfish (*Ictalurus punctatus*, Burleson and Smatresk, 1990), on all gill arches and also in the orobranchial cavity in the tambaqui (*Colossoma mesopotamicus*, Sundin *et al.*, 2000). It required complete denervation of all gill arches to eliminate the hypoxic bradycardia and the cardiac responses to internal and external NaCN in the pacu, *Piaractus mesopotamicus*.

Studies using isolated gill arches demonstrated a very important difference between fish and the air-breathing vertebrates. In mammals the O_2 chemoreceptors, due their location, are sensitive to the oxygen content of the blood, but in fish there are chemoreceptor populations that monitor preferentially or exclusively the blood and others that respond to O_2 levels in the water. The O_2 chemoreceptors involved in producing the hypoxic bradycardia appear to sense the change in the external water coursing over the gills in many teleosts (tench, *Hemitripterus americanus*, Saunders and Sutterlin, 1971; trout, *Oncorhynchus* sp., Daxboeck and Holeton, 1978; Smith and Jones, 1978; cod, *Gadus morhua*, Fritsche and Nilsson, 1989; Adriatic sturgeon, *Acipenser naccarii*, McKenzie *et al.*, 1995). In some teleosts, however, the receptors are oriented such that they sense changes in O_2 tension in the blood passing through the gills (traíra, *Hoplias malabaricus*, Sundin *et al.*, 1999) while in others they sense changes in both water and blood (tambaqui, *Colossoma macropomum*, Sundin *et al.*, 2000; and pacu, *Piaractus mesopotamicus* Leite *et al.*, 2007). In this last study, the magnitude of the bradycardia resulting from intravenous NaCN injection was not reduced by section of IXth or the complete denervation of the first gill arch. This could indicate that the first gill arch does not possess internally oriented O_2 chemoreceptors that participated in the hypoxic bradycardia, or that the receptors on the other arches are capable of producing a full response.

The pseudobranch, innervated by the VIIth and the IXth cranial nerve (Nilsson, 1984), is also an important potential site for O_2 chemoreception (Randall and Jones, 1973). That is the reason probably, that many other studies on O_2 chemoreception did not eliminate the ventilatory response to hypoxia by gill denervation (Hughes and Shelton, 1962; Davis, 1971; Saunders and Sutterlin, 1971; Bamford, 1974; Jones, 1983). In trout the pseudobranch is responsive to decreases in O_2 (Laurent and Rouzeau, 1972). In other studies where the pseudobranch was denervated, the respiratory response to hypoxia was completely abolished (the bowfin, *Amia cavia*, McKenzie *et al.*, 1991; the channel catfish, *Ictalurus punctatus*, and the gar, *Lepisosteus osseus*, Smatresk, 1989). However, both traíra (Sundin *et al.*, 1999) and tambaqui (Sundin *et al.*, 2000; Milsom *et al.*, 2002), which do not possess pseudobranchs, showed increases in V_{AMP} but no bradycardia in response to hypoxia following complete gill denervation, apparently arising from externally oriented receptors.

So, results from fish species studied to date suggest that chemoreceptors do not share common locations and distribution among species (Saunders and Sutterlin, 1971; Smatresk *et al.*, 1986; Burleson and Smatresk, 1990b; McKenzie *et al.*, 1991; Burleson and Milsom, 1993; Sundin *et al.*, 1999, 2000; Milsom *et al.*, 2002). Different populations of receptors monitor the blood or the water, or some of them may be sensing the O_2 levels in both. Some of these groups trigger just cardiac and others just respiratory reflexes in response to changes in oxygen tension. Until now, we do not have a clear picture that can link O_2 receptor distribution to phylogenetic or life history traits. The overall similarities between tambaqui and pacu, and the relatively small differences, suggest that it may prove rewarding to continue to try to map locations and distribution as a function of phylogeny, habitat and life history. Two important factors remain unknown: the possibility of synergistic relationships between the different population of O_2-receptors, so that the effects of denervation of specific areas could be masked by the residual population, and the short-term plasticity due to the previous history of exposure to hypoxia that may lead to changes in responsiveness and even in functionality of some areas.

Mechanoreceptors

The respiratory muscles in fish contain length and tension receptors, in common with other vertebrate muscles, and the gill arches bear a number of mechanoreceptors with various functional characteristics. De Graaf and Ballintijn (1987) and De Graaf (1990) described slowly adapting position receptors on the gill arches and phasic receptors on the gill filaments and rakers of the carp. They interpreted their function as maintenance of the gill sieve and detection of and protection from clogging or damaging material. Mechanical stimulation of the gill arches is known to elicit the 'cough' reflex in fish and a reflex bradycardia (Taylor, 1992). These mechanoreceptors will be stimulated by the ventilatory movements of the gill arches and filaments. Stimulation of branchial mechanoreceptors by increasing rates of water flow may be the trigger for the cessation of active ventilatory movements during 'ram ventilation' in fish (Johansen, 1971; Randall, 1982).

Evidence for the involvement of baroreceptors in vasomotor control in fish was long contentious and it has been proposed that the evolution of barostatic control of the heart may be associated with the evolution of air-breathing because the gills of fish are supported by their neutral buoyancy in water. Ventilation of the gills generates hydrostatic pressures which fluctuate around, but predominantly above ambient. Arterial blood pressures in the branchial circulation of fish and the pressure difference across the gill epithelia are relatively low, despite the fact that the highest systolic pressures are generated in the ventral aorta, which leaves the heart to supply the afferent branchial arteries. Consequently, the need for functional baroreceptors in fish is not clear. However, there is now ample evidence that many teleost fish exhibit heart rate changes in response to blood pressure variations (e.g., Sandblom and Axelsson, 2005).

Increased arterial pressure has been shown to induce a bradycardia in both elasmobranchs, and teleosts. However, in both cases the increase in pressure required to cause a significant reduction in heart rate was relatively high (10-30 mmHg) and dogfish seem not to control arterial pressure following withdrawal of blood (Barrett and Taylor, 1984a). However, in teleosts injection of adrenaline, which raised arterial pressure, caused a bradycardia, abolished by atropine (Randall and Stevens, 1967), while low frequency oscillations in blood pressure, similar to the Mayer waves in mammals, were abolished by injection of the a-adrenoreceptor antagonist yohimbine (Wood, 1974). These data imply active regulation of vasomotor tone and the balance of evidence indicates that functional arterial baroreceptors may exist in the branchial circulation of teleost fishes.

The branchial branches of cranial nerves IX and X provide the afferent arm for the reflex changes in ventilation and heart rate following stimulation of the gill arches or increases in arterial pressure. Central stimulation of branchial nerves in both elasmobranchs (Lutz and Wyman, 1932a; Young et al., 1993) and teleosts (Mott, 1951) caused a bradycardia. However this could have stimulated mechanoreceptor and/or chemoreceptor afferents. Afferent information reaching the brain in the IXth and Xth cranial nerves is also known to influence the respiratory rhythm, with fictive breathing rate slowing in teleosts and increasing in elasmobranchs following transection of the branchial nerves or paralysis of the ventilatory muscles (Johansen, 1971; Taylor, 1992). Central stimulation of branchial branches of the vagus in the dogfish with bursts of electrical pulses entrained the efferent activity in neighbouring branchial and cardiac branches (Taylor, 1992). The entrained activity in the cardiac vagus drove the heart at rates either slower or faster than its intrinsic rate (see below).

The branchial branches of cranial nerves IX and X, supplying the gill arches of fish, project to the sensory nuclei lying dorsally and laterally above the sulcus limitans of His immediately above the equivalent motor nuclei in the medulla. These have an overlapping, sequential rostro-caudal distribution in the brainstem. Other than this generalization, the central projections of the sensory afferents contributing to cardioventilatory control in fish have not been identified (Burleson *et al.*, 1992).

Air-breathing Fish

The sites and afferent innervation of O_2-sensitive chemoreceptors that stimulate gill ventilation and air breathing have been studied in various species of gar and in the bowfin. They are found diffusely distributed in the gills and pseudobranch, innervated by cranial nerves VII, IX and X (Smatresk *et al.*, 1986; Smatresk, 1988; McKenzie *et al.*, 1991b). In gar and bowfin, gill denervation (with pseudobranch ablation in the latter case) almost completely abolished air-breathing in normoxia and abolished responses to hypoxia and NaCN (Smatresk, 1988; McKenzie *et al.*, 1991b), indicating that such responses are indeed dependent on afferent (O_2 chemoreceptor) feedback (Shelton *et al.*, 1986; Smatresk, 1990a, 1994). Smatresk *et al.* (1986) suggested that in gar there is central integration of input from internally and externally oriented receptors whereby internal receptors set the level of hypoxic drive and external receptors set the balance between air-breathing and gill ventilation.

There is no experimental evidence for baroreceptor responses in air-breathing fish. Most air-breathing fish supply their various air-breathing organs from the systemic circulation. Lungfish and all of the tetrapods have distinct pulmonary arteries and veins in association with true lungs having highly permeable surfaces; the lungfish, *Protopterus*, has a diffusion distance of 0.5 μm over the ABO, which is similar to the mammalian lung (Munshi, 1976). However, possibly because they retain gills, lungfish have similar, relatively low blood pressures in the respiratory and systemic circuits and may as a consequence not have a functional requirement for baroreceptor responses.

CENTRAL CONTROL OF CARDIORESPIRATORY INTERACTIONS

Cardiac Vagal Tone

As described previously, the heart in all fish except cyclostomes and in all tetrapods is supplied with inhibitory parasympathetic innervation via the vagus nerve. The inhibitory effect is mediated via muscarinic cholinoreceptors associated with the pacemaker and atrial myocardium. The heart in vertebrates typically operates under a degree of inhibitory vagal tone that varies with physiological state and environmental conditions. Heart rate in the dogfish varied directly with PO_2; hypoxia induced a reflex bradycardia, a normoxic vagal tone was released by exposure to moderate hyperoxia, and extreme hyperoxia induced a secondary reflex bradycardia, possibly resulting from stimulation of venous receptors; all of these affects were abolished by injection of the muscarinic cholinergic blocker atropine (Taylor, 1992). In addition, cholinergic vagal tone, assessed as the proportional change in heart rate following atropinization, increased with increasing temperature of acclimation. These data indicate that variations in the degree of cholinergic vagal tonus on the heart serve as the predominant mode of nervous cardioregulation in elasmobranchs and that the level of vagal tone on the heart varies with temperature and oxygen partial pressure. A similar reliance of cardiac vagal tone on inputs from peripheral receptors has been identified in mammals (see above). An exception to this rule is the sturgeon which exhibited no change in normoxic heart rate following atropinization (McKenzie et al., 1995).

In the teleost fish the heart receives both a cholinergic vagal supply and an adrenergic sympathetic supply. Available data on the extent of vagal tone on the teleost heart give a wide range of values revealing species differences, and the effects of different environmental or experimental conditions. In the trout, vagal tone on the heart, although higher than in the dogfish at all temperatures, decreased at higher temperatures, but the cardioacceleration induced by adrenaline injection into atropinized fish increased with temperature (Wood et al., 1979). An inhibitory vagal tonus was significantly greater in warm acclimated than in cold-acclimated eels, *Anguilla anguilla*, and blocking vagal function with benzetimide reduced a nearly complete temperature compensation (Seibert, 1979). These data indicate that adaptation of heart rate to temperature in the eel was largely mediated by the parasympathetic nervous system. Further evidence for temperature related changes in heart rate being determined centrally was provided by work on Antarctic fishes which indicated that the very low resting heart rates in normoxia are attributable to high levels of vagal tone (Axelsson *et al.*, 1992; Taylor, 1992).

Cardiorespiratory Synchrony

In fish, there is a close matching of the rates of respiratory water flow and cardiac output, according to their relative capacities for oxygen (the ventilation/perfusion ratio), that is thought to optimize respiratory gas exchange over the functional countercurrent at the gills (Taylor, 1992). The matching of the flow rates of water and blood over the counter-current at the gills of fish, according to their relative capacities for oxygen, is essential for effective respiratory gas exchange (Piiper and Scheid, 1977). Both water and blood flow have been shown to be pulsatile over the gills (e.g., in unrestrained cod (Jones *et al.*, 1974)). The pumping action of the heart generates a pulsatile flow of blood, which in fish is delivered directly down the ventral aorta to the afferent branchial vessels. To optimize respiratory gas exchange this pulsatile blood flow should probably be synchronized with the respiratory cycle, which typically consists of a double pumping action; with a buccal pressure pump alternating with an opercular or septal suction pump to maintain a constant but highly pulsatile water flow throughout the respiratory cycle. The flow is maximal early in the respiratory cycle and declines during the last two thirds of a cycle (Hughes 1960; Hughes and Shelton, 1962). Thus, the supposed functional significance of cardiorespiratory synchrony relates to the importance of continuously matching relative flow rates of water and blood over the counter current at the gill lamellae.

When cod were cannulated and released into large holding tanks of normoxic seawater they showed periods of 1:1 synchrony (Jones *et al.*, 1974; Taylor, 1992). The importance of these observations is that they measured dorsal aortic blood flow, which was markedly pulsatile in phase with variation in buccal pressure, confirming a role for cardiorespiratory synchrony (CRS) in the generation of concurrent flow patterns of ventilation and perfusion over the gills. CRS has been reported in both resting dogfish (Taylor, 1992) and hypoxic trout (Randall, 1967). Cardiac vagotomy or injection of atropine abolished CRS in the dogfish (Taylor, 1992), while in the sculpin, *Myoxocephalus scorpius*, injection of atropine raised mean heart rate in normoxia and abolished a hypoxic bradycardia and cardiac vagotomy abolished heart rate variability (Campbell *et al.*, 2004). Both these observations confirm the dependence of beat-to-beat variability of heart rate on tonic vagal control.

Close beat-to-beat temporal relationships between heart beat and ventilation, or cardiorespiratory synchrony (CRS), has long been hypothesized for fish (see review by Taylor, 1985). Recordings of differential blood pressure and gill opacity in the dogfish revealed a brief period of rapid blood flow through the lamellae early in each cardiac cycle (Satchell, 1960), and as the R wave of the ECG tended to occur at or near the mouth-opening phase of the ventilatory cycle, this could result in coincidence of the periods of maximum flow rate of blood and water during each cardiac cycle (Shelton, 1970; Shelton and Randall, 1970). The improvement in gill perfusion and consequent oxygen transfer resulting from pulsatile changes in transmural pressure and intralamellar blood flow, described by Farrell *et al.* (1980), may be further improved by synchronization of the pressure pulses associated with ventilation and perfusion. Cardiorespiratory synchrony may, therefore, increase the relative efficiency of respiratory gas exchange (i.e., maximum exchange for minimum work).

However, synchrony is often absent from recordings and the absence of synchrony, or even consistent close coupling, as opposed to a drifting phase relationship, is most often attributable to changes in heart rate, which is more variable than ventilation rate (Taylor and Butler, 1971; Taylor, 1985). This may be reliably interpreted in the dogfish as variations in cardiac vagal tone. A decrease in vagal tone on the heart, such as that recorded during exposure to moderately hyperoxic water, caused heart rate to rise towards ventilation rate (Barrett and Taylor, 1984), suggesting that when vagal tone was relatively low a 1:1 synchrony could occur. When cannulated dogfish were allowed to settle in large tanks of running, aerated seawater at 23° C they showed 1:1 synchrony between heartbeat and ventilation for long periods (Taylor, 1985). This relationship was abolished by atropine, confirming the role of the vagus in the maintenance of synchrony. Whenever the fish was spontaneously active or disturbed the relationship broke down due to a reflex bradycardia and acceleration of ventilation, so that the 2:1 relationship between ventilation and heart rate characteristic of the experimentally restrained animal was re-established. Thus, it is possible that the elusiveness of data supporting the proposed existence of cardiorespiratory synchrony in dogfish was due to experimental procedures that increase vagal tone on the heart.

Elasmobranchs

The heart in the dogfish operates under a variable degree of vagal tone (see above). This implies that the cardiac vagi will show continuous efferent activity. Recordings from the central cut end of a branchial cardiac branch of the vagus in decerebrate, paralysed dogfish revealed high levels of spontaneous efferent activity, which could be attributed to two types of unit (Taylor and Butler, 1982; Barrett and Taylor, 1985a, c). Some units fired sporadically and increased their firing rate during hypoxia. Injection of capsaicin into the ventilatory stream of the dogfish, which was accompanied by a marked bradycardia, powerfully stimulated activity in these non-bursting units recorded from the central cut end of the cardiac vagus (Jones *et al.*, 1995). Consequently, we suggested they may initiate reflex changes in heart rate, as well as playing a role in the determination of the overall level of vagal tone on the heart, which as stated previously seems to vary according to oxygen supply. Other, typically larger units fired in rhythmical bursts which were synchronous with ventilatory movements (Taylor, 1992). We also hypothesized that these units, showing respiration related activity which was unaffected by hypoxia, may serve to synchronize heartbeat with ventilation (Taylor and Butler, 1982).

The separation of efferent cardiac vagal activity into respiration-related and non-respiration-related units was discovered to have a basis in the distribution of their neuron cell bodies in the brainstem. Extracellular recordings from CVPN identified in the hindbrain of decerebrate, paralysed dogfish by antidromic stimulation of a branchial cardiac branch revealed that neurons located in the DVN were spontaneously active, firing in rhythmical bursts which contributed to the respiration related bursts recorded from the intact nerve (Barrett and Taylor, 1984, 1985c). Neurons located ventrolaterally outside the DVN were either spontaneously active, firing regularly or sporadically but never rhythmically, or were silent. Thus the two types of efferent activity recorded from the cardiac nerve arise from the separate groups of CVPN, as identified by neuroanatomical studies (Taylor, 1992).

Activity recorded from the central cut end of the cardiac vagus, or centrally from CVPN, in the decerebrated, paralysed dogfish is likely to be centrally generated. In the intact fish stimulation of peripheral receptors will affect patterns of activity. All of the spontaneously active CVPN from both divisions and some of the silent CVPN fired in response to mechanical stimulation of a gill arch, which implies that they could be entrained to ventilatory movements in the spontaneously breathing fish (Taylor, 1992). Support for this idea was provided by phasic electrical stimulation of the central cut end of a branchial branch of the vagus in the decerebrated dogfish (Young et al., 1993). This entrained the efferent bursting units recorded from the central cut end of the ipsilateral branchial cardiac branch, presumably due to stimulation of mechanoreceptor afferents (Taylor, 1992). The firing rates of the non-bursting units recorded from the branchial cardiac were also increased, suggesting that chemoreceptor afferents were being stimulated as well.

Consequently, normal-breathing movements in the intact fish may indirectly influence cardiac vagal outflow, and subsequently heart rate, by stimulating branchial mechanoreceptors. Thus, the typical reflex bradycardia in response to hypoxia may arise both directly, following stimulation of peripheral chemoreceptors and indirectly, via increased stimulation of ventilatory effort, which by stimulating branchial mechanoreceptors may increase vagal outflow to the heart. This is reminiscent of, but opposite in kind to, the hypoxic response in the mammal, where stimulation of lung stretch receptors causes an increase in heart rate (Daly and Scott, 1962).

These data support a previous conclusion that synchrony in the dogfish was reflexively controlled, with mechanoreceptors on the gill arches constituting the afferent limb and the cardiac vagus the efferent limb of a reflex arc (Taylor and Butler, 1982). However, the spontaneous, respiration-related bursts recorded from the branchial cardiac nerve continued in decerebrated dogfish, after treatment with curare which stopped ventilatory movements, suggesting that they originated in the brainstem. Direct connections between bursting CVPN and RVM are possible in the dogfish hindbrain, as both are located in the DVN with an overlapping rostro-caudal distribution (see earlier). As the bursts are synchronous, the innervation of CVM is likely to be excitatory rather than inhibitory as described for the mammal and it is equally possible that a direct drive from a central pattern generator operates both on the RVM and the CVPN (Taylor, 1992).

These data from elasmobranchs suggest that cardiorespiratory synchrony, when present, is due primarily to central interactions generating respiration-related activity in CVPN located in the DVN, which are then effective in determining synchronous heart beating when overall cardiac vagal tone, attributable primarily to activity in CVPN located outside the DVN, is relatively low in normoxic or hyperoxic fish. Synchrony will be reinforced in the spontaneously breathing fish by rhythmical stimulation of branchial mechanoreceptors.

Confirmation that the heart may beat at a rate determined by bursts of efferent activity in the cardiac vagi was obtained by peripheral electrical stimulation of these nerves in the prepared dogfish. Although continuous vagal stimulation normally slows the heart, it proved possible to drive the denervated heart at a rate either lower or somewhat higher than its intrinsic rate with brief bursts of stimuli, delivered down one branchial cardiac vagal branch (Taylor *et al.*, 2006). At a rate several beats higher than its intrinsic one, the heart responded to alternate bursts of electrical pulses so that it began beating at half the rate of the bursts. Interestingly, similar results were obtained from a mammal. In the anaesthetized dog, electrical stimulation of the vagus nerve towards the heart with brief bursts of stimuli, similar to those recorded from efferent cardiac vagal fibers, caused heart rate to synchronize with the stimulus, beating once for each vagal stimulus burst over a wide frequency range (Levy and Martin, 1984).

Teleosts

Work on teleosts has stressed the importance of inputs from peripheral receptors in the genesis of cardiorespiratory synchrony. Randall (1966) recorded efferent nervous activity from the cardiac branch of the vagus in the tench that was synchronized with the mouth-opening phase of the breathing cycle. It was suggested that this activity maintains synchrony between heartbeat and breathing movements and that both a hypoxic bradycardia and synchrony were mediated by reflex pathways. Randall and Smith (1967) described the development of an exact synchrony between breathing and heart beat in the trout during progressive hypoxia. In normoxia heart rate was faster than ventilation; hypoxia caused an increase in ventilation rate and a reflex bradycardia which converged to produce a 1:1 synchronization of the two rhythms. Both the bradycardia and synchrony were abolished by atropine. In addition, Randall and Smith

(1967) were able to demonstrate 1:1 synchronization of hypoxic heart rate with pulsatile forced ventilation, which was clearly generated by reflex pathways, presumably arising from mechanoreceptors on the gills, because the spontaneous breathing efforts of the incubated fish were out of phase with imposed changes in water velocity and were without effect on heart beat. Current work on pacu (Taylor *et al.*, 2007) has revealed that respiration related, bursting activity is only recorded from the cardiac vagus when fish are hyperventilating or coughing, implying that the bursts arise reflexively, following stimulation of branchial mechanoreceptors. Recordings from the central cut end of the cardiac branch of the vagus on either side of the lightly anaesthetized normoxic animal revealed spontaneous activity that increased on bolus injection of cyanide. The spontaneous occurrence of a cough or bout of increased ventilatory effort was accompanied by bursts of activity in the cardiac vagus, with a consequent bradycardia (Fig. 17.2). In fish rendered hypoxic by the cessation of the flow of water irrigating the gills, a period of spontaneously increased ventilatory amplitude was accompanied by respiration related bursts of activity in the cardiac vagus, which were not apparent in the inactive, normoxic fish and appeared to recruit the heart (Fig. 17.3).

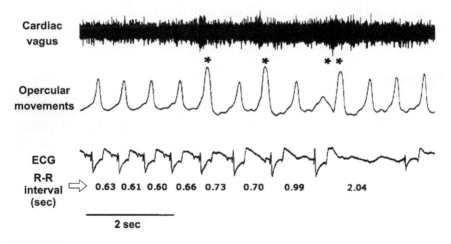

Fig. 17.2 *Piaractus mesopotamicus* Spontaneous electrical activity recorded from the central cut end of the right cardiac vagus (upper trace); ventilation (opercular movements) and ECG, of a lightly anaesthetized (benzocaine) fish. Increased ventilation amplitude (*) was associated with a burst of activity in the cardiac nerve and an increased cardiac interval. A subsequent cough (**) caused a brief cardiac arrest. Adapted from Taylor *et al.* (2007).

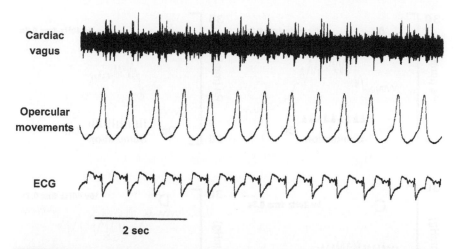

Cardiac
vagus

Opercular
movements

ECG

2 sec

Fig. 17.3 *Piaractus mesopotamicus* Spontaneous electrical activity of the central cut end of the right cardiac vagus (upper trace); ventilation (opercular movements) and ECG and of a hypoxic lightly anaesthetized fish (benzocaine). Ventilation amplitude was increased from its normoxic level, phasic, respiration related activity was recorded from the central cut end of cardiac vagus and this appeared to entrain the heart. Adapted from Taylor *et al.* (2007.)

So, in the hypoxic pacu, phasic, respiration-related activity was recorded from the cardiac vagus and this appeared to entrain the heart (Fig. 17.3). Stimulating the cardiac branchial branch of the vagus peripherally can also entrain the heart over a wide range of frequencies, both below and above the intrinsic rate (Fig. 17.4). It proved possible to either slow or increase the heart rate from for example 70 to 120 beats a minute (Taylor *et al.*, 2006b). The injection of atropine abolished all effects of vagal stimulation on heart rate. These data were interpreted as signifying that the heart was responding to a reflex arising from stimulation of branchial mechanoreceptors (Taylor *et al.*, 2006). This possibility was tested by central stimulation of respiratory branches of the vagus nerve, glossopharyngeal and also the facial nerve. All these nerves, when phasically stimulated, to simulate stimulation of gill mechanoreceptors by respiratory movements, entrained of the heart over a wide range of frequencies (Leite *et al.*, 2007). During progressive hypoxia, caused by the cessation of the water flow for the forced irrigation of the anaesthetized animal, heart rate decreased while ventilation rate and amplitude both increased. There was a phase of cardiorespiratory synchrony in normoxia that was maintained into moderate hypoxia as ventilation rate increased (Fig. 17.5). Below a critical PO_2 of about

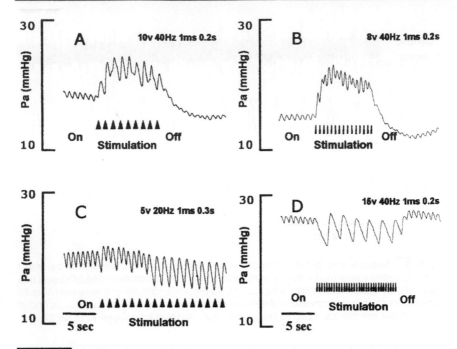

Fig. 17.4 *Piaractus mesopotamicus* Phasic stimulation of the right cardiac nerve towards the heart. The heart was entrained by the bursts of stimuli over a wide range both slower than the intrinsic rate (A & C) and faster than the intrinsic rate (B). At very high stimulation rates the heart was entrained to a fractional rate such as 1:5 (D). Adapted from Taylor *et al.* (2007).

40 mmHg the developing bradycardia and increasing rate of ventilation caused the ratio between fR and fH to increase. Prolonged cessation of water flow caused a concomitant marked increase in activity in the cardiac vagus, accompanied by a marked bradycardia.

All of these data imply that respiration related activity in the cardiac vagus of teleost fish is generated reflexively and does not arise by feed-forward control in the CNS, as described above for elasmobranchs. It is interesting in this regard that heart rate was observed to rise immediately upon the onset of ram ventilation in the trout, implying a reduction in vagal tone (Taylor, 1992). As this can be attributed to the effect of cessation of activity both in the central respiratory pattern generator and in the respiratory apparatus it implies that respiratory activity to some extent generates cardiac vagal tone. This is the obverse of the situation in mammals, where cessation of ventilation, for example during SLN stimulation, increases vagal tone on the heart (Jordan and Spyer, 1987).

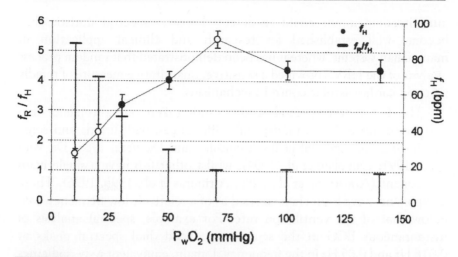

Fig. 17.5 *Piaractus mesopamicus* submitted to progressive hypoxia (140, 70, 50, 30, 20, 10 mmHg O_2). The hypoxic bradycardia starts about the critical tension point for this species (= 40 mmHg). The open symbols denote statistical difference from the normoxic fH. The relationship between fR/fH is about 1 in normoxia and remains at 1 down to a P_wO_2 of 70 mmHg.

Thus, we are left with an apparent conflict of evidence on the mode of generation of cardiorespiratory synchrony, which in elasmobranchs may be centrally generated in inactive, normoxic, or hyperoxic fish when cardiac vagal tone is low; while in teleosts it appears during hypoxia perhaps as a result of stimulation of branchial mechanoreceptors during forced ventilation and is generated reflexively by increased vagal tone. The differences between these two groups of fish may be real and it is of interest that branchial denervation increases fictive ventilation rate in elasmobranchs but decreases it in teleosts.

Power Spectral Analysis

One-to-one CRS is not the only possible temporal interaction between heart rate and ventilation rate. For many years researchers desperately seeking after piscine CRS reported complex and often drifting phase relationships between fH and fV (Satchell, 1960; Taylor and Butler, 1971). These relationships were often subject to relatively complex analysis (e.g., Hughes, 1972), but the techniques of power spectral analysis were not available to fish physiologists of that generation. Interest in heart rate variability (HRV) research increased once it became clear that oscillations in cardiac interval generated by sympathetic, parasympathetic and circadian inputs could be detected. Power spectral statistics has become well established for research and clinical application in mammalian systems, where it has been demonstrated that random process analysis of the HRVS provided a sensitive, quantitative measure of rapidly reacting cardiovascular control mechanisms.

The few power spectral studies undertaken on non-mammalian vertebrates have shown interspecific differences, with lizards and some fish having dual spectral peaks (Gonzalez and De Vera, 1988; De Vera et al., 1991; Campbell et al., 2004), whilst other fish have a single main component (Armstrong et al., 1988; Altimiras et al., 1995, 1996). These components have been characterized as of a relatively low frequency, well below that of the ventilation rate. For example, spectral analysis of instantaneous ECG in the sculpin produced dual spectral peaks at 0.018 Hz and 0.05 Hz in the frequency domain, equivalent to periodicities of 55 and 20 s, respectively, in the time domain. As ventilation occurred every 3.3 ± 0.2 s (0.30 Hz), it was about six times faster that the highest-frequency component (0.05 Hz). It was concluded that this component probably did not represent centrally generated respiration-related activity, analogous to the RSA described in recordings of HRV in mammals, and consequently that the sculpin did not show CRS (Campbell et al., 2004).

However, one very important point that has historically been overlooked in the vertebrate literature is that RSA is only possible when heart rate is at least twice as fast as ventilation (i.e., respiration) rate. Two cardiac intervals (heartbeats) are required for each breath in order to have a characteristic waxing and waning of heart rate with each respiratory cycle. In other words, at the very minimum, there must be a longer heartbeat associated with expiratory phase and a shorter one associated with inspiratory phase. This minimum two-to-one relationship is known as the Nyquist frequency. Therefore, if ventilation rate is greater than one-half the heart rate, another phenomenon occurs that is termed aliasing: It is no longer possible to meaningfully calculate RSA because there are less than two beats per breath, and any possible synchrony between respiration period and IBI must be evaluated over a longer duration than the single breath. If one knows exactly what the cardiac frequency and the ventilation frequency are for an animal, then one can determine the exact aliasing frequency as the difference between the sampled signal (heart rate) and the sampling frequency.

Aliasing of components is sometimes called folding-back and the point at which the frequency component is folded back is given by:

$$[f_{bk}] = f_s - f_a$$

where f_{bk} is the fold-back frequency, f_s is the sampling frequency and is the f_a actual frequency (Campbell *et al.*, 2006). If this is applied to the data from Campbell *et al.* (2004) it reveals that HRV in the sculpin included a component that appeared to be synchronous with the respiratory rhythm (Taylor *et al.*, 2006) so that, contrary to previous analysis of these data, a respiratory coupling may indeed contribute to HRV in fish and perhaps other lower vertebrates. Campbell *et al.* (2004) noted that as rates of oxygen uptake in the sculpin fell during recovery from surgery, a time domain index of HRV (standard deviation of successive R-R intervals, SDRR) mirrored this change more closely than the progressive reduction in heart rate. Re-examination of their data, using anti-aliasing techniques has led them to conclude that HRV could be an important component of the mechanisms optimizing respiratory gas exchange (Taylor *et al.*, 2006). The present reanalysis of their data implies that what was actually involved in optimizing the regulation of oxygen uptake on recovery was the establishment of CRS. As stated above, CRS was observed directly in the dogfish and cod when they were allowed to settle, unrestrained in large tanks of aerated seawater (Taylor, 1992). So it appears that CRS may be characteristic of an inactive state in some fish.

Concluding Remarks

Heart rate in fish is controlled by activity in the autonomic nervous system that originates in the CNS, either from central interactions or reflexively from stimulation of peripheral receptors sensitive to O_2 or CO_2 levels or to mechanoreceptors on the gills. The parasympathetic, vagus nerve is the sole nervous regulator of heart rate in elasmobranchs and predominates in teleosts. There is accumulating evidence for cardiorespiratory interactions, including synchrony, that, in elasmobranchs, originates primarily from central interactions between the respiratory rhythm generator and CVPN and is characterized by relatively low levels of vagal tone on the heart, but in teleosts is driven reflexively, by stimulation of peripheral chemoreceptors and mechanoreceptors, and is characterized by increased levels of vagal tone on the heart.

Acknowledgements

EWT was supported by a visiting Professorship funded by FAPESP, CACL has a studentship from CNPq, HAC is a postdoctoral associate funded by the NERC in the UK, II is funded by his University in Thailand and TW by the Danish Research Council.

References

Agnisola, C., D.J. Randall and E.W. Taylor. 2003. The modulatory effects of noradrenaline on vagal control of the heart in the dogfish, Squalus acanthius. Physiological and Biochemical Zoology 76: 310-320.

Altimiras, J., A. Aissaoui and L.Tort. 1995. Is the short-term modulation of heart rate in teleost fish physiologically significant? Assessment by spectral analysis techniques. Brazilian Journal of Medical and Biological Research 28: 1197-1206.

Altimiras, J., A.D.F. Johnstone, M.C. Lucas and I.G. Priede. 1996. Sex differences in the heart rate variability spectrum of free-swimming Atlantic salmon (Salmo salar L.) during spawning season. Physiological Zoology 69: 770-784.

Altimiras, J., A. Aissaoui, L. Tort and M. Axelsson. 1997. Cholinergic and adrenergic tones in the control of heart rate in teleosts. How should they be calculated? Comparative Biochemistry and Physiology A118: 131-139.

Armstrong, J.D. 1986. Heart rate as an indicator of activity, metabolic rate, food intake and digestion in pike, Esox lucius. Journal of Fish Biology 29: 207-221.

Augustinsson, A., R. Fange, A. Johnels and E. Ostlund. 1956. Histological, physiological and biochemical studies on the heart of two cyclostomes, hagfish (Myxine) and lamprey (Lampetra). Journal of Physiology (London) 13: 257-276.

Axelsson, M. 1988. The importance of nervous and humoral mechanisms in the control of cardiac performance in the Atlantic cod Gadus morhua at rest and during non-exhaustive exercise. Journal of Experimental Biology 137: 287-303.

Axelsson, M., F. Ehrenström and S. Nilsson. 1987. Cholinergic and adrenergic influence on the teleost heart in vivo. Experimental Biology 46: 179-186.

Axelsson, M., A.P. Farrell and S. Nilsson. 1990. Effects of hypoxia and drugs on the cardiovascular dynamics of the Atlantic hagfish, Myxine glutinosa. Journal of Experimental Biology 151: 297-316.

Axelsson, M., W. Davison, M.E. Forster and A.P. Farrell. 1992. Cardiovascular responses of the red-blooded Antarctic fishes Pagothenia bernacchii and P. borchgrevinki. Journal of Experimental Biology 167: 179-201.

Bamford, O.S. 1974. Oxygen reception in the rainbow trout (Salmo gairdneri). Comparative Biochemistry and Physiology 48: 69-76.

Barrett, D.J. and E. Taylor. 1984. Changes in heart rate during progressive hypoxia in the dogfish, Scyliorhinus canicula L.: evidence for a venous oxygen receptor. Comparative Biochemistry and Physiology 78: 697-703.

Barrett, D.J. and E.W. Taylor. 1985a. Spontaneous efferent activity in branches of the vagus nerve controlling heart rate and ventilation in the dogfish. Journal of Experimental Biology 117: 433-448.

Barrett, D.J. and E.W. Taylor. 1985b. The location of cardiac vagal preganglionic neurones in the brainstem of the dogfish. *Journal of Experimental Biology* 117: 449-458.

Barrett, D.J. and E.W. Taylor. 1985c. The characteristics of cardiac vagal preganglionic motoneurones in the dogfish. *Journal of Experimental Biology* 117: 459-470.

Burleson, M.L. and W.K. Milsom. 1993. Sensory receptors in the first gill arch of rainbow trout. *Respiration Physiology* 93: 97-110.

Burleson, M.L. and N.J. Smatresk. 1990a. Evidence for two oxygen sensitive chemoreceptor loci in channel catfish. *Physiological Zoology* 63: 208-221.

Burleson, M.L. and N.J. Smatresk. 1990b. Effects of sectioning cranial nerves IX and X on cardiovascular and ventilatory reflex responses to hypoxia and NaCN in channel catfish. *Journal of Experimental Biology* 154: 407-420.

Burleson, M.L., N.J. Smatresk and W.K. Milsom. 1992. Afferent inputs associated with cardioventilatory control in fish. In: *Fish Physiology*, W.S. Hoar, D.J. Randall and A.P. Farrell (eds.). Academic Press, New York, Vol. 22B, pp. 390-426.

Butler, P.J. and E.W. Taylor. 1971. Response of the dogfish (*Scyliorhinus canicula* L.) to slowly induced and rapidly induced hypoxia. *Comparative Biochemistry and Physiology* A39: 307-323.

Butler, P.J. and E.W. Taylor. 1975. The effect of progressive hypoxia on respiration in the dogfish (*Scyliorhinus canicula*) at different seasonal temperatures. *Journal of Experimental Biology* 63: 117-130.

Butler, P.J., E.W. Taylor and S. Short. 1977. The effect of sectioning cranial nerves V, VII, IX and X on the cardiac response of the dogfish *Scyliorhinus canicula* to environmental hypoxia. *Journal of Experimental Biology* 69: 233-245.

Butler, P.J., E.W. Taylor, M.F. Capra and W. Davison. 1978. The effect of hypoxia on the level of circulating catecholamines in the dogfish, *Scyliorhinus canicula*. *Journal of Comparative Physiology* 127: 325-330.

Cameron, J.S. 1979. Autonomic nervous tone and regulation of heart rate in the goldfish, *Carassius auratus*. *Comparative Biochemistry and Physiology* C63: 341-349.

Cameron, J.S. and S.E. Brown. 1981. Adrenergic and cholinergic responses of the isolated heart in the goldfish *Carassius auratus*. *Comparative Biochemistry and Physiology* C70: 109-116.

Campbell, H.A., E.W. Taylor and S. Egginton. 2004. The use of power spectral analysis to determine cardio-respiratory control in the short-horned sculpin *Myoxocephalus scorpius*. *Journal of Experimental Biology* 207: 1969-1976.

Campbell, H.A., J.Z. Klepacki and S. Egginton. 2006. A new method in applying power spectral statistics to examine cardio-respiratory interactions in fish. *Journal of Theoretical Biology* 241: 410-419.

Daly, M. de B. and M.J. Scott. 1962. An analysis of the primary cardiovascular reflex effects of stimulation of the carotid body chemoreceptors in the dog. *Journal of Physiology* (London). 162: 555-573.

Davis, J.C. 1971. Circulatory and ventilatory responses of rainbow trout (*Salmo gairdneri*) to artificial manipulation of gill surface area. *Journal of Fisheries Research Board Canada* 28: 1609-1614.

Daxboeck, C. and G.F. Holeton. 1978. Oxygen receptors in the rainbow trout, *Salmo gairdneri*. *Canadian Journal of Zoology* 56: 1254-1256.

De Graaf, P.J.F. 1990. Innervation pattern of the gillarches and gills of the carp (*Cyprinus carpio*). *Journal of Morphology* 206: 71-78.

De Graaf, P.J.F. and C.M. Ballintijn. 1987. Mechanoreceptor activity in the gills of the carp II. Gill arch proprioceptors. *Respiration Physiology* 69: 183-194.

De Vera, L. 1991. The heart rate variability signal in rainbow trout (*Oncorhynchus mykiss*). *Journal of Experimental Biology* 156: 611-617.

Falck, B., C. Mecklenburg, H. Myhrberg and H. Persson. 1966. Studies on adrenergic and cholinergic receptors in the isolated hearts of *Lampetra fluviatilis* (Cyclostomata) and *Pleuronectes platessa* (Teleostei). *Acta Physiologica Scandinavica* 68: 64-71.

Farrell, A.P. 1984. A review of cardiac performance in the teleost heart — Intrinsic and humoral regulation. *Canadian Journal of Zoology* 62: 523-536.

Farrell, A.P. 1991. From hagfish to tuna — A perspective on cardiac-function in fish. *Physiological Zoology* 64: 1137-1164.

Farrell, A.P. and D.J. Jones. 1992. The heart. In: *Fish Physiology*, W.S. Hoar, D.J. Randall and A.P. Farrell (eds.). Academic Press, New York, Vol. 12A, pp. 1-73.

Farrell, A.P., S.S. Sobin, D.J. Randall and S. Crosby. 1980. Intralamellar blood flow patterns in fish gills. *American Journal of Physiology* 239: R428-R436.

Fritsche, R. and S. Nilsson. 1989. Cardiovascular responses to hypoxia in the Atlantic cod, *Gadus morhua*. *Experimental Biology* 48: 153-160.

Gannon, B.J. and G. Burnstock. 1969. Excitatory adrenergic innervation of the fish heart. *Comparative Biochemistry and Physiology* 29: 765-773.

Gonzalez, J.G. and L.D. de Vera. 1988. Spectral-analysis of heartrate variability of lizard, *Gallotia galloti*. *American Journal of Physiology* 254: R242-R248.

Green, C.W. 1902. Contributions to the physiology of the California hagfish *Polistrotremas stoutii* II. The absence of regulative nerves for the systemic heart. *American Journal of Physiology* 6: 318-324.

Holmgren, S. 1977. Regulation of the heart of a teleost, *Gadus morhua*, by autonomic nerves and circulating catecholamines. *Acta Physiologica Scandinavica* 99: 62-74.

Hughes, G.M. 1960. The mechanism of gill ventilation in the dogfish and skate. *Journal of Experimental Biology* 37: 11-27.

Hughes, G.M. 1972. The relationship between cardiac and respiratory rhythms in the dogfish, *Scyliorhinus canicula* L. *Journal of Experimental Biology* 57: 415-434.

Hughes, G.M. and G. Shelton. 1962. Respiratory mechanisms and their nervous control in fish. *Advances in Comparative Physiology and Biochemistry* 1: 275-364.

Jean, A. 1993. Brainstem control of swallowing: Localization and organization of the central pattern generator for swallowing. In: *Neurobiology of the Jaws and Teeth*, A. Taylor (ed.). Macmillan, London, pp. 294-321.

Johansen, K. 1971. Comparative physiology: Gas exchange and circulation in fishes. *Annual Review of Physiology* 33: 569-612.

Johansen, K. and W. Burggren. 1980. Cardiovascular function in the lower vertebrates. In: *Hearts and Heart-like Organs*, G.H. Bourne (ed.). Academic Press, New York, Vol. 1, pp. 61-117.

Jones, D.R. and D.J. Randall. 1978. The respiration and circulatory systems during exercise. In: *Fish Physiology*, W.S. Hoar and D.J. Randall (eds.). Academic Press, New York, Vol. 7, pp. 425-501.

Jones, J.F.X., M. Young, D. Jordan and E.W. Taylor. 1993. Effect of capsaicin on heart rate and fictive ventilation in the decerebrate dogfish (*Scyliorhinus canicula*). *Journal of Physiology* (London) 473: 236P.

Jones, D.R. 1983. Ontogeny and phylogeny of the oxygen response. *Proceedings of the Physiological Society of New Zealand* 3: 79-81.

Jones, D.R., B.L. Langille, D.J. Randall and G. Shelton. 1974. Blood flow in dorsal and ventral aortas of the cod, *Gadus morhua. American Journal of Physiology* 226: 90-95.

Jordan, D. and K.M. Spyer. 1987. Central neural mechanisms mediating respiratory-cardiovascular interactions. In: *Neurobiology of the Cardiorespiratory System*, E.W. Taylor (ed.). Manchester University Press, Manchester, pp. 322-341.

Laurent, P. and J.D. Rouzeau. 1972. Afferent neural activity from the pseudobranch of teleosts. Effects of PO_2, pH, osmotic pressure and Na^+ ions. *Respiration Physiology* 14: 307-331.

Laurent, P., S. Holmgren and S. Nilsson. 1983. Nervous and humoral control of the fish heart: Structure and function. *Comparative Biochemistry and Physiology* A76: 525-542.

Leite, C.A.C., C.D.R Guerra, L.H. Florindo, T.C. Belao, F.T. Rantin and E.W. Taylor. 2007. Centrally derived activity in respiratory branches of the vagus and its role in generating cardiorespiratory interactions in the Neotropical fish, the pacu, *Piaractus mesopotamicus*: a neuranatomical and neurophysiological study (In Preparation).

Levy, M.N. and P. Martin. 1984. Parasympathetic control of the heart. In: *Nervous Control of Cardiovascular Function*, W.C. Randall (ed.). Oxford University Press, Oxford.

Liem, K.F. 1989. Respiratory gas bladders in teleosts: Functional conservatism and morphological diversity. *American Zoologist* 29: 333-352.

Lutz, B.R. and L. Wyman. 1932. Reflex cardiac inhibition of branchio-vascular origin in the elasmobranch, *Squalus acanthias. Biological Bulletin* 62: 10-16.

McKenzie, D.J., M.L. Burleson and D.J. Randall. 1991. The effects of branchial denervation and pseudobranch ablation on cardio-ventilatory control in an air-breathing fish. *Journal of Experimental Biology* 161: 347-365.

McKenzie, D.J., M.L. Burleson and D.J. Randall. 1991. The effects of branchial denervation and pseudobranch ablation on cardioventilatory control in an air-breathing fish. *Journal of Experimental Biology* 161: 347-365.

McKenzie, D.J., E.W. Taylor, P. Bronzi and C.L. Bolis. 1995. Aspects of cardioventilatory control in the Adriatic sturgeon (*Acipencer naccarii*). *Respiration Physiology* 100: 45-53.

Milsom, W.K. 2002. Phylogeny of CO_2/H^+ chemoreception in vertebrates. *Respiratory Physiology and Neurobiology* 131: 29-41.

Morris, J. and S. Nilsson. 1994. The circulatory system. In: *Comparative Physiology and Evolution of the Autonomic Nervous System*, S. Nilsson and S. Holmgren (eds.). Switzerland, Hardwood Academic Publishers, pp. 193-246.

Milsom, W.K., S.G. Reid, F.T. Rantin and L. Sundin. 2002. Extrabranchial chemoreceptors involved in respiratory reflexes in the neotropical fish *Colossoma macropomum* (the tambaqui). *Journal of Experimental Biology* 205: 1765-1774.

Mott, J. 1951. Some factors affecting the blood circulation in the common eel (*Anguilla anguilla*). *Journal of Physiology* (London) 114: 387-398.

Munshi, J.S.D. 1976. Gross and fine structure of the respiratory organs of air-breathing fishes. In: *Respiration in Amphibious Vertebrates*, G. M. Hughes (ed.). Academic Press, New York, pp. 73-104.

Nilsson, S. 1983. *Autonomic Nerve Function in the Vertebrates*. Springer-Verlag, Berlin.

Nilsson, S. 1984. Innervation and pharmacology of the gills. In: *Fish Physiology*, W.S. Hoar and D.J. Randall (eds.). Academic Press, Orlando, Vol. 10A, pp. 185-227.

Nilsson, S. and S. Holmgren. 1994. *Comparative Physiology and Evolution of the Autonomic Nervous System*. Harwood, Academic Publishers, Chur, Switzerland.

Perry, S.F. and S.G. Reid. 2002. Cardiorespiratory adjustments during hypercarbia in rainbow trout (*Oncorhynchus mykiss*) are initiated by CO_2 external receptors on the first gill arch. *Journal of Experimental Biology* 205: 3357-3366.

Pettersson, K. and S. Nilsson. 1979. Nervous control of the branchial vascular resistance of the Atlantic cod *Gadus morhua*. *Journal of Comparative Physiology* 129: 179-183.

Piiper, J. and P. Scheid. 1977. Comparative physiology of respiration: functional analysis of gas exchange organs in vertebrates. In: *International Review of Physiology Respiratory Physiology, II*. J.G. Widdicombe (ed.). Baltimore: University Park Press, Vol. 14, pp. 219-253.

Randall, D.J. 1966. The nervous control of cardiac activity in the tench (*Tinca tinca*) and the goldfish (*Carassius auratus*). *Physiological Zoology* 39: 185-192.

Randall, D.J. 1968. Functional morphology of the heart in fishes. *American Zoologist* 8: 179-189.

Randall, D.J. 1982. The control of respiration and circulation in fish during exercise and hypoxia. *Journal of Experimental Biology* 100: 275-288.

Randall, D.J. 1990. Control and co-ordination of gas exchange in water breathers. In: *Advances in Comparative and Environmental Physiology*, R.G. Boutilier (ed.). Springer-Verlag, Berlin, Vol. 6, pp. 253-278.

Randall, D.J. and D.R. Jones. 1973. The effect of deafferation of the pseudobranch on the respiratory response to hypoxia of the trout (*Salmo gairdneri*). *Respiration Physiology* 17: 291-301.

Randall, D.J. and J.C. Smith. 1967. The regulation of cardiac activity in fish in a hypoxia environment. *Physiological Zoology* 40: 104-113.

Randall, D.J. and E.D. Stephens. 1967. The role of adrenergic receptors in cardiovascular changes associated with exercise in salmon. *Comparative Biochemistry and Physiology* 21: 415-424.

Randall, D.J. and E.W. Taylor. 1991. Evidence of a role for catecholamines in the control of breathing in fish. *Reviews in Fish Biology and Fisheries* 1: 139-158.

Sandblom, E. and M. Axelsson. 2005. Baroreflex mediated control of heart rate and vascular capacitance in trout. *Journal of Experimental Biology* 208: 821-829.

Satchell, G.H. 1960. The reflex co-ordination of the heart beat with respiration in dogfish. *Journal of Experimental Biology* 37: 719-731.

Satchell, G.H. 1971. *Circulation in Fishes*. Cambridge University Press, Cambridge.

Saunders, R.L. and A.M. Sutterlin. 1971. Cardiac and respiratory response to hypoxia in the searaven, *Hemipterus americanus*, an investigation of possible control mechanism. *Journal of Fisheries Research Board Canada* 28: 491-503.

Seibert, H. 1979. Thermal adaptation of heart rate and its parasympathetic control in the European eel, *Anguilla anguilla* (L.). *Comparative Biochemistry and Physiology* C64: 275-278.

Shelton, G. 1970. The regulation of breathing. In: *Fish Physiology*, W.S. Hoar and D.J. Randall (eds.). Academic Press, New York, Vol. 4, pp. 293-359.

Shelton, G. and D.J. Randall. 1970. The relationship between heart beat and respiration in teleost fish. *Comparative Biochemistry and Physiology* 7: 237-250.

Shelton, G., D.R. Jones and W.K. Milsom. 1986. Control of breathing in ectothermic vertebrates. In: *Handbook of Physiology*, A.P. Fishman, N.S. Cherniak, J.G. Widdicombe and S.R. Geiger (eds.). American Physiological Society, Bethesda. Section 3. *The Respiratory System, Vol. 2, Part 2, Control of Breathing*, pp. 857-909.

Short, S., P.J. Butler and E.W. Taylor. 1977. The relative importance of nervous, humoral and intrinsic mechanisms in the regulation of heart rate and stroke volume in the dogfish, *Scyliorhinus canicula*. *Journal of Experimental Biology* 70: 77-92.

Smatresk, N.L. 1989. Chemorelfex control of respiration in an air-breathing fish. In: *Chemoreceptors and Chemoreflexes in Breathing Cellular and Molecular Aspects*, S. Lahiri, R.E. Foster II, R.O. Davies and A.I. Pack (eds.). Oxford University Press, London, pp. 29-52.

Smatresk, N.J. 1988. Control of the respiratory mode in air-breathing fishes. *Canadian Journal of Zoology* 66: 144-151.

Smatresk, N.J. 1990. Respiratory defense reflexes in an air-breathing fish, *Lepisosteus oculatus*. *American Zoologist* 30: 67A.

Smatresk, N.J. 1994. Respiratory control in the transition from water to air breathing in vertebrates. *American Zoologist* 34: 264-279.

Smatresk, N.L., M.L. Burleson and S.Q. Azizi. 1986. Chemoreflexive responses to hypoxia and NaCN in longnose gar: evidence for two chemoreceptive loci. *American Journal of Physiology* 251: 116-125.

Smith, F.M. and P.S. Davie. 1984. Effects of sectioning cranial nerves IX and X on the cardiac response to hypoxia in the coho salmon, *Oncorhynchus kisutch*. *Canadian Journal of Zoology* 62: 766-768.

Smith, F.M. and D.R. Jones. 1978. Localization of receptors causing hypoxic bradycardia in trout (*Salmo gairdneri*). *Canadian Journal of Zoology* 56: 1260-1265.

Sundin, L., S.G. Reid, A.L. Kalinin, F.T. Rantin and W.K. Milsom. 1999. Cardiovascular and respiratory reflexes: the tropical fish, traira (*Hoplias malabaricus*) O_2 chemoresponses. *Respiration Physiology* 116: 181-199.

Sundin, L., S.G. Reid, F.T. Rantin and W.K. Milsom. 2000. Branchial receptors and cardio-respiratory reflexes in a neotropical fish, the tambaqui (*Colossoma macropomum*). *Journal of Experimental Biology* 203: 1225-1239.

Taylor, E.W. 1985. Control and co-ordination of gill ventilation and perfusion. *Symposium of the Society for Experimental Biology* 39: 123-161.

Taylor, E.W. 1992. Nervous control of the heart and cardiorespiratory interactions. In: *Fish Physiology*, W.S. Hoar, D.J. Randall and A.P. Farrell (eds.). Academic Press, New York, Vol. B12, pp. 343-387.

Taylor, E.W. 1994. The evolution of efferent vagal control of the heart in vertebrates. *Cardioscience* 5: 173-182.

Taylor, E.W. and P.J. Butler. 1971. Some observations on the relationship between heart beat and respiratory movement in the dogfish (*Scyliorhinus canicula* L.). *Comparative Biochemistry and Physiology* A39: 297-305.

Taylor, E.W. and P.J. Butler. 1982. Nervous control of heart rate: activity in the cardiac vagus of the dogfish. *Journal of Applied Physiology* 53: 1330-1335.

Taylor, E.W., S. Short and P.J. Butler. 1977. The role of the cardiac vagus in the response of the dogfish, *Scyliorhinus canicula* to hypoxia. *Journal of Experimental Biology* 70: 57-75.

Taylor, E.W., D. Jordan and J.H. Coote. 1999. Central control of the cardiovascular and respiratory systems and their interactions in vertebrates. *Physiological Reviews* 79: 855-916.

Taylor, E.W., M.A.D. Al-Ghamdi, Y.I. Ihmied, T. Wang and A.S. Abe. 2001. The neuranatomical basis of central control of cardiorespiratory interactions in vertebrates. *Experimental Physiology* 86: 781-786.

Taylor, E.W., H.A. Campbell, J.J. Levings, M.J. Young, P.J. Butler and S. Egginton. 2006. Coupling of the respiratory rhythm in fish with activity in hypobranchial nerves and with heart beat. *Physiological and Biochemical Zoology*. (In Press).

Taylor, E.W., C.A.C. Leite, C.D.R. Guerra, L.H. Florindo, T. Belao and F.T. Rantin. 2007. Vagal control of the heart and of cardiorespiratory interactions in the Neotropical fish, the pacu, *Piaractus mesopotamicus*: a neuranatomical and neurophysiological study. *Journal of Experimental Biology*. (Submitted).

Withington-Wray, D.J., B.L. Roberts and E.W. Taylor. 1986. The topographical organisation of the vagal motor column in the elasmobranch fish, *Scyliorhinus canicula* L. *Journal of Comparative Neurology* 248: 95-104.

Withington-Wray, D.J., E.W. Taylor and J.D. Metcalfe. 1987. The location and distribution of vagal preganglionic neurones in the hindbrain of lower vertebrates. In: *Neurobiology of the Cardiorespiratory System*, E.W. Taylor (ed.). Manchester University Press, Manchester, pp. 304-321.

Wood, C.M. 1974. Mayer waves in the circulation of a teleost fish. *Journal of Experimental Biology* 189: 267-274.

Wood, C.M., P. Pieprzak and J.N. Trott. 1979. The influence of temperature and anaemia on the adrenergic and cholinergic mechanisms controlling heart rate in the rainbow trout. *Canadian Journal of Zoology* 57: 2440-2447.

Xiang, H., E.W. Taylor, N.M. Whiteley and D.J. Randall. 1994. Modulation of noradrenergic action by neuropeptide Y in dogfish. *Physiological Zoology* 67: 204-215.

Young, M.J., E.W. Taylor and P.J. Butler. 1993. Central electrical stimulation of the respiratory nerves of the anaesthetised, decerebrate dogfish, *Scyliorhinus*, and its effect on fictive respiration. *Journal of Physiology* (London). 459: 104P.

Young, J.Z. 1933. The autonomic nervous system of selachians. *Quarternary Journal of Microscopical Science* 15: 571-624.

Young, J.Z. 1950. *The Life of Vertebrates*. Clarendon Press, Oxford.

Index

Color Section

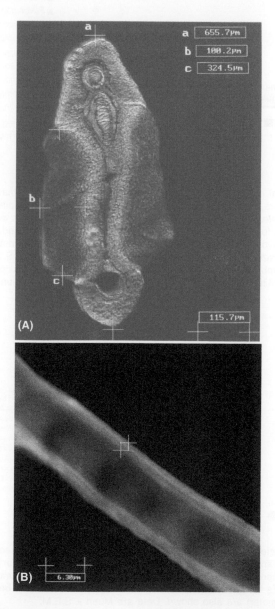

Fig. 9.2 (A) Transverse section of a sea filament, viewed by confocal laser-scan microscopy. The monolateral surfaces of two contiguous lamellae are shown. (B) Measure of gas diffusion distance in a fresh lamella. Adapted from M. Saroglia *et al.* (2000, 2002).

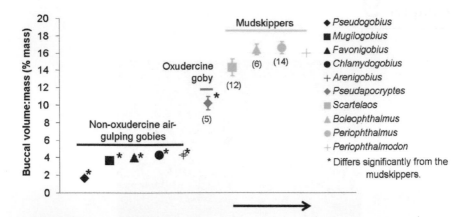

Chapter 14

Fig. 14.8 Buccal chamber volumes, expressed as percent of body mass, for the four mudskipper genera, *Pseudapocryptes*, and five non-oxudercine air-gulping gobies. Data for *Pseudapocryptes*, *Scartelaos*, *Boleophthalmus*, and *Periophthalmus* were determined in this study, data for *Periophthalmodon* are from Aguilar *et al.* (2000), and *Pseudogobius*, *Mugilogobius*, *Favonigobius*, *Chlamydogobius*, and *Arenigobius* are from Gee and Gee (1991). (Sample size is noted in parentheses when available. Mean ± S.E.M., * denotes the buccal chamber volume is significantly different than mudskipper volumes.) Oxudercines (right half of Fig.) are in order of their increased specialization for amphibious life (arrow).

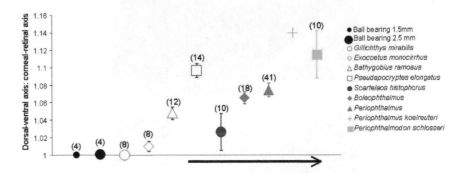

Fig. 14.9 Diameter ratios for the dorsal-ventral: corneal-retina lens axes determined for the mudskippers and *Pseudapocryptes* in this study, together with data for *Periophthalmus koelreuteri* (Karsten, 1923), two non-oxudercine gobies (*Bathygobius ramosus*, *Gillichthys mirabilis*), and the flying fish (*Exocoetus monocirrhus*). Ratios for ball bearings (1.5 mm and 2.5 mm diameter) are also shown. Data are Mean ± S.E.M.

Chapter 15

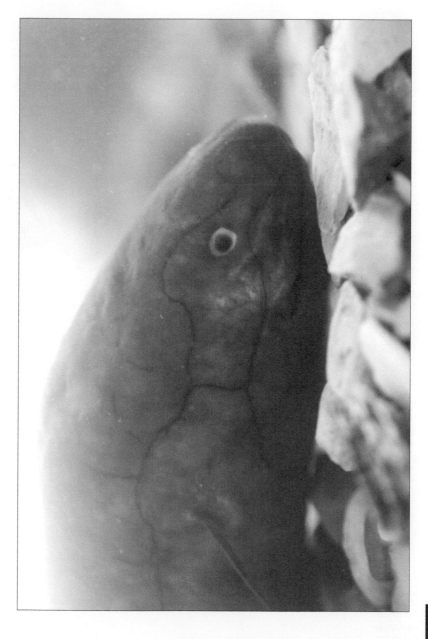

Fig. 15.1 Photo of the South American lungfish, *Lepidosiren paradoxa*.

Printed and bound by CPI Group (UK) Ltd, Croydon, CR0 4YY

23/10/2024

01778263-0008